EXPLANATION, CAUSATION AND DEDUCTION

THE UNIVERSITY OF WESTERN ONTARIO
SERIES IN PHILOSOPHY OF SCIENCE

A SERIES OF BOOKS
IN PHILOSOPHY OF SCIENCE, METHODOLOGY,
EPISTEMOLOGY, LOGIC, HISTORY OF SCIENCE,
AND RELATED FIELDS

Managing Editor

ROBERT E. BUTTS

Dept. of Philosophy, University of Western Ontario, Canada

Editorial Board

JEFFREY BUB, *University of Western Ontario*

L. JONATHAN COHEN, *Queen's College, Oxford*

WILLIAM DEMOPOULOS, *University of Western Ontario*

WILLIAM HARPER, *University of Western Ontario*

JAAKKO HINTIKKA, *Florida State University, Tallahassee*

CLIFFORD A. HOOKER, *University of Newcastle*

HENRY E. KYBURG, JR., *University of Rochester*

AUSONIO MARRAS, *University of Western Ontario*

JÜRGEN MITTELSTRASS, *Universität Konstanz*

JOHN M. NICHOLAS, *University of Western Ontario*

GLENN A. PEARCE, *University of Western Ontario*

BAS C. VAN FRAASSEN, *Princeton University*

VOLUME 26

FRED WILSON

Philosophy Department, University of Toronto, Canada

EXPLANATION, CAUSATION AND DEDUCTION

D. REIDEL PUBLISHING COMPANY

A MEMBER OF THE KLUWER ACADEMIC PUBLISHERS GROUP

DORDRECHT / BOSTON / LANCASTER

Library of Congress Cataloging in Publication Data

Wilson, Fred, 1937–
  Explanation, causation, and deduction.

  (The University of Western Ontario series in philosophy of science; v. 26)
  Bibliography: p.
  Includes indexes.
  1. Explanation (Philosophy) 2. Logic. 3. Causation.
I. Title. II. Series.
BD237.W54   1985   162   85-10797
ISBN 90-277-1856-3

Published by D. Reidel Publishing Company,
P.O. Box 17, 3300 AA Dordrecht, Holland

Sold and distributed in the U.S.A. and Canada
by Kluwer Academic Publishers,
190 Old Derby Street, Hingham, MA 02043, U.S.A.

In all other countries, sold and distributed
by Kluwer Academic Publishers Group
P.O. Box 322, 3300 AH Dordrecht, Holland

D. Reidel Publishing Company is a member of the Kluwer Group.

All Rights Reserved
© 1985 by D. Reidel Publishing Company
No part of the material protected by this copyright notice may be reproduced or
utilized in any form or by any means, electronic or mechanical,
including photocopying, recording or by any information storage and
retrieval system, without written permission from the copyright owner

Printed in The Netherlands

To Gustav Bergmann

TABLE OF CONTENTS

PREFACE	ix
CHAPTER 1 / THE DEDUCTIVE MODEL OF EXPLANATION: A STATEMENT	1
1.1. Explanation and Deduction	2
1.2. The Humean Account of Laws	10
1.3. The Evidential Worth of Law-Assertions	23
1.4. That Some Explanations Are Better than Others	43
1.5. That Technical Rules of Computation Are Laws	66
CHAPTER 2 / THE REASONABILITY OF THE DEDUCTIVE MODEL	73
2.1. Why Ought the Deductive Model Be Accepted?	75
2.2. Are There Reasoned Predictions Which Are Not Explanations?	86
2.3. Is Correlation Less Explanatory than Causation?	106
2.4. Is Causation Inseparable from Action?	118
2.5. Are There Explanations Without Predictions?	136
2.6. Explanation and Judgment	159
CHAPTER 3 / EXPLANATIONS AND EXPLAININGS	173
3.1. Explanations in the Context of Communication	173
3.2. Formalist Criticisms of the Deductive Model	187
3.3. Explanations and Explanatory Content	231
3.4. Narrative and Integrating Explanations	251
3.5. Are Laws Evidence for, or Part of, Explanations?	263
3.6. Can We Know Causes Without Knowing Laws?	290
CONCLUSION	343
NOTES	345
Notes to Chapter 1	345
Notes to Chapter 2	351

Notes to Chapter 3	353
Notes to Conclusion	365
BIBLIOGRAPHY	367
NAME INDEX	375
SUBJECT INDEX	377

PREFACE

The purpose of this essay is to defend the *deductive-nomological model* of explanation against a number of criticisms that have been made of it.

It has traditionally been thought that scientific explanations were causal and that scientific explanations involved deduction from laws. In recent years, however, this three-fold identity has been challenged: there are, it is argued, causal explanations that are not scientific, scientific explanations that are not deductive, deductions from laws that are neither causal explanations nor scientific explanations, and causal explanations that involve no deductions from laws. The aim of the present essay is to defend the traditional identities, and to show that the more recent attempts at invalidating them fail in their object. More specifically, this essay argues that a Humean version of the deductive-nomological model of explanation can be defended as (1) the correct account of scientific explanation of individual facts and processes, and as (2) the correct account of causal explanations of individual facts and processes.

The deductive-nomological model holds that to explain an event E, say that a is G, one must find some *initial conditions* C, say that a is F, and a *law* or *theory* T such that T and C jointly entail E, and both are essential to the deduction. Thus, if the law T is 'All F are G', or, in symbols, '$(x)(Fx \supset Gx)$', then the *argument, assuming its premises are true, and that the general premiss is a law*,

$T: (x)(Fx \supset Gx)$
$C: Fa$
$E: Ga$

will be an *explanation* of a's being G, and equally, if a's being F is observed prior to its being observed to be G, it will be a *prediction* of a's being G. Thus, to defend the deductive-nomological model of explanation is also to defend the *symmetry thesis*, that every explanation is a potential prediction an that every prediction – that is, every *reasoned* prediction, every prediction based on laws – is an explanation.

The criticisms we shall be examining proceed largely by way of *counter-example*. One group attacks the model by way of the symmetry thesis, purporting to discover explanations which are not predictions and (reasoned) predictions which are not explanations. Another group attacks the model on formalist grounds, giving examples which satisfy the requirement of law-deduction but which do not seem to be explanations, or showing how, formally, any event seems to be able to explain any other event. A third group attacks the model on grounds that it fails to do justice to the pragmatics of explanation. A fourth group attacks the model on grounds that there are explanations, and, specifically, causal explanations, which do not involve laws or deduction from laws. Interestingly enough, as we shall see, the last three groups are not unrelated.

For a *matter-of-fact generalization* 'All F are G' to serve as a premiss in an explanatory argument, it must be a *law*. A necessary condition for a matter-of-fact generalization to be a law, is that it be *true*. This is not sufficient for a generalization to be a law, however, since there are true matter-of-fact generalizations, e.g., "all the coins now in my pocket are copper", which are, as one says, accidental generalities, i.e., not laws. In order to discuss the criticisms indicated above, it is necessary for this essay to take some stand on the issue of what constitutes a law. The position adopted is that of Hume (*Treatise of Human Nature*) and Mill (*System of Logic*) that what distinguishes those generalizations that count as laws and those that do not is *contextual* or *pragmatic*. Lawfulness is not a syntactic or semantic feature of generalizations but rather is, as Rescher (*Scientific Explanation*) puts it, a matter of *imputation*. A generalization is accepted as a law, or, as I shall say, is *law-asserted* just in case that *we are prepared to use it for purposes of explanation, prediction, and the support of contrary-to-fact conditionals*. And one is *justified* in law-asserting a generalization just in case that *the evidence for its truth has been collected in accordance with the rules of the scientific method*. This is the position proposed by Hume and defended by Mill, and, more recently, Bergmann (*Philosophy of Science*), Nagel (*The Structure of Science*), and Rescher.

The present essay does not undertake a full-scale defence of this Humean account of laws. It has, of course, come under attack recently, both by those, e.g., Armstrong (*Universals and Scientific Realism*), who speak of primitive necessary connections among properties – a theme that goes back to Aristotle – and by those, e.g., Lewis (*Counterfactuals*), who propose to analyze lawfulness in terms of possible worlds. But as a single essay cannot do everything, many of the points raised by these critics are not dealt with; the aim is not so much to look at objections to the Humean account of laws as to look at objections to the deductive-nomological model of explanation. For this reason, while we adopt the Humean account of lawfulness, its defence is only sketched in, just enough for our purposes with no full-scale discussion attempted.

In adopting the Humean account of laws, we introduce not merely the *objective conditions* of validity and true premises for an argument *to count as an explanation* but also the *subjective condition* of the general premiss being supported by scientific evidence *for an argument to be acceptable as an explanation*. It is subjective because the evidence available varies from time to time and from person to person, so that a generalization may not be law-assertible at one time or for one person but law-assertible at another time or for another person. It therefore varies from subject to subject; and, unlike validity and truth, is not objective, independent of the subject.

The thesis we are defending, then, is that *law-deduction is necessary and sufficient for scientific explanations*, bearing in mind that what renders a generalization a law is not a matter of either syntax or semantics nor, indeed, of truth alone, but also evidential context. Included in this thesis are the subtheses that there are no scientific explanations which are not potentially predictions, and no (law-based) predictions which are not explanations. Furthermore, also included is the subthesis that all causal explanations, insofar as they are genuine explanations, are scientific and therefore both deductive from laws and potentially predictions.

Several gambits will be used in replying to the critics of the deductive-nomological model as we attempt to show how their proposed counter-examples really do fit with that model.

Thus, one of the immediate consequences of the Humean-contextualist account of laws is that an argument may be acceptable as explanatory in one evidential context but not in another. We shall make use of this point in several of our replies to critics of the deductive-nomological model of explanation.

Another of the immediate consequences of the Humean account of laws is that, since we observe only a sample and not the population, and since a generalization is about a population and not the sample, we never have *conclusive* evidence that a generalization we assert or law-assert is *true*. It may well be, then, that a generalization the law-assertion of which is supported by available data is in fact false. Thus, an argument may well be *subjectively acceptable as an explanation* but still *not an explanation* if the *objective condition* of true premises is not fulfilled We shall also make use of this point in replying to critics of the deductive-nomological model of explanation.

Deductive-nomological explanations interest us because they provide knowledge that permits prediction and control. That is, we have certain pragmatic interests in obtaining such knowledge. We also are sometimes interested in such explanations for their own sake; i.e., to speak with Veblen, we are interested in them out of idle curiosity. Now, one explanation may *better satisfy* these *cognitive interests* – pragmatic interests and idle curiosity – than does another. This permits us to distinguish *better and worse* among explanations according to the extent to which they satisfy the cognitive interests that motivate us to obtain explanations. We shall make extensive use of this point when we reply to critics of the deductive-nomological model. In particular, we shall discover that many of them fail to distinguish, what we shall insist are distinct, an argument *being a non-explanation* from an argument *being a weak explanation*. Those who claim to find predictions which are not explanations are often guilty of this confusion.

It should be noted, by the way, that while our cognitive interests, that is, our subjective values, determine a scale of better and worse among explanatins, once that scale is give then whether or not an explanation falls at a higher or lower place on this scale is an objective matter.

If we have two explanations available, and one is better than another on the scale of worth defined by our cognitive interests, then, other things being equal, it would be *wrong* to use the weaker, less desirable, explanation as an *act of explaining something to someone*. It is important to distinguish (1) and explanation, i.e., an argument, (2) the acceptability of an argument as an explanation, and (3) the use of an acceptable explanation in an act of explaining. In particular, an argument may be acceptable as an explanation but for all that not appropriate for use in an act of explaining. Critics of the deductive-nomological model often confuse an argument's *being inappropriate for use in an act of explaining* with its *being a non-explanation*.

Some critics of the model tend to rely upon an assumption that any deductive-nomological explanation can have a law-premiss no more complicated in logical form than 'All F and G'. This is often the case with those who claim to have found explanations which are not predictions. These criticisms can be shown to be invalid once it is noted that *there are laws* in such explanations, and that these laws do *not* violate the symmetry thesis, but that the logical form of the explanatory argument is more complex than that of a syllogism in *Barbara*. The same point also serves to rebut criticisms based on the idea that some explanations somehow rely upon 'intuitive judgment' rather than scientific reason to determine their acceptability; and is relevant, too, to criticisms based on the idea that there are explanations which are causal but which do not involve laws, or law-deduction.

Finally, the distinction between better and worse among explanations enables one to distinguish laws into various kinds, e.g. process laws, causal laws, correlations. Criticis of the deductive-nomological model are often guilty of confusing the concept of explanation with one of the special cases, very often that of causal explanation. However, while a law that gives a correlation may not yields a causal explanation, and may be weaker, or worse as an explanation, than a causal explanation, it does *not* follow, as too many wrongly infer, that correlations have *no* explanatory power, that arguments with correlations as their law-premisses are non-explanatory.

The Humean version of the deductive-nomological model that we propose to defend is laid out in Chapter 1. This chapter is essentially exposition, rather than defence. It begins by stating briefly the thesis that explanation of individual events is by deduction from generalities and initial conditions. However, not all generalities yield explanations: for the latter, the general premiss must be a law. The thesis we are defending is that the Humean account of laws suffices. On this account, the characteristic of being lawlike is *contextual*: a generality is lawlike just in case that we are prepared to use it to support the assertion of counterfactual conditionals. It is then argued that some generalities are more *worthy* of being treated as lawlike than others; a generality is worthy of being treated as lawlike just in case that evidence for it has been gathered, or at least advanced, in terms of the norms of scientific research. This involves a crucial distinction between the objective and the subjective (or contextual) conditions that a deduction must satisfy before it can count as, or reasonably be used as, an explanation. Among the relevant contextual elements are matters of evidence. It is not the point of this essay to analyze the logic of confirmation. (Some aspects of this problem — or set of problems — are discussed in my *Reasons and Revolutions*, currently being prepared for publication.) But clearly, the topic is important enough for what we are about that the issues cannot be wholly avoided, and so a brief discussion of these matters is included. It is suggested that the basic patterns of confirmation are those described by Hume and John Stuart Mill. But, using some ideas of Peirce, it is also suggested that these patterns are compatible with ideas more recently presented by Kuhn. However, no *full* discussion of these points is attempted.

Mill's account of the Principle of Causation leads directly to the idea of what Bergmann has called "process knowledge". It has been suggested by

Bergmann that this notion provides an ideal or standard of excellence for the evaluation of explanations of individual facts and processes. This notion of better and worse explanations is developed in Chapter 1, and will play an important role in what follows. Again, the thrust of the discussion at this point is expository; defence of the notion will be given in Chapter 2.

John Stuart Mill's views on confirmation have often been condemned as foolish. Chapter 1 attempts to show that Stuart Mill's position on the justification of the Principle of Causality and, more generally, on the *confirmation of laws about laws* is a position that is at least plausible, sufficiently plausible, anyway, to permit us to rely upon them for our limited purpose.

That purpose is the defence of the Humean version of the deductive-nomological model of explanation. This defence begins in Chapter 2. This chapter begins by outlining the positive reasons for accepting the model as one that – relative to certain cognitive interests – *ought* to be adopted. It then proceeds to discuss various ways in which the deductive-nomological model has been challenged. It argues that, contrary to some, there are no reasoned predictions (i.e., predictions based on deductions from laws) that are not explanations. It argues that, contrary to some, correlations provide explanations just as do causal relations. It argues that, contrary to some, the concept of causation is independent of that of action. It argues that, contrary to some, there do not exist explanations that are not also predictions. Throughout these arguments the distinction between process laws and imperfect knowledge that falls short of this ideal plays a crucial role: in general the argument is that the objections to the deductive-nomological model that we discuss appear successful only if this distinction is ignored. The chapter closes with a criticism of the views of those who hold that 'judgment' plays a role in explanation that is not allowed for by those who defend the deductive-nomological model.

Chapter 3 develops the contrast between explanations and acts of explaining. In this context, it examines and rejects a number of recent formalist criticisms of the deductive-nomological model. The chapter then goes on to discuss whether laws are part of, rather than merely presupposed by, certain sorts of scientific explanation, and whether causes can be known, e.g., by 'judgment', without our also knowing laws. It is argued that laws are indeed part of scientific explanations, and that in knowing a cause one *ipso facto* knows a law.

Generally speaking, aspects of the deductive-nomological model are developed only to the extent that they are needed for replying to critics. For example, the defence of the criteria for better and worse scientific explanations of individual fact is merely sketched. Again, since our concern is with explanations of *individual* facts and events, another sort of deductive explanation, that of laws by theories, receives very little attention. Both these topics — criteria for better and worse, and the nature of theoretical explanation — receive a more detailed discussion in my forthcoming *Reasons and Revolutions*.

In this connection it is perhaps worth noting a proposal concerning explanations advanced by a number of recent authors such as Michael Friedman ('Explanation and Scientific Understanding', 1974) and Philip Kitcher ('Explanatory Unification', 1981). While these authors disagree about what constitutes unification (cf. Kitcher, 'Explanation, Conjunction and Unification', 1976), they do agree that "to explain is to unify" — Kitcher cites E. M. Forster's injunction: "Only connect" — and argue that models developed in accordance with this injunction ought to replace the now discredited covering law or deductive-nomological model.

Now it seems obvious that in a sense I do not agree with this position, since the whole thrust of the present essay is to argue that the deductive-nomological model has in fact not been discredited. That being so, there is no need to replace it.

On the other hand, the injunction is not without merit: there is indeed a real point to the idea that explanation consists in unification. But of course, the deductive-nomological model already acknowledges this point. This model asserts that to explain an event is to subsume it and another, independent event, under a law that *connects* the two. It is part of the very idea of the deductive-nomological model that explanation of individual events consists in unifying those events by means of laws.

There are, of course, degrees of unifying power. The connection between events that some laws establish is tighter than the connection established by other laws. What we argue below is that, so far as concerns individual events, the ideal of explanation is process knowledge. When one explains an event in terms of process knowledge, that event is located as part of the ongoing development of a system over time, a development for which the process knowledge enables one, on the basis of a knowledge of the present state of the system, to predict the state of the system at *all* future times and at *all* past times. Process knowledge fills in *all details* of a process. Thus, when an event is explained by means of process knowledge it is *gaplessly connected to all other events in an ongoing process*. In the case of individual events, then, the injunction "Only connect" is fulfilled to the greatest degree possible when process knowledge is attained.

Kitcher, Friedman, and others who adopt their position are thus mistaken in their view that defending the deductive-nomological model is somehow incompatible with defending the thesis that explanation consists in unification.

The notion that "to explain is to unify" is, however, one that has its greatest impact at the level not of explanation of individual facts but at the level of *theoretical explanation*, the explanation of laws. Our cognitive interest in scope, as well as our interest in theories that can guide research into hitherto undiscovered laws, both lead us to seek theories that unify bodies of law. Such laws about laws (to use Mill's phrase) explain the laws they unify, and, insofar as they apply to areas where the specific laws are not yet known, they predict, and guide researchers to the discovery of, new laws. It is hard to over-estimate the importance of such unifying laws. Yet a thorough discussion of the issues would be out of place in the present essay, which has as its main focus of concern the explanation of individual facts rather that the explanation of laws. To this extent, then, a pursual of these aspects of the issues raised by Kitcher and Friedman is out of place.

On the other hand, the issue cannot entirely be avoided, if only because the laws about laws, the unifying laws that explain other laws, can themselves at times be used to explain individual facts. The discussion of Mill in Chapter 1 gives an account of how certain sorts of generic over-hypotheses can unify more specific hypotheses. According to Kitcher ('Explanatory Unification', p. 519), "The problem of explanation is to specify which set of arguments we ought to accept for explanatory purposes given that we hold certain sentences to be true". The constraints on such arguments are constraints on the laws that appear in these explanatory arguments; and these constraints are themselves factual, that is, lawful. These conditions "jointly imposed by the presence of nonlogical expressions in the pattern and by the filling instructions" as well as "conditions on logical structure" (*ibid.*, p. 518) are precisely the constraints imposed by the generic laws about laws that Mill discussed. The view that we sketch, then, of how theoretical explanation proceeds by unification is close indeed to that of Kitcher.

Kitcher does not develop his position at any length. He does not attend to the two specific forms of theoretical unification that have played so great a role in the actual history of science, to wit, unification by means of a composition law, and unification by reducing one area (the macro-area) to another (the micro-area). Nor does Kitcher develop the idea that unifying theories are indispensable tools for guiding research. For a detailed analysis of a generic law and its role in concept formation, one might look at my essay 'Is Operationism Unjust to Temperature?'; and for their role in research two other essays 'Kuhn and Goodman: Revolutionary vs. Conservative Science', and 'A Note on Hempel on the Logic of Reduction', are relevant. Some aspects of the latter issue are also dealt with in Section 3.3, below. The whole complex of issues is discussed in some detail in another forthcoming essay of mine, *Empiricism and Darwin's Science*.

Another topic that is certainly worth discussing is that of statistical explanations; but this too is almost completely ignored. This in spite of the fact that it is of considerable relevance to what we are about. I can only plead by way of excuse that one cannot deal *at once* with *all* the objections to the deductive-nomological model. The objections I have considered all fit together, so it is they that I have discussed, rather than challenges to the deductive-nomological model based on so-called 'statistical explanations'. If the reader wishes some very brief indication of my views on the latter, he might glance at my review of N. Rescher's *Scientific Explanation*.

A number of critics of the deductive-nomological model who shall be discussed have been heavily influenced by the doctrine of so-called 'Oxford philosophy' that 'meaning is use'. I have critically evaluated the latter doctrine in a monograph, *Meaning Is Use*, currently being prepared for publication. Certain parts of the present work supplement and elaborate points about scientific explanation made more briefly in this latter monograph.

It is my belief that many objections to the deductive-nomological model depend upon an inadequate analysis of certain examples of scientific theories and of certain historical developments of science. For that reason, I have ventured *detailed* analyses where appropriate. In particular, an extended analysis of geometrical optics occurs in Chapter 3. Much of this is straightforward physics, or history of science, yet I am convinced that only by presenting these straightforward things in a deliberately philosophical context can many objections to the deductive-nomological model be adequately met.

It is also my view that it is sometimes important to know not only *that* a conclusion follows but also *how* it follows. Sometimes a knowledge of the logic of the deduction is important to understanding its philosophical implications. This is particularly so with respect to some points of Davidson and of Kim that are discussed in Chapter 3. I have therefore tried to present such deductions in sufficient detail to make clear what is going on. The result is no doubt more symbolic logic than is comfortable for many. My only excuse is that experience with students and colleagues over the years has convinced me that clear understanding can be attained only by a patient working-through of the details.

I would also like to thank the following authors for permission to quote from the works cited:

S. Bromberger: 'Why-Questions'. First appeared in *Mind and Cosmos: Essays in Contemporary Science and Philosophy*, Robert Colodny, editor. Copyright by the University of Pittsburgh Press, 1966. Used by permission.

A. W. Collins: 'Explanation and Causality'. *Mind* 75 (1966). Used by permission of the editor of *Mind*.

D. Davidson: 'Causal Relations'. First published in *Journal of Philosophy* 64 (1967). Reprinted in D. Davidson: *Actions and Events*, Oxford University Press, 1980. Quoted from the latter by permission of the Oxford University Press.

S. Toulmin: *Philosophy of Science*. Hutchison, 1953. Used by permission.

G. H. von Wright: 'On the Logic and Epistemology of the Causal Relation'. First appeared in *Logic, Methodology and Philosophy of Science IV*, P. Suppes *et al.*, editors. North-Holland Publishing Co., 1973. Used by permission.

Portions of the following publications have been incorporated into the present essay:

'Hume and Ducasse on Causal Inference from a Single Experiment', *Philosophical Studies* 35 (1979), pp. 305–9. Copyright 1979, D. Reidel Publishing Co.

'Goudge's Contribution to Philosophy of Science', in L. W. Sumner, J. G. Slater, and F. Wilson, (eds.), *Pragmatism and Purpose: Essays in Honour of T. A. Goudge*, University of Toronto Press, Toronto, 1980. Copyright 1980, University of Toronto Press.

'Mill on the Operation of Discovering and Proving General Propositions', *Mill News Letter* 17 (1982), pp. 1–14. Copyright 1982, University of Toronto Press.

Editors and publishers are thanked for permission to include this material.

A number of key ideas in this essay were developed at a Working Conference on Causality held at Dalhousie University, July–August 1973. All the participants should be thanked for the stimulation they provided. The organizer, David Braybrooke, deserves special thanks. So, too does the Canada Council which partially funded the conference.

The draft of the book was completed while I was on leave from the University of Toronto in 1979–80, made possible by a leave fellowship granted by the Social Sciences and Humanities Research Council of Canada.

CHAPTER 1

THE DEDUCTIVE MODEL OF EXPLANATION:
A STATEMENT

The method for answering the question, 'What is a scientific explanation?' seems to be the following. We look, *first*, at what everyone agrees are in fact cases of scientific explanation. The *second* step is to find features of the examples in terms of which one can evaluate other purported explanations as to whether or to what extent they are really explanations. The *third* step is to justify those evaluative criteria as *justified* norms. This last step is crucial. Philosophy of scientific is not merely descriptive, even if in its first step it is merely that. It is also, or ought also to be a normative discipline, laying down norms for what constitutes a *good* explanation, an explanation that *ought* to be accepted.

The result towards which we shall argue will be that explanation is *deduction from laws*. This is the *deductivist* position. This position has been seriously attacked of late. Our articulation of the deductivist position will therefore involve responding to various aspects of these attacks. Because this is the purpose the three steps above will not be followed in their natural order. On the contrary, they will, so to speak, be done simultaneously, with order of exposition determined by the strategy of defending the deductivist position.

The examples to which we will appeal will generally come from physics. We shall thus look at Newton's explanation of the behaviour of the solar system as an example of the explanation of individual facts, and at Newton's explanation of Kepler's and Galileo's laws as examples of the explanation of laws. There are other possible examples. It is perhaps unfortunate that recently discussion has centred about the internal happenings of automobiles in cold weather. It is understandable why such an example was first introduced: for purposes of illustration. The attack has, however, tended to take it, rather than Newton, as providing the paradigm of scientific explanation. The results have been unfortunate. In order that these unfortunate results be avoided we shall also have to look at these other perhaps less appropriate examples.

Popper has stated the thesis excellently as follows:

To give a *causal explanation* of an event means to deduce a statement which describes it, using as premises of the deduction one or more *universal laws*, together with certain singular statements, the *initial conditions*.[1]

2   CHAPTER 1

Elsewhere he tells us that "To my knowledge, [this] theory of causality . . . was first present in my book, *Logik der Forschung* (1935) — now translated as *The Logic of Scientific Discovery* (1959)."[2] But actually, the deductivist thesis is much older than Popper. Indeed, it is a cornerstone of empiricism, and has been recognized as such by empiricists ever since Hume. Popper's report on his knowledge of his predecessors may indeed be a correct report of Popper's knowledge, but to repeat, the thesis is a standard one of empiricists. It is, for example, clearly stated almost in Popper's terms in J. S. Mill's *System of Logic* (1858):

> An individual fact is said to be explained, by pointing out its cause, that is, by stating the law or laws of causation, of which its production is an instance . . . a law or uniformity in nature is said to be explained, when another law or laws are pointed out, of which the law itself is but a case, and from which it could be deduced.[3]

1.1. EXPLANATION AND DEDUCTION

An explanation of certain facts is a set of sentences, including a sentence about the fact or facts (individual or general) to be explained. The sentences are *about* facts, including the fact or facts to be explained. The sentences which *are* the explanation constitute an explanation of the facts because sentences in the explanation are about those facts.

To speak about sentences being *about* facts introduces a semantical feature into the notion of explanation. Deduction is a matter of syntax. That means deduction cannot be the whole story about explanation. But, of course, no one has ever denied that point. It is therefore rather useless to attack the deductivist position by pointing to this non-deductive element.[4] *No one* has ever denied a semantical dimension to the language used in explanations. But some have denied explanations are deductive, or have insisted that they are more than deductive. It was against these that the defenders of the deductivist position wished to argue. That was why they emphasized deduction as the defining feature of scientific explanation. Certainly, the emphasis was not merely a matter of perversity, as some of the more hysterical of the attacks seem to suggest.

As for why one defines an explanation in terms of a set of sentences, one can say easily that it is a stipulation; not a "mere stipulation", but one which will prove useful both in describing our examples and in defending the claims about the norms of good explanation which we shall make. The defence of the stipulation consists in showing its utility, by actually using it.

Two points are important about explanations. (1) What a sentence is about is independent of who utters or asserts it, once the linguistic conventions have been laid down. These conventions are sufficient to establish what it is about, i.e., to establish its meaning.[5] (2) Whether a sentence is true or false does not depend upon who uses the sentence, once the linguistic conventions have been laid down. But the conventions are not sufficient to establish that truth or falsity, so far as sentences about facts are concerned: truth or falsity also depends upon whether what the sentence is about exists or not, and whether it exists or not depends on the way the world is, on the way what language is about is, and not on the language, not on the linguistic conventions.[6] (3) Whether a set of sentences constitutes a deductively valid argument depends only on the logical form of the set, that is, upon the syntax of the sentences. The syntax of a sentence depends on the linguistic conventions which establish the meaning of the sentence. So logical form is a matter of the meaning of the sentence. Thus, validity depends only on the meaning of the sentence, and not on who happens to use it, or on the occasion on which it might be used.[7]

Another point will be useful for later reference. Consider a conversation between A and B.

>A asks: Why is a an F?

This question is, of course, not: is a an F? but rather: a is an F and why is that so? That is, the way A asks the question makes it clear he is prepared to assert that a is an F, that in this context he is certain about a's being F.

>B explains: Because a is also a G.

B is the explainer, A the explainee. B engages in the activity of explaining (something to A). The tool B uses in this activity is an explanation. This explanation is a sequence of sentences. This sequence consists of two simpler sentences connected with 'because', namely, 'Fa because Ga'. The sequence is not completely uttered by B; only A utters 'Fa'. 'Fa' can still be counted as part of the explanation, however, without any difficulty. Nothing in our definition of 'explanation' requires the whole explanation to be uttered by the explainer. 'Fa' is actually uttered only by the explainee, but precisely just because he has already uttered it, the explainer is not required to utter it a second time. That is, *part* of the explanation is already available in the context, so the explainer need not repeat it.

It must be recognized that the *activity* of explaining is a transaction between B and A. The explanation itself is not a transaction, but, as we have defined it, a sequence of sentences. In this we disagree with Collins,[8] who defines 'explanation' to be the transaction. That is his privilege and ours, to use words as we please. Only, when he find certain things are true of explanations in his sense which are not true of explanations in the sense of this essay, he cannot criticize this essay for failing to do justice to those features of explanation. For, his concept is not the concept of the deductivist position. What will be argued below, however, is that the explanation, in the sense in which we are proposing to use that term, is a crucial ingredient in the transaction, and therefore worth singling out and abstracting from the process. (To abstract is not to falsify.) It is crucial from a normative point of view. One might suggest that Collin's preference for defining 'explanation' in terms of the transaction results from his abandonment of a normative for a merely descriptive point of view. But that would be to charge him with abandoning philosophy of science for the psychology and sociology of communication. I will leave it to the reader to judge the justice of the suggestion.

Let me now make some assertions about explanations. Then I will turn to defending them. There are reasons justifying this procedure. The first is that it will make clear what exactly I propose to defend. The second is that it will enable me to dispose of certain obviously irrelevant objections, before turning to those which go to (or towards) the substance of the theses to be defended. I therefore begin to assert.

Consider first the explanation of individual facts or events, including sequences of events (processes).

I shall take it completely without argument that a scientific explanation of individual facts and processes is one which shows that, in some sense of 'must', since event E_1 occurred, the event E_2 (which is to be explained) *must* have occurred. This is where one must be purely descriptive. All one can say here is simply that this is as a matter of fact what is meant by 'scientific explanation'. As far as I can tell, both the older deductivist tradition and its more recent critics agree on this much. It is only at this point where they begin to disagree. The 'must' is philosophically problematic, and requires explication. The critics in fact leave this notion unexplicated. They simply take it as primitive. In their hands, this problematic aspect of explanation remains problematic. In contrast, the deductive model proposes such an explication. We shall examine the defence of this explication below. At this point the explication will merely be stated.

Explanations of individual facts and events must be deductive. The premises must be true, and must include at least one sentence expressing a law,

THE DEDUCTIVE MODEL OF EXPLANATION 5

which sentence is essential for the deduction (if it was dropped, the remaining premises would not entail the conclusion). The sentence expressing the law, conjoined with sentences about individual facts (the initial conditions) entail sentences about the facts to be explained (the explanandum).[9] Presupposing the law being true and the deductive validity of the argument, then given the initial conditions, the explanandum *must* occur. That is, the deductive model explicates (it is claimed) the problematic 'must' with which we began. Now, I do not want to claim the law statement is always explicit. For example, it is not in our *A* and *B* transaction above. But it is always implicit. Or so I propose to argue. Scriven for one disagrees.[10] So does Collins.[11] Part of the argument (especially that of Scriven) simply turns on pointing out the statement of lawful generality does not occur explicitly. That is, of course, irrelevant. All that is being claimed here is that the law is implicit. Or, if it is not, then one does not have a scientific explanation, in *the reasonable* sense of 'scientific explanation'. Why the deductive model provides *the reasonable* sense will be argued in Chapter 2.

The crucial ingredient in the scientific explanation, upon the explication proposed by the deductive model, is the law, a matter-of-fact generality. When science explains, it explains via laws. It follows we can speak of the aim of science as that of explanation or as that of knowledge of laws, since the knowledge of laws permits one to explain (supposing one also knows the initial conditions).

Knowledge of laws which permit explanation is also knowledge which permits prediction. We explain the water boiling by pointing to its having been heated (initial conditions) and citing the law that water, when heated, boils. Noticing the water being heated we can use the law to deduce and thereby predict that it will boil. Upon the deductive model, explanation and prediction are symmetric: where one can predict, there one has explained, and where one explains, there one could have predicted. I take it as obvious that 'could have predicted' here means: could have predicted were one able to have had knowledge of the initial conditions. Unfortunately, some critics have not taken this obvious qualification seriously, as we shall see below. Failure to take it seriously initiates a good many criticisms of the deductive model.

Besides permitting explanation and prediction, knowledge of laws enables us to control. Thus, knowing that heating is sufficient for boiling, if we want boiling water (e.g., for tea) then we can intelligently interfere in the system through turning on the stove. This action has the causal consequence that heating of the water occurs. And, since water, when heated, boils, this in turn has the causal consequence that the water boils. Knowing laws enables us to know what the initial conditions must be if certain desired events are to occur (e.g., the water boiling). Such knowledge is necessary for purposes of control. But not sufficient. For, it is also necessary for control that there be technology available that enables us through our actions to bring about the relevant initial conditions. Newton told us a great deal about the solar system, enough for explanation and prediction, but we still cannot control it: technology is hardly up to the task — except perhaps we now have it within our means to remove one planet from the system by blowing it up. At any rate, just as we had to qualify the idea that Knowledge of laws permits prediction, so we must equivalently qualify the idea that knowledge of laws permits control.

But, just what is this notion of a *law* that we have been using? We have claimed that the *basic* idea of a law is that it is expressed by a statement of the sort

(1.1)   All F are G

where (1.1) is to be understood as a *true, synthetic, matter-of-fact generality*. It's being a generality is secured by the quantifier 'all'. It's being matter-of-fact is secured by insisting that 'F' and 'G' be either observation concepts or concepts which are explicitly defined in terms of the observational base. Now, the only known explication of the analytic/synthetic distinction is that involving the customary truth-table apparatus, and that requires there be no (basic) connectives other than the truth-functional. This means the form of (1.1) will be

(1.2) $(x)(Fx \supset Gx)$

And, indeed, generalities being of this form will guarantee that, if the initial condition, say,

(1.2a) Fa

obtains, then the event

(1.2b) Ga

obtains. Given (1.2a), then, upon (1.2), the event (1.2b) *must* occur: it cannot be otherwise. So (1.2) explains why (1.2b) occurs, given the occurence of (1.2a). And, obviously, if one knows (1.2a) obtains but not whether (1.2b) does, then (1.2) enables one to predict the occurrence of (1.2b). Again, it is a matter of the *must*. This *must* is a matter of logical deduction. (1.2) and (1.2a) logically imply or entail (1.2b) – in the sense of 'entail' explicated in the usual way by the techniques of formal logic. Since (1.2) and (1.2a) entail (1.2b), then, given that the former are true, then the latter *must* also be true.

A sound argument (logically valid, true premisses) of the sort "(1.2), (1.2a), ∴ (1.2b)" constitutes an *objectively acceptable* explanation or prediction, that is, an objectively acceptable reason justifying why one event or fact *must* occur, given that some other event or fact has occurred. Yet, in fact, we do not tend to count as providing acceptable explanations or reliable predictions many sound arguments of the cited sort. Only a subset of the set of all true generalities of the form (1.2) are taken as providing the needed general premiss for an acceptable explanation or reliable prediction. We distinguish true generalities of the form (1.2) into the class of "accidental generalities" and "laws of nature", and only the latter provide acceptable explanations and reliable predictions. Thus, to use a classic example,[12] there is a park bench on the Boston Common on which only Irishmen have sat. So far as we know, then, we have the true generality

(1.3)   Only Irishman sit on this bench.

Still, we count this as an "accidental generality", one that obtains by "accident" and not as a matter of "law". For, we would not accept it as providing reliable predictions: we would not, on the basis of (1.3), venture with much confidence the prediction that

(1.3a)   The next man who sits on this bench will be an Irishman.

There would be no confidence in the prediction simply because you know that you yourself, who are not an Irishman, could falsify both the prediction and the generality by simply sitting on the bench, though, of course, you do not choose to so act, so that (for all you know) the generality (1.3) does in fact happen to be true – but only as a matter of "accident".

Similarly, we would not tend to count as acceptable the explanation

(1.4a)   This coin is copper because it is now in my pocket.

based on the generality

(1.4)   All the coins now in my pcoket are copper.

even though this generality does happen to be true. In contrast, we would accept the explanation

(1.5a) This powder is soluble because it is sugar.

based on the equally true generality

(1.5) Sugar is soluble.

The difference between (1.5) and (1.4) can be brought out by noting that in connexion with (1.5) we are quite prepared to assert the contrary-to-fact subjunctive conditional

(1.5b) If this powder [which is quartz] were sugar then it would dissolve.

whereas (1.4) does not in the same way support the assertion of the subjunctive conditional

(1.4b) If this coin [which is made of silver] were now in my pocket then it would be copper.

And in similar fashion, we would count as reliable the prediction

(1.5c) If this sample of sugar is put in tea then it will dissolve.

based on (1.5) where we would not count as reliable the prediction (1.3a) based on (1.3). Nor does (1.3) support our asserting the contrary-to-fact subjunctive conditional

(1.3b) If I [who am not Irish] were to sit on this bench then I would be an Irishman.

in the way (1.5) does support asserting (1.5b). In general, we may say that *we tend to accept a generality of the form (1.2) as providing an acceptable explanation only if, of course, we have reason to believe it to be true, but further, only if the generality can support the assertion of subjective conditionals and can yield reliable predictions.*[13]

Now, a number of philosophers have taken this as meaning that (1.2) does not after all give the logical form of "natural laws". They have argued that lawlike generalities somehow assert a sort of "natural necessity", and it is this which enables them to support asserting subjective conditionals and reliable predictions. Instead of (1.2), laws will have a form of the sort

(a) $N(x)(Fx \supset Gx)$

where 'N' represents "natural necessity", or of the sort

(b) $(x)(Fx \rightarrowtail Gx)$

where '$\rightarrowtail$' represents "causal implication".

The difference between (a) and (b) is not crucial. If the modal operator is taken to be basic then (a) may be used to define the special implication of (b). Or the latter may be used to define the former. In any case, these philosophers hold there is a special *non-truth-functional* feature in the logical form of law-statements that is not present in the logical form of non-law-statements.[14] This means that the truth-conditions for law-statements differ from those of non-law statements. The non-truth-functional element in the law-statements represents a special sort of nonlogical connection that (supposedly) obtains objectively in the facts of the world. If the position is to be made intelligible, a semantics must be provided for the non-truth-functional component of law-statements, hooking that linguistic feature to the objectively existing nomological connection. Unfortunately, some philosophers who insist upon introducing non-truth-functional components are content simply to give an axiomatization for the signs, without giving any semantics.[15] But, as always, syntax without semantics,[16] or, what is the same, implicit definition,[17] yields no philosophical insight,[18] and for that reason can safely be ignored. Others[19] are content to argue that there somehow must be a feature in the world for that non-truth-functional component to represent,[20] for otherwise we would be unable to draw the law/accidental generality distinction as we do. But some who argue this way then never go on to tell us how to identify when this objective nomological connection is present, with the consequence that *for practical purposes* there will be no distinction of logical form between law and non-law statements.[21] But others of this group do recognize the semantics must be spelled out, that we must be able to identify the objective nomological connection if it is claimed it is such a connexion that enables us to draw the law/non-law distinction as we do. With metaphysical and epistemological sophistication, Aristotle argued for the presence of such a special nomological connection and for our knowledge of it. And, much more recently, Scriven, too, has argued we can and do judge which such a connection is objectively present. We shall look at these views of Aristotle and of Scriven later. Our conclusion then will be that they are in fact mistaken. Nor should this surprise us. For, if non-truth-functional components of the sort required by (a) and (b) are introduced into one's language, then one can no longer explicate the analytic-synthetic distinction. Or, at least, the only viable explication of that distinction will no longer be available, since it depends upon there being only truth-functional connectives. If, therefore, we are to defend the analytic-synthetic distinction — as I think we must — then we should reject any claims that an objective nomological connection exists, and any claims that one must introduce into one's language a non-truth-functional component to represent it. We should, therefore, insist that the logical form of lawlike statements is given by (1.2), that there is no logical or ontological difference in form between lawlike generalities and accidental generalities.

But we do, of course, recognize such a distinction: there is a clear difference between (1.5) on the one hand and (1.3) and (1.4) on the other, a difference we marked by the suggestion that a true generality is a law only if we take it as yielding reliable predictions and as supporting the assertion of subjunctive conditionals: and that is simply to say we do draw a distinc- between *propter hoc* and *post hoc*. If we are to hold (1.2) is the logical form of laws then we cannot hold this distinction is one of logical form. What, then, is the nature of this distinction? The answer to this question was first given, I believe, by Hume, in his account of causation found in his *Treatise on Human Nature*.[22]

1.2. THE HUMEAN ACCOUNT OF LAWS

Hume has often been criticized for his views on causation. But very many of these criticisms are in fact directed at views which are not found in the *Treatise*. Thus, for example, Ewing[23] describes a "regularity" theory of causality, which holds there is nothing more to causal assertions than assertions of *de facto* regular connection. This theory is then criticized for being unable to distinguish causal associations from those that occur by chance, in other words, for being unable to distinguish laws from accidental generalities. The difficulty is that Hume never held this "regularity" theory. The *Treatise* fully recognizes the distinction Ewing's "regularity" theory denies, between *post hoc* and *propter hoc*. As it points out, "An object may be contiguous and prior to another, without being consider'd as its cause. There is a NECESSARY CONNEXION to be taken into consideration; and that relation is of much greater importance, than any of the other two above-mentioned."[24] And most of the rest of Book I, Part III is devoted in one way or another to an investigation of the nature of this necessary connection. My purpose here is simply to state the account of the *Treatise* in such a way that it is clear such criticisms as that of Ewing of "Humean" theories of causation at least are not applicable to the account of the *Treatise*.

I propose to state this account, but not to argue for it much further than I have done here, for, that would take us too far afield,[25] though it should be noted that our future arguments against Scriven and Aristotle would be part of any full case in favour of the Humean view. And further, as will soon be clear, simply to see Hume's view accurately is already to see that many standard criticisms do not apply. In stating Hume's view, since it is for our purposes and not the *explication du texte*, I shall not restrict myself to the language of Hume: I shall not hesitate to translate, where appropriate, Hume's way of speaking into language more familiar to the twentieth century. And in this connection I shall try to separate Hume's account of causation from other theories which he holds, specifically his account of impressions and ideas, his account of relations, and his account of concept formation. No doubt these are of interest, but equally without doubt the account of causation is logically independent of these other theories. Of these theories all we shall retain are two points. The first is the doctrine about observable particulars (impressions) that they are logically independent of each other.[26] As Wittgenstein's *Tractatus* shows us, such logical atomism is not logically connected to the psychological atomism of associationist psychology. And the second point to be retained is that empirical concepts are either primitive or defined, and that the primitive concepts designate sensible properties of observable particulars. Which is, of course, the logical core of Hume's principle " ... that all our ideas are deriv'd from correspondent impressions."[27] Criticisms of Hume's account of causation are sometimes directed not so much at that account *per se* as at the associationist psychology that provided the broader context and terminology for the account as Hume himself stated it in the *Treatise*.[28] Stating the account of causation apart from this context will enable us to see the irrelevance of these criticisms. Which is not to say those who offer these criticisms have an adequate understanding of the associationism we shall more or less put to one side. Prichard for one, isolated provincially as he was for all his life at Oxford, had neither understanding of what psychology was, nor any interest in finding out.

Fundamental to understanding Hume's account of causation is the distinction that can be drawn with respect to our beliefs between the propositional attitude and the propositional content. If one person believes Toronto is west of Hamilton and a second disbelieves Toronto is west of Hamilton, the propositional content of both beliefs is the same, namely, the proposition that Toronto is west of Hamilton. But the propositional attitudes in the two beliefs are different. In one case the attitude is that of believing or asserting, in the other it is that of disbelieving or denying. The propositional content is true or false. Its truth-value depends only on the facts it is about. In particular it does not depend upon what the propositional attitude one has with respect to that content. This independence is important since it is precisely this independence which enables truth and falsity to function as *standards* justifying the attitude one has. The attitude of believing or asserting with respect to a proposition is *objectively justified* if and only if that proposition is true. And the attitude of disbelieving or denying is objectively justified if and only if the proposition is false. Clearly, if truth-value depended on attitude the former could not provide an objective standard for evaluating the latter.

The first definition of "cause" is this: "an object precedent and contiguous to another, and where all the objects resembling the former are plac'd in a like relation of priority and contiguity to those objects, that resemble the latter."[29] In other words, one event causes a second where the two are subsumable under a matter-of-fact generality. Particular instances of causation are to be understood in terms of *de facto* regularities. This means any proposition stating a causal regularity has the logical form "$(x) (Fx \supset Gx)$", where the descriptive predicates are logically independent of each other. This renders in the language of Russell the Humean point[30] that cause and effect are logically independent of each other.

It follows that so far as propositional content is concerned the assertion of a causal regularity involves no difference in logical form from the assertion of an accidental generality. Both are of the form "$(x) (Fx \supset Gx)$", from which it follows that *objectively there is no logical difference between causal regularities and accidental generalities*. Objectively, then, there is no difference of a logical sort between *post hoc* and *propter hoc*. Yet, such a distinction is to be drawn, as Hume recognizes,[31] so he raises the question, "What is our idea of necessity, when we say that two objects are necessarily connected together."[32] What he finds is that when two sorts are causally connected, upon the appearance of an object of the one sort "the mind is *determin'd* by custom to consider its usual attendant" and that it is this "*determination*, which affords me the idea of necessity."[33] This yields the second definition of 'cause' as "An object precedent and contiguous to another, and so united with it in the imagination, that the idea of the one determines the mind to form the idea of the other, and the impression of the one to form a more lively idea of the other."[34] If a propositional content of the form "$(x) (Fx \supset Gx)$" is such that *as a matter of psychological fact* it is used to support assertions of subjunctive conditionals ("the idea of the one determines the mind to form the idea of the other") and to make predictions ("the impression of the one [determines the mind] to form a more lively idea of the other") *then* the assertion of that proposition is the assertion of a causal generality. A generality is lawlike just in case it is in fact used to predict and to support subjunctive conditionals; otherwise it is a statement of "mere regularity". The connection between lawlike generalities and subjunctive conditionals was noted above. Chisholm once put the point very neatly:

Can the relevant difference between law and nonlaw statements be described in familiar terminology without reference to counterfactuals. without the use of modal terms such as 'causal necessity", "necessary condition", "physical possibility", and the like, and without use of metaphysical terms such as "real connections between matters of fact"? I believe that no one has shown the relevant difference *can* be so described.[35]

Hume agrees that the law/nonlaw distinction cannot be made without reference to counterfactuals, and without reference to a notion of "causal necessity". But he denies the need to introduce non-empirical "real connections". The latter would introduce an *objective* difference between lawful and nonlawful regularities, and such a difference is denied by the Humean. Rather than put "causal necessity" into the objective facts, the Humean puts it on the subjective side, and identifies it with a preparedness to assert counterfactuals. Far from denying the connection between laws and counterfactuals the Humean uses the connection to *define* lawlikeness. Or, more accurately, *he takes the preparedness to use a generality to support the assertion of subjunctive conditionals as the defining characteristic of a lawlike or "causally necessary" generality*. Thus, a generality is causal or not depending upon its psychological context.[36] Lawlikeness is a matter of the propositional attitude which obtains with respect to the generality. The generality is lawlike if and only if it is not merely believed or asserted, but asserted in a certain more specific way, namely, with a preparedness to take risks with it, that is, to predict and to assert subjunctive conditionals. We may call this the law-assertion attitude.

Kneale[37] has criticized a view of Popper's that is somewhat akin to the Humean account of natural laws or causes. Popper[38] once proposed to analyze contrary-to-fact conditionals in terms of universal material implications of the form "$(x) (Fx \supset Gx)$", where 'F' and 'G' are not lists of proper names but rather unrestricted descriptive predicates. Given the connection between laws and subjunctive conditionals, it follows Popper is proposing to transcribe natural laws into universal material implications. Kneale argued against such transcriptions by means of two examples. The first[39] turns on the logical point that "$(x)(Fx \supset Gx)$" is logically equivalent to the statement of non-existence of things which are both F and not-G: $\sim(\exists x)(Fx \& \sim Gx)$. Kneale lets '$Fx$' be '$x$ is a chain reaction of plutonium' and 'Gx' be 'x is outside a strong steel shell containing heavy hydrogen'. He then considers the statement that "there never has been and never will be a chain reaction of plutonium within a strong steel shell containing heavy hydrogen", which upon Popper's account, will be the statement of law that "No chain reaction of plutonium occurs within steel shells containing heavy hydrogen". Now, it is at least conceivable that the statement of non-existence is true, but in order to accept that statement we do not need to accept what Popper's account would entail, that there is a law of nature excluding such a combination of events. This is clear since the non-existence of such combinations can clearly be accounted for, on the one hand, in terms of the improbability of such a combination occurring without human planning, and on the other, in terms of the prevalence of the belief that such an event would have disastrous consequences. The Humean response here is that not every statement of non-existence is equivalent to a law of nature, and that, in this case in particular, the generality "No chain reactions etc." is not a law of nature. The Humean can so reply where Popper cannot because he distinguishes, as Popper does not, between universal material implications as such and universal material implications that are laws of nature. The Humean will claim that "No chain reactions etc." is not a law because one has not adopted with respect to it the law-assertive attitude.

In his second example[40] Kneale supposes a musician, lying on his deathbed, composes in his imagination an intricate tune he is too feeble to write down or even speak. In his last moment the thought comes to him that "No human being has ever heard or ever will hear this tune", meaning by "this tune" a certain complex pattern of sounds which could be described in general terms using unrestricted descriptive predicates. Here it is clear the musician means to assert no law of nature. So not every universal material implication is a law of nature. This tells against Popper. But it does not touch the Humean, who will agree with Kneale that the musician's assertion was not an assertion of law. In fact the Humean will argue, a law was not asserted simply because the musician's attitude was just that of assertion, rather than that of law-assertion.

No doubt, however, Kneale would not take the Humean to be really siding with him as against Popper. For, so far as the objective facts are concerned, Popper and the Humean agree: there is nothing more to laws than regularities, universal material implications. Kneale would not agree with Hume that lawlikeness is a subjective matter, not objective, a feature of the logical form of causal propositions. Probably he would disagree for the same reason rationalists tend to disagree with a similar thesis in ethics. The thesis in ethics that value is not a matter of propositional content but rather of the psychological attitude is often called "emotivism". Hume's account of causation may therefore perhaps not unreasonably be characterized as an emotivist account of causation. And, as with those who defend an emovitist theory in ethics, the immediate question raised is this: does not emotivism, when it denies the existence of an objective standard, entail that whether the attitude is adopted is not something that permits of justification? Or, in other words, does not Hume's account reduce causal reasoning to irrationalism?

Now, the reply to this, in Hume's case, as in ethics, is to challenge the presupposition of the objector's question, and ask what might one *reasonably* mean by 'justification" in this context? In ethics, if the emotivist's arguments that objectivism is false are accepted, then it is no serious objection to his account that he leaves no room for an objective justification of value judgments. Of course he has left no room: he has just finished arguing that such justification is not possible. And if it is not possible, it is not reasonable to insist upon it. Whatever justification amounts to in ethics the one thing that cannot *reasonably* be demanded is objective justification. And in the case of Hume's account of causation the same sort of response is called for. *Given* Hume's argument that, objectively considered, all causal assertions are assertions of constant conjunction, *then* it is not reasonable to demand an objective justification for the adoption of the law-assertion attitude towards some generalities, rather than others.[41] What Hume must do is give an alternative account of what justifies adopting the law-assertion attitude. This, of course, he does.

Hume clearly recognizes that there are cases where the law-assertion attitude holds and it is not justified. His discussion of credulity,[42] his discussion of the often adverse effects of education,[43] his discussion of the role of imagination,[44] his discussion of unphilosophical probability,[45] to which a whole chapter is devoted (I, III, xiii), all make evident that the *Treatise* draws a distinction between these cases, where the attitude is unjustifiably held, and cases where it is justifiably held, where its adoption is in accordance with the "Rules by which to judge of causes and effects" which appear in their own chapter (I, III, xv)[46] with that very title.

An assertive attitude is objectively justified if and only if the proposition in question is true. But this holds equally for laws and for "mere regularities": provided the latter are true they may justifiably be asserted. So truth, while *sufficient* to justify the attitude of assertion or mere assertion, is only *necessary* to justify the attitude of law-assertion. Thus, if a generality is false one is objectively unjustified in holding towards it the law-assertion attitude. This will be so even if the other necessary conditions of justification (whatever they may be) are all fulfilled. And it will be so even if we have *all possible reason* to believe that that necessary condition of truth is fulfilled.

A generality is a statement about a total population. Normally all one ever observes is a sample. Between sample and population there is a logical gap. This gap is such that properties which are regularly associated or constantly conjoined in the observed sample may not be constantly conjoined in the population. Hume argues for the existence of this logical gap,[47] which he elsewhere expresses as the principles "That there is nothing in any object, consider'd in itself, which can afford us a reason for drawing a conclusion beyond it" and "That even after the observation of frequent or constant conjunction of objects, we have no reason to draw any inference concerning any object beyond those of which we have had experience,"[48] that is, reason drawn from those objects "consider'd in themselves" as suggested by the preceding principle, rather than "never any sort of reason," for, after all, Hume does go on to give us the "Rules by which to judge of causes" where he sketches the conditions under which one *can* reasonably infer from a sample to a population. That properties are constantly conjoined in a sample is a necessary condition for their being constantly conjoined in the population. A necessary condition for justifiably making a law-assertion is that the generality asserted be true. Given the *logical* gap between sample and population it is *not possible* to *know* whether this necessary condition is fulfilled simply by observing that regularity obtains in the sample. The *best* we can do is know a necessary part of this necessary condition obtains, namely, that the regularity holds in the sample. This is the *best* we can do, short of omniscience. And since it is the best we *can* have, we must make do with it. *If we observe a regularity holds in a sample we thereby have every objective reason it is (at that point)*[49] *possible to have to justify one in believing the regularity holds in the population.* Subjectively the only and best objective evidence a regularity obtains overall is that we have observed it to obtain among the facts we already know. So, subjectively we may be justified in asserting a generality when objectively the assertion is not justified.[50] Still, if we have done the best we can do, if we assert only when we are subjectively justified, we cannot be blamed for not having done more, even where we are objectively unjustified. Fallibility is not a vice.

We have still not distinguish causal from "mere" regularities. The remarks

THE DEDUCTIVE MODEL OF EXPLANATION

just made drawing attention to our fallibility, apply equally to both sorts of regularity. An observed constant conjunction is a necessary condition for our being subjectively justified in adopting the law-assertion attitude towards a generality, but it is not a sufficient subjective condition.

Let me try to bring out what is relevant by commenting on an argument of Ducasse.[51] According to Hume, wherever one asserts a causal connection, there one asserts a generality. Ducasse argues that in fact this is false and therefore Hume's account is inadequate. Ducasse represents Hume as holding that a causal assertion involves nothing more than the mere assertion of a matter-of-fact regularity. This is, I've argued, a serious misrepresentation of Hume's view. But as Hume's account holds the assertion of a generality is part of what is involved, Ducasse's argument, if sound, still counts against the position. Alas, it is not sound.

Ducasse describes a situation. In the situation a causal assertion is made. From his description Ducasse draws certain conclusions which have the anti-Humean import that the causal assertion in the situation does not involve the assertion of a generality. But Ducasse's description is by no means complete. By filling in certain details which may quite reasonably be assumed to be part of the situation, it is possible for the Humean to deny that Ducasse's conclusions may be inferred from the situation he describes. The reasonable filling out of Ducasse's description enables the Humean consistently to maintain that the causal assertion made in the situation does, after all, involve the assertion of a generality.

Ducasse's argument is this:

> I bring into the room and place on the desk a paper-covered parcel tied with string in the ordinary way and ask the students to observe closely what occurs. Then, proceeding slowly so that observation may be easy, I put my hand on the parcel. The end of the parcel the students face at once glows. I then ask them what caused it to glow at that moment, and they naturally answer that the glowing was caused by what I did to the parcel immediately before.[52]

He concludes: In this case it is clear that what the spectators observed, and what they based their judgment of causation upon, was not repetition of a certain act of mine followed each time by the glow, but *one single case* of sequence of the latter upon the former. The case, that is to say, does not conform to Hume's definition of causation as constant conjunction but is nevertheless judged by unprejudiced observers to be a case of causation.

Davidson has recently accepted this argument of Ducasse.[53] He accepts Ducasse's conclusions we have in his example a *singular* causal statement, that this statement entails no statement of law, i.e., that no law is meant or known when the singular causal judgement is made. He does believe, however, that no singular causal statement is true unless there is a law to be known. On the basis of this point he proposes a sort of reconciliation between Ducasse and Hume.

> The reconciliation depends . . . on the distinction between knowing there is a law 'covering' two events and knowing what the law is: in my view, Ducasse is right that singular causal statements entail no law; Hume is right that they entail there is a law.[54]

We shall have more to say about Davidson's views later.[55] Right now I merely wish to make clear that the Humean we are talking about (and the historical Hume is in this respect a Humean) maintains against Davidson that the singular causal judgement made in Ducasse's example not merely entails there is a law, but entails a definite law, that, in other words, Ducasse's students do assert a definite generality when they assert of Ducasse that his then touching the box caused it to glow. The Humean, if he is wise, should agree with Ducasse (and Davidson) that in this case one does (correctly) judge a causal relation obtains. But he can and should deny Ducasse's conclusion. He does this by arguing that what the students *infer* from what is observed is that a *generality* is true. Or, rather, they deduce from (a) *two* events and (b) background knowledge that a certain kind of event is sufficient for another kind (relative to a field).

Let H = hand on box, G = box glowing. The field F^{55a} is objects of this boxy sort, which are able to glow, etc. Let this particular box be a. Then we have *two* events: a at t_1 when H and G are absent and a at t_2 when both are present.

On the basis of background knowledge we can list a set of conditions the presence or absence of which would be relevant to the presence of G in a in field F. Thus we have wires (W), antenna (A), internal structure (I) (which we know is there though we may not know its details). On this basis we can use the familiar Millian Method of Difference, which is the Sixth of Hume's "Rules by which to judge of causes". We apply this Method according to the data:

	G	H	W	A	I
a at t_1	a	a	a	a	p
a at t_2	p	p	a	a	p

and infer that in this context being H is sufficient for being G, that, relative to the field F, $(x)(Hx \supset Gx)$ obtains. Here, the context is defined by the field *F and the background knowledge* that implies there is a sufficient condition for G and that it is one of or built up out of H, W, A and I.

On the basis of this knowledge and these observations we deduce being H at t is sufficient for being G at t (relative to F). In other words, we do know a causal relation in this context, but this involves a generality which we have discovered in the context. We use events a, t_1 and a, t_2 to arrive at the law. We then use this law in asserting a causal relation between $G(a, t_2)$ and $H(a, t_2)$. We use the events to infer the law, and then use the law to explain the events.

This reply to Ducasse depends upon the perfectly reasonable assumption that the students enter the situation with a certain amount of *background knowledge*. This knowledge permits a number of hypotheses to be formulated about the circumstance which confronts them. They then reason from observational data which they *come to acquire* to the conclusion that, of the possible hypotheses, one alone is consistent with that data. The uneliminated hypothesis is, of course, a generality and as such asserts *more than* what is in the observational data available at the end of the situation. The point is not that we here overcome the logical gap between observed sample and total population, but simply that, of the possible hypotheses, this one alone is subjectively justified.

Hume notes carefully that there are many situations in which the mind is confronted with a set of alternative hypotheses.[56] *The choice between these may be made by collecting relevant observational evidence, or by some other principle. If it is on some other basis than observational evidence then the resulting law-assertion is unjustified.* Hume cites the principle that we choose as worthy of law-assertion the hypothesis we want to be true, quoting Cardinal de Retz on the principle that the wish is the father of the belief, "that there are many things in which the world wishes to be deceiv'd."[57] Where the world wishes to be deceived it can avoid trying to gather together the *evidence* relevant to reasonably deciding between the possible hypotheses.[58] In the chapter "of unphilosophical probability"[59] a number of such unreasonable principles are mentioned. Those who desire a *reasonable decision* among contrary hypotheses must go out and *actively collect* additional observational evidence that will permit a decision to be made. Very often the data does not make itself as readily available as it did to Ducasse's students. The data which is given is often acquired only with great difficulty. As Hume puts it, directly after stating the "Rules by which to judge of causes":

There is no phenomenon in nature, but what is compounded and modify'd by so many different circumstances, that in order to arrive at the decisive point, we must carefully separate what is superfluous, and enquire by new experiments, if every particular circumstance of the first experiment was essential to it. These new experiments are liable to a discussion of the same kind; so that the utmost constancy is requir'd to make us persevere in our enquiry, and the utmost sagacity to choose the right way among so many that present themselves.[60]

One makes a reasonable decision among alternative possible hypotheses when one has *actively* sought out such data as would permit the data to make such a decision and render exactly one of the hypotheses subjectively worthy of law-assertion.

The set of possible hypotheses is determined by background knowledge. This knowledge will be more or less generic. Ultimately there is one major premise, "the supposition, that the future resembles the past",[61] which Hume includes as his Rule 4, that "The same cause always produces the same effect, and the same effect never arises but from the same cause."[62] This is, of course, J. S. Mill's "Axiom of the uniformity of the course of Nature", which is "the ground of induction."[63] *In the beginning* this background knowledge is acquired by the mechanism of association. Hume introduces considerations from psychology at this point, specifically, from learning theory. His suggestion is that learning *initially* is a matter of what we now call classical conditioning. But given his reference to the idea that the wish is often father of the belief, it is clear he knows reinforcement plays its role in learning also; the same is made evident also by his later systematic references to utility. But, minimally, observation of a constant conjunction gives rise to the habit of expecting the effect, given the cause:

... after a frequent repetition, I find, that upon the appearance of one of the objects, the mind is determin'd by custom to consider its usual attendant, and to consider it in a stronger light upon account of its relation to the first object. 'Tis this impression ... or determination, which affords me the idea of necessity.[64]

The observation of a constant conjunction causes[65] one to acquire the law-assertion attitude with respect to the generality that extends the same association to the whole population.

But this passive acquisition of law-assertive habits later comes to be replaced by the active adjustment of assertions to deliberately-sought-out data. Thus, suppose we have a class K of objects, and that we are interested in the property G and what causes its presence in objects of sort K. We might very well have as a piece of background knowledge that it is some one thing

THE DEDUCTIVE MODEL OF EXPLANATION   21

of genus $\mathscr{F}$ that is necessary and sufficient for G being present in a K; and further that the only $\mathscr{F}$s in Ks are F, F' and F''. For example, G may be a type of disease had by persons K, and we know that G is caused by some species or other of germ, that is, some species under the genus $\mathscr{F}$, without knowing just which species F, F', F'' of that genus we have discovered in Ks is actually the cause of that disease. One might now perform the following experiment to determine whether F is the cause of G in Ks. One locates two individuals, a_1 and a_2, which enable one to conclude F is sufficient for G. The set-up will yield these data

	F	F'	F''	G
a_1	F	T	T	F
a_2	T	T	T	T

This eliminates F' and F'' as possible conditioning properties of G, and *given the background knowledge* that the cause of G is one among these properties it follows that F is sufficient for G, that '$(x)(Fx \supset Gx)$' is worthy of the law-assertive attitude, at least so long one is talking only about Ks. a_1 and a_2 need not be two individuals. They could be the same individual at different times. The experimenter brings about a change in a from being not-F to being F, and ensures it changes in no other relevant respect. If a then changes from not-G to G it may be concluded F causes G. The similarity of the above table to one we used in our discussion of Ducasse is obvious and relevant. But the experiment need not involve the active intervention by the experimenter; a_1 and a_2 might be quite distinct Ks. If the experimenter now proceeds to locate a third individual a_3 such that

	F	F	F'	F''	G
a_3		T	F	F	T

he may conclude that G is not just caused by F but that it is caused only by F. In terms of one example, he can conclude the disease is present only if germs of species F are present. Thus, he concludes that G is sufficient for F. So, his research, guided by his background knowledge, determines that among the three possible hypotheses

(2.1) $(x)(Kx \supset (Fx \equiv Gx))$
        $(x)(Kx \supset (F'x \equiv Gx))$
        $(x)(Kx \supset (F''x \equiv Gx))$

only the first is subjectively worthy of the law-assertive attitude.

But induction is open. It always remains possible that a proposition, however worthy in terms of evidence available that it may be to adopt the law-assertive attitude, may turn out to be objectively unworthy of that attitude. It may turn out that, after an initial period of success with the proposition that being F is the only and invariable cause of K's being G, our experimenter discovers a K or group of Ks for which this does not hold. In terms of our example, what our scientist discovers is that germs F are not, after all, necessary and sufficient for the disease G in Ks, that, in spite of a consistent connection in the past between the presence of the germs and the occurrence of the disease, there is a group of Ks in the members of which the germs do not produce the disease. He then forms the hypothesis that the class of Ks may be divided non-enumeratively into two exhaustive subclasses, those for which F is necessary and sufficient for G and those for which it is not. This hypothesis is to the effect that *there is* some characteristic f such that, for all Ks, F is necessary and sufficient for G if and only if f is present:

$$(\exists f)(x) \{Kx \supset (fx \equiv (Fx \equiv Gx))\}$$

The 'non-enumerative' is required since there is trivially always such an f, namely, the characteristic of being identical with one or another of the individual Ks in which F is necessary and sufficient for G. If F was necessary and sufficient for G in all and only the individuals a_1 and a_2 and these were both Ks, then the characteristic defined by

$$x = a_1 \vee x = a_2$$

would be an f satisfying the above formula. So we are placing restrictions upon the properties over which the variable f ranges, and the hypothesis of our scientist would more accurately be formulated as something like

(2.2) $\quad (\exists f) \{ \mathscr{C}f \,\&\, (x) [Kx \supset (fx \equiv (Fx \equiv Gx))] \}$

where '$\mathscr{C}$' designates the limitations we are placing upon the character f. On the basis of this hypothesis our scientist now proceeds to try to discover just which $\mathscr{C}$ it is that makes this hypothesis true. He tries to discover a characteristic C of genus $\mathscr{C}$ for which the instantiation

(2.3) $\quad (x) [Kx \supset (Cx \equiv (Fx \equiv Gx))]$

of the hypothesis holds. (Here, we mean the latter formula instantiates the hypothesis in the way in which 'ga' is an instantiation of '$(\exists x)(fx \,\&\, gx)$'

THE DEDUCTIVE MODEL OF EXPLANATION 23

given that 'fa'.) If the experimenter discovers the appropriate individuals then the logic of experiment, i.e., Mill's Methods, will lead him to knowledge of the factor C. The logic will be slightly more complicated than in the example where F was inferred to be the only and invariable cause of G, but not essentially different: the same principle of elimination will be the operative logical tool. With these data and the principles of experimental logic, our scientist will discover, let us say, the physiological basis which renders certain persons immune to the disease he has been startled to discover is not invariably consequent upon the presence of the germs which cause that disease. By such methodical acquisition of knowledge, we come to know the conditions which limit the range of applicability of imperfect knowledge. The methodical acquisition of evidence leads us to knowledge which is better than the knowledge we previously had.

1.3. THE EVIDENTIAL WORTH OF LAW-ASSERTIONS

There are two aspects to this notion of 'better'. One is relevant here, and the other is worth noting because subsequently much indeed shall be made of it.

After our initial research it was reasonable, on the basis of the evidence then available, to adopt the law-assertive attitude with respect to (2.1). It turned out, however, that (2.1) was, as a matter of objective fact, false. Objectively, then, it was not — never was — worthy of the law-assertive attitude. Our belief in (2.1) was radically defective as knowledge: it never was knowledge. If we suppose (2.3) to be in fact true, then our adoption of the law-assertive attitude towards it consequent upon our further research turns out not merely to be justified by available evidence but also justified relative to the objective facts. In this sense, then, the knowledge we have when we law-assert (2.3) is better than the knowledge, or rather, the purported knowledge we had when we asserted (2.1).

On the other hand, however, we must recognize a sense in which the law-assertion of (2.1) was not so bad after all. Though false, it was not, as one says, wholly false. The problem is to make clear what is involved in this idea. It must be granted at the beginning that the idea of "degrees of falsity" makes no sense.[66] A proposition is either true or false, and that is all there is to it. And in this sense (2.1) is just false. What sense, then, attaches to the idea it is not "wholly false"? The idea involved seems to be a comparison of (2.1) and (2.3). What (2.3) does is show that, while (2.1) is literally false, it nonetheless holds "under certain conditions." More strongly, (2.3) states exactly what those conditions are. In the light of (2.3) what we can say is that the person who asserted (2.1) was asserting unconditionally what was objectively worthy only of conditional assertion. In this sense, then, the law-assertion of (2.1) was never wholly unworthy.

Mill put this point rather neatly when he explained that it is experience which teaches us which regularities hold unconditionally, and which hold only under certain conditions, and that it is experience which further teaches us just what those conditions are:

> ... it is experience itself which teaches us that one uniformity of sequence is conditional and another unconditional .... Though a fact may, in experience, have always been followed by another fact, yet if the remainder of our experience teaches us that it might not always be so followed, or if the experience itself is such as leaves room for a possibility that the known cases may not correctly represent all possible cases, the hitherto invariable antecedent is not accounted the cause; but why? Because we are not sure that it *is* the invariable antecedent.[67]

The distinction between knowing that a regularity holds only under certain conditions and knowing exactly what those conditions are is of considerable importance, as we shall see. It is an important ingredient in the idea that one piece of knowledge can be "better" than another. The distinction, and its role, can be illustrated by (2.2) and (2.3). Both these generalities assert that (2.1) holds "Under certain conditions." But (2.3) makes a stronger assertion than (2.2). (2.2) merely asserts that *there is* a relevant factor of a certain *generic* sort the presence of which is the condition for (2.1) holding. (2.2) does not say *specifically* what that factor is. In contrast, (2.3) does say specifically what that factor is. Our scientist's research was guided by the hypothesis (2.2). This hypothesis, if it be true, guarantees that he will be able to discover specific law of sort (2.3). (2.2) asserts that something like (2.3) is there to be discovered.[68] The gathering of evidence aims at discovering just which specific character of sort $\mathscr{C}$ it is that conditions the causing of G by F. The scientist proceeded on this further research, not because (2.2) was false, as was the case with (2.1), but rather because he wanted better, more specific knowledge than (2.2) provides, a better sort of knowledge that (2.2) itself says is there to be known.

Our scientist adopts the law-assertive attitude towards (2.3). This he does partly on the basis of its predictive successes. But that need not be the only reason; more strongly, if it were the only reason then one would question the worthiness of that attitude in that evidential context. However, suppose the law-assertive attitude is appropriate to (2.2). Observational data which eliminate $\mathscr{C}$s other than C as conditioning properties lead to the belief that C is the only property that could satisfy the conditions (2.2) lays down.

In that case the law-assertive attitude appropriate to (2.2) would turn out to be appropriate also to (2.3). We see once again, as we saw in our discussion of Ducasse, that background knowledge plays a major role in justifying the adoption of law-assertive attitudes.

But what would justify adopting the law-assertive attitude towards (2.2)? As we set out our tale of research, the discovery of the falsity of (2.1) led to the formulation of the hypothesis (2.2). However, the failure of (2.1) to successfully predict hardly entails the truth of (2.2) or renders the latter worthy of law-assertion. Moreover, it is hard to see how (2.2) could directly be put to the test. Predictive success will, of course, tend to confirm (2.2). It predicts a law of the sort (2.3). So the discovery of the latter will tend to confirm (2.2). Or, rather, given $\mathscr{C} G$, (2.3) entails (2.2), so any instance which confirms the former also tends to confirm the latter.[69] But if these instances do not really justify adopting the law-assertive attitude towards (2.3), they equally cannot justify adopting it toward (2.2). And further, they could hardly justify adopting (2.2) as a guiding hypothesis prior to the discovery of (2.3). The search for the data that lead to the acceptance of (2.3) can be said to be reasonable in its aim only if the guiding hypothesis which defines that aim is itself rationally acceptable, and not merely some flight of fancy. What, then, justifies law-asserting (2.2)?

This question is complicated by the fact that (2.2) is generic rather than specific. Asking for data to confirm it prior to the discovery of (2.3) is exactly like asking for data to confirm '$(\exists x)(fx)$', e.g., 'There are abominable snowpersons', prior to the discovery of some particular a which is f, some particular abominable snowperson. Now, the discovery of the falsity of (2.1) suggested the hypothesis (2.2) because, in spite of the fact (2.1) was literally false, it nonetheless still seemed to have limited or conditional validity. We were able to make this judgement even without knowing what those limiting conditions specifically were. That judgement is, of course, simply the adoption of (2.2). Our question, then, is this: are there any principles which establish that law-asserting (2.2) not only *seems* reasonable but actually *is* reasonable? Are there any principles by means of which one can establish F is conditionally relevant to G without knowing specifically what those conditions are?

It turns out that in fact techniques have been developed for testing hypotheses of the sort (2.2). These techniques enable one to test the hypothesis (2.2) without actually identifying the factor C. Thus, these techniques enable one to discover a conditioned regularity between F and G without knowing specifically what the condition is.

Let us restrict ourselves to *K*s, and write '(*x*) (*Fx* ⊃ *Gx*)' as '*F* ⊃ *G*'. Finally, let us indicate that we know there is a factor, while not knowing specifically what it is, by the predicate variable '*f*'. Then (2.2) will appear as

(2.2*) $f \equiv (F \equiv G)$

We are interested in establishing whether this obtains. Clearly, the methods of elimination we used in the previous cases cannot be applied directly since we do not know what the conditioning property *F* is. What we must do is set up experimental situations that take account of the presence and the absence of the unknown factor *f*. In order to establish the relevance of *F* to *G* given *f*, we need a pair of individuals a_1 and a_2 which *are f* and which yield the data of the following table

		F	G
(T$_1$)	a_1	F	F
	a_2	T	T

as in the case of testing (2.1). The difficulty is to find the a_1 and the a_2 which are *f*, since we do not know specifically what character *f* is, and so cannot identify individuals as being or not being of that kind. The problem of finding data to verify (2.2*) is equivalent to finding data to verify

(2.2**) $F \equiv (f \equiv G)$

since the latter is logically equivalent to the former. This indicates we may treat *F* as the conditioning property and $(f \equiv G)$ as the conditioned property, and a set of data analogous to those of (T$_1$) would confirm (2.2**).

		F	$(f \equiv G)$
(T$_2$)	a_1	F	F
	a_2	T	T

Now this data would confirm that *F* is relevant to *G* given *f* provided we knew the absence of the conditioned proeprty in a_1 was due to the absence of *G*, and its presence in a_2 was due to the presence of *G*. If we knew this, we would have the data summarized in (T$_1$) and could then affirm *F* as the conditional cause of *G*. Now, $(f \equiv G)$ is absent when *G* is absent and present

THE DEDUCTIVE MODEL OF EXPLANATION

when G is present just in case f is present. So we would confirm the relevance of F to G if our data looked like

$$
\begin{array}{cccc}
 & & F & (f \equiv G) \\
(\mathrm{T}_3) & a_1 & F & \mathrm{T\ F\ F} \\
 & a_2 & T & \mathrm{T\ T\ T}
\end{array}
$$

But, again, we do not know what f is, and so cannot identify its presence. We must take into account cases where f is absent when F is absent and cases where f is absent when F is present:

$$
\begin{array}{cccc}
 & & F & (f \equiv G) \\
 & a_1' & F & \mathrm{T\ F\ F} \\
 & a_1'' & F & \mathrm{F\ F\ T} \\
(\mathrm{T}_4) & a_2' & T & \mathrm{T\ T\ T} \\
 & a_2'' & T & \mathrm{F\ T\ F}
\end{array}
$$

(T_4) suggests we proceed as follows. We should select a group in which F is absent, *and in which, although not knowing what f is, we may reasonably expect it to be present in some of the group, but absent in the rest.* Similarly, we should select a group in which F is present and in which we may expect f to be present in some, absent in the rest. What we require is that the number in each group in which f is present be the same. We expect f to be present in the same propertion in the a_1 group as in the a_2 group. *If it turns out that G is present in the a_1 group (at a_1'') in the same proportion as it is present in the a_2 group (at a_2''),* then we may conclude that F conditionally causes G. What we have to do is ensure the groups a_1 and a_2 satisfy the indicated conditions. If the assumption is granted that f is randomly distributed in the Ks then it is easy to select the groups, namely, randomly, and if they are large enough then the probability will be that the proportion of fs in each group will correspond to the proportion of fs in the population as a whole. This principle of randomization in fact turns out to work fairly well.[69a]

The *principle of randomization* may be stated thus: since, in random procedures, every member of a population has an equal chance of being selected, members with certain distinguishing characteristics – male and female, Republican and Democrat, extravert and introvert, high and law intelligence, and so on and on – will, if selected, probably be counterbalanced in the long run by the selection of other members with the "opposite" quantity or quality of the characteristics. We might say that this is a practical principle that indicates what happens. We will not say it is a law of nature. It is simply a statement of what usually happens when random procedures and used.[70]

Randomization is a way of using the methods of elimination, ensuring their automatic working even when one (or more) of the properties they are working upon remains unknown.

Systematic research can thus proceed even where some of the relevant variables are unknown. But what makes it reasonable to put (2.2) to such a test? One must answer in the same way one answered with respect to (2.1): background knowledge. (2.2*) must be thought of as competing with, say,

$$f \equiv (F' \equiv G)$$
$$f \equiv (F'' \equiv G)$$

just as (2.1) had similar competitors. We have this group of hypotheses which research proceeds to put to the test. That just this group is relevant will be justified by background knowledge of the sort

(3.1) $\quad (\exists f)(\exists h) \{ \mathscr{C}f \, \& \, \mathscr{F}h \, \& \, (x)\,[Kx \supset (fx \equiv (hx \equiv Gx))] \}$

and

$(h)(\mathscr{F}h \supset : h = F \vee h = F' \vee h = F'')$

We indicated how, knowing (2.2) one could proceed to the law-assertion of (2.3). We have now indicated how (2.2) could come reasonably to be law-asserted. It turns out that in order to law-assert (2.2) we must law-assert (3.1). And once again, we must ask the question: what evidential basis would render (3.1) worthy of law-assertion? Well, one way would be for it to follow from some more comprehensive hypothesis. Thus, the disease G may itself be a species of a kind $\mathscr{G}$. For other members of this genus we may well have discovered also that for these other kinds of disease there are conditions $\mathscr{C}$ the absence of which confers immunity. These discoveries will have confirmed the hypothesis

(3.2) $\quad (g) \, \{ \mathscr{G}g \supset (\exists f)(\exists h) \, [\mathscr{C}f \, \& \, \mathscr{F}h \, \& \, (x)\,[kx \supset (fx \equiv (hx \equiv gx))]$

Since (3.2) has been confirmed for $\mathscr{G}$s other than G, we will have reason for predicting it will hold for G. (3.2) and $\mathscr{G}G$ predict that (3.1) obtains. That is, on the basis of law-asserting (3.2), we can law-assert (3.1). The observational data which establish that Fs are the only $\mathscr{F}$s causing G, now justify the law-assertive attitude being as appropriate to (2.2) as to (3.1). In this way, the law-assertion of the background knowledge leading to the discovery of (2.3) can in its turn be justified on the basis of the law-assertion of other background knowledge.

But, once more one must ask, what justifies the law-assertive attitude with respect to (3.2)? One can retreat further up a hierarchy of laws to a more and more generic level. Ultimately one comes to the most generic law of all: for every event there is a cause. As Hume indicates, this is the ultimate principle justifying adopting the law-assertive attitude. Again the question can be raised, however. At the level of (2.3) mere predictive success did not justify the law-assertive attitude. Reference had to be made to background knowledge, some more generic hypothesis – in the case of (2.3) reference was made to (2.2). But as one proceeds to the top of the generic hierarchy such background theory ceases to be available, for the simple reason that at the top of the generic hierarchy there would be no more generic principle to which appeal could be made. However, upon reflection this difficulty disappears. As one proceeds up the generic hierarchy the need to rely upon background knowledge to justify the law-assertive attitude becomes less necessary, and it suffices to rely upon predictive success alone. Hume is perhaps not as clear on this as one would like, but certainly Mill is clear enough that the methods of elimination, the logic of experiment, requires background theory, that the ultimate background theory is the principle of causation, and that for this principle induction by simple enumeration suffices to justify adopting the law-assertive attitude.

Stuart Mill explicitly adopts Hume's identification of causal belief with association.[71] According to Hume the truth of a causal belief is a matter of *de facto* regularity (Hume's first "definition" of "cause"). What makes the belief one such that the generality is *propter hoc* rather than *post hoc* is that an association obtains between the idea of the antecedent and that of the consequent (Hume's second "definition" of "cause"). An observed constant conjunction results in the coming to be of the association of ideas. This does not account for the difference between prejudice and justified causal beliefs, so Hume goes on to state his "Rules" for judging of causes. Stuart Mill similarly insists that observed association is not sufficient ground or evidence for (causal) belief.[72] The tendency of the mind to move from observed conjunction to causal belief – its generalizing tendency – must be disciplined.[73] The mind eventually arrives at rules which enable it reasonably to judge of causes: the mind " ... presently discovers that the expectations which are least often disappointed are those which correspond to the greatest and most varied amount of antecedent experience .... In other words, it considers the conditions of right inference from experience; and by degrees arrives at principles or rules, more or less accurate, for inductive reasoning."[74] Nagel, in his criticisms of Carnap's views on "inductive logic",[75] has recently similarly emphasized the role of variety in evidence, where 'variety' here means 'variety of *kinds* of individuals'. Stuart Mill states in detail the relevant rules in his *System of Logic*,[76] and from there they have descended to contemporary logic texts as "Mill's Methods", though of course Bacon had previously stated them, as had Hume in his "Rules".[77] In any case, however, where

> ... belief is really grounded on evidence, it is grounded in the ultimate result, on the constancy of the course of nature .... Whatever it is that we believe, that justification of the belief must be, that unless it were true, the uniformity of the course of nature would not be maintained .... What we call evidence, whether complete or incomplete, always consists of facts or events tending to convince us that some ascertained general truths or laws of nature must have proved false, if the conclusion which the evidence points to is not true.[78]

"Mill's Methods" are the methods of eliminative induction. Consider again the case of necessary and sufficient conditions, and again restrict ourselves to Ks. There is a property G for which there is a set, say, F, F', and F'', of possible conditioning properties. An individual a_i such that G is present in it and F_i is absent eliminates F_i as a necessary, and therefore as a necessary and sufficient condition of G. An individual a_j such that G is absent and F_j is present eliminates F_j as a sufficient, and therefore as a necessary and sufficient condition. Suppose we observe a set of individuals that eliminate all possible conditioning properties save (say) F. All the examined individuals will be such that for each in which G is present F will also be present, and for each in which G is absent F will also be absent. What this means is that of the set of all possible hypotheses [1], '$(x) (Fx \supset Gx)$', alone is both confirmed and uneliminated.[79] ([1] is (1.1), given our restriction to the class of Ks.) *If* certain assumptions are made then it is possible to conclude that this uneliminated hypothesis is true.[80] One needs to assume that there is a necessary and sufficient condition for G and that it is one of the Fs. More explicitly it is necessary to assume that [A] At least one F is a necessary and sufficient condition of G; and that [B] At most one F is a necessary and sufficient condition of G. Under assumptions [A] and [B], together with the observed individual facts, it follows that [1] alone of the set of possible hypotheses is true.

As we have seen, this sort of inference is not unusual in scientific practice. We considered the case of diseases, where we had a number of species of "symptoms", which fall within a certain genus. About these there is the hypothesis that they are caused by germs. The task of the researcher is to isolate for each disease the unique species of germ which is the cause of that disease. Here the working hypothesis is [C]: For each specific symptom g of genus $\mathscr{G}$, there is a unique species f of germs of genus $\mathscr{F}$ such that for any human the presence of f is necessary and sufficient for the presence of g. ([C] is (3.2) except for the consideration about $\mathscr{C}$'s, which we are here dropping because it would only complicate the point we wish to make.) If we now have a particular disease G such that [D] G is of genus $\mathscr{G}$, then we can deduce [E] there is a unique species f of genus $\mathscr{F}$ such that f causes G in humans. The medical researcher is interested in the bacteriological causes of disease G. The law [E] tells him that there is a set $\mathscr{F}$ of such possible causes. Extensionally, $\mathscr{F}$ is a set: F, F', F''. [E] states that exactly one member of this set is necessary and sufficient for G. Or, in brief, [E] asserts the conjunction of [A] and [B]. Research then proceeds to eliminate the false hypotheses of the set and to isolate that definite species of germ that causes the disease. Various experiments are arranged. Individuals are examined. For each of these individuals F alone is present and absent according as G is present or absent. So upon the assumption [H] that F is of genus $\mathscr{F}$ we deduce that [1] is the correct hypothesis: the germ F is the cause of G.

Represent the statements of individual fact by [I]. The truth of these is known by observation. Assume we know [D] and [H] by observation also. The inferences we have gone through depend upon the necessary truth of

[J] $[E] \supset ([H] \supset ([I] \supset [1]))$

and of

[K] $[C] \supset ([D] \supset [E])$

Now, if the consequent of [J] is confirmed then, by what philosophers have come to call the converse consequence condition of confirmation, we have thereby confirmed [E]: the necessity of [J] ensures this. To confirm the consequent of [J] one must confirm both [I] and [1]. [I] is confirmed by observation. And, since [I] are instances of [1] we have in [I] confirming instances of [1]. We assume [H] is also confirmed by observation. Whence the consequent of [J] is confirmed and thereby so is [E]. And with this confirmation of [E] we have also confirmed [C], given the necessity of [K] and the converse consequent condition. Thus, the actual isolation of the species of germ F confirms the hypothesis [C] which guided our research. Antecedently to the isolation of the cause of disease G the hypothesis [E] can be rendered antecedently plausible by other confirmations of [C]. We may suppose we have had various other diseases G', G'', of the genus $\mathscr{G}$ for which science has succeeded in isolating the relevant causes, the relevant species of germ. These successes will have tended to confirm [C]. We assume [D] to be confirmed. Then, by the necessity of [K] and what we have come to call the (special) consequence condition of confirmation, it follows we have confirmed [E].

The consequence and converse consequence conditions of confirmation are nowadays controversial.[80a] Or at least, purely formal versions of confirmation theory tend to call them into question. Given the plausibility of these conditions – the obvious fact that they *do* play a role in science – I myself should say that this calls into question the idea of formalizing confirmation theory. For what we are about, however, it suffices to note that Stuart Mill accepts both these not implausible conditions.[81] Or at least, his discussion makes it clear that he accepts as cases of confirmation what *we* should characterize as cases instancing the consequence and converse consequence conditions. I emphasize this last point because I think at least part of the reason why his position is thought foolish is a failure to attempt such translation of his terminology as could reasonably be suggested as necessary.

Consider the proposition [α] John has the property of being fathered uniquely. In order to verify [α] it is necessary to examine a variety of individuals (*not*: variety of kinds). Many identificatory hypotheses will be conceivable. The relevant evidence will have to be considered in all its its instances, and the false hypotheses will have systematically to be eliminated. Contrast this to the case where [α] is generalized to [β] everyone has the property of being fathered uniquely. In order to confirm [β] one would look at its instances. One such instance is [α], and there will be obviously similar instances about individual persons Bill, Jones, Zavier, Leo Straus, Raymond Lull, etc., etc. This pattern with respect to the confirmation of [β] is not unreasonably described as one of enumerating confirming instances. To set about confirming [α] is to set about examining a variety of instances; to set about confirming [β] is to set about enumerating instances. [E] ascribes a property to G in the way [α] ascribes a property to John. (Assume, again, each F_i is known to be $\mathscr{F}$.) Confirming [E] requires one to examine a variety of *kinds* of individuals. For, confirmation of [E] requires elimination of hypotheses, hypotheses which identify the bacteriological cause of the disease with different species of germ. The various identificatory hypotheses are eliminated until [1] alone remains. [C] generalizes from [E] over all diseases of the genus $\mathscr{G}$, in the way in which [β] generalizes from [α] over all individual persons: G has the property of being caused by a unique $\mathscr{F}$, G' has the property of being caused by a unique $\mathscr{F}$, G'' has the property ..., and so one concludes that [C] all $\mathscr{G}$ have the property of being caused by a unique $\mathscr{F}$. In this case confirmation is reasonably described as one of confirmation by simple enumeration. In the case of [E] and [α] one moves from an uneliminated identificatory hypothesis via the converse consequence condition to [E] and [α]. And then, again via the converse consequence condition, one moves to [C] and [β]. The logical forms involved are rather complex (we know since Russell)[82] but what makes the "variety vs. simple enumeration" description appropriate can, I think, be seen when one recognizes that [E] and [α] are confirmed only if one member of an exclusive disjunction is true, while [C] and [β] are confirmed only if every member of an indefinite conjunction is true.

We begin with a set of possible hypotheses, of which [1] was a member. These were gathered together as relevant possibilities by [E]. This role of [E] was justified by virtue of its being an instance of [C]. Now, [C] itself will be a member of a set

[S]
All $\mathscr{G}$ have a unique bacteriological cause (=[C])
All $\mathscr{G}$ have a unique viral cause
All $\mathscr{G}$ have a unique psychosomatic cause
. . .

These hypotheses will have been gathered together into a set of relevant possible hypotheses by an over-hypothesis of the same sort as [E]. The confirmation of [E] for G, of the hypothesis for G', of the hypothesis for G'', . . . , thus not only yield the confirmation of [C], but also serve to eliminate the members of [S] other than [C]. In other words, the enumeration which confirms [C] is not just a simple enumeration after all, but one in which the eliminative mechanism is also operating. It operates through the presence of the [E]-sort over-hypothesis. The confirmation of this over-hypothesis will involve not just simple enumeration but also considerations involving variety. Furthermore, the hypothesis of the [E]-sort that gathers the set [S] together will justifiably play this role by virtue of being an instance of a more general hypothesis of the same sort as [C]. The confirmation of the [E]-sort hypothesis, through an examination of a variety of kinds, will tend to confirm the [C]-sort hypothesis which generalizes it. The confirmation of the latter will be by simple enumeration. Unless, of course, it too, is one of a family of relevant hypotheses. One will reach an hypothesis whose confirmation is appropriately described as by simple enumeration alone only at the very top of the hierarchy. One will have to move up the species-genus hierarchy, from G, to $\mathscr{G}$, to the supreme genus.

What will be the appropriate hypothesis at the top of this generic hierarchy? Surely none other than the *Principle of Causality:*[83] Every event has a unique cause.

This principle may be expressed in more detail as this: For each species *g* of event, there is a unique species *f* such that for any event *f* is necessary and sufficient for *g*. In order to avoid complete trivialization qualifications must be added, corresponding to the '$\mathscr{F}$' in [C], which place restrictions upon the *f*. Hume recognizes the need for these qualifications in his first three rules, and so does John Stuart Mill. The basic idea is that the operation of causes is a matter of *continuous* causal processes. The idea of science (of which, more presently)[84] is that with respect to each event one can explain it by what Bergmann has called "process knowledge". This is Russell's idea[85] of replacing our quotidien language of "cause" with that of law or functional connection, where the functions satisfy conditions of continuity and differentiability with respect to time, so that processes can be represented by systems of differential equations. This idea — essentially exemplified by Newton's explanation of the behaviour of the solar system — is, I believe, the idea Hume[86] and Stuart Mill[87] were attempting to articulate, though as Russell's discussion makes clear, in this they failed. But such a failure should hardly surprise us, since a clear statement, such as Russell's, of the logical structure of process knowledge could not be given until the logical structure of continuous and differentiable functions had been clarified, a task completed only with the monumental work of Russell himself. So I suggest it is not unreasonable to read Mill as intending the Principle of Causality to state the principle of determinism, that each event can be explained by subsumption under a process law, that for each event there is a process law which covers it.

So understood, the Principle of Causality is indeed an empirical statement, as Mill claimed, and one for which confirming evidence can be sought, as Mill also claimed. And his description of its confirmation is that this Principle is confirmed by simple enumeration alone, and not through variety, as with laws lower in the generic hierarchy. It is Mill's argument that " ... we are justified in ... holding induction by simple enumeration to be good for proving this general truth [the Principle of Causality], the foundation of scientific induction, and yet refusing to rely upon it for any of the narrower inductions".[88] His position is that induction by simple enumeration " ... though a valid process, ... is a fallible one, and fallible in very different degrees ... ",[89] with " ... the precariousness of the method of simple enumeration ... in an inverse ratio to the largeness of the generalization

THE DEDUCTIVE MODEL OF EXPLANATION   35

... ",[90] and since the Principle of Causality " ... stands at the head of all observed uniformities in point of universality ... " he concludes that it is also at the head of all observed uniformities " ... in point of certainty".[91] Simple enumeration [92] serves to justify our relying upon the Principle of Causality to justify all inferior inductions: "When we have ascertained that the particular conclusion must stand or fall with the general uniformity of the laws of nature – that it is liable to no doubt except the doubt whether every event has a cuase – we have done all that can be done for it".[93]

If certainty does attach to the Principle of Causality the last quote makes clear that, for Stuart Mill, so also does doubt attach to that Principle. He quite explicitly recognizes that this Principle is an empirical generalization,[94] making an assertion about a population, where the evidence justifying its assertion is knowledge of a sample only.[95] *Any* inference from sample to population is open to some doubt, in the sense that it is possible on the basis of the evidence that the conclusion is false. Von Wright [96] has argued in detail that several attempts to "justify" induction are attempts to bridge this logical gap between sample and population, between a subset of conjuncts and the total set of conjuncts of an indefinitely long conjunction. But, as von Wright argues, such an aim is self-contradictory. Which was, of course, Hume's point, and it is still correct. Stuart Mill himself recognizes the point. " ... [The] very circumstance that complicated processes of induction are sometimes necessary, shows that cases exist in which this regular order of succession is not apparent to our unaided apprehension. If, then, the processes which bring these cases within the same category with the rest require that we should assume the universality of the very law which they do not at first sight appear to exemplify, is not this a *petitio principii*?" [97] In terms of our example, if [C] is required if we are to conclude [1], then how can we speak of [1] as providing confirming evidence for [C]? To which the immediate answer is that so long as [1] is uneliminated, the observed individuals do constitute confirming instances of it and therefore (as we would say, by the converse consequence condition) they also tend to confirm [C].[98] And to which the final answer is that the inference, by simple enumeration, to the most universal law, the Principle of Causality, *is* fallible. "In matters of evidence, as in all other human things, we neither require, nor can attain, the absolute. We must hold even our strongest convictions with an opening left in our minds for the reception of facts which contradict them ... ".[99] It is a fact of our humanity that, unlike the Deity, we are not omniscient, but that, like any Diety save Descartes', we are unable to do the impossible. To try to bridge the logical gap, to which Hume so forcefully directed our attention, between sample and population, is to deny the simple fact that we are after all, human beings. We are neither divine, nor even semi-divine, whatever claims to the contrary have been made by Aristotle, Kant, or Whewell. The humane spirit of the Enlightenment enabled men to recognize for the first time this fact about themselves. Mill was the chief representative in his age of that spirit of enlightenment, and his reply to the narcissistic pretensions that we can overcome our humanity is simply to re-affirm that spirit and recall to us the fact our our fallibility.

To say we are fallible is to say we might be mistaken. It is not to say we *are* mistaken. For is it to say we have reason to believe we are mistaken. Nor even is it to say we cannot be certain and reasonably certain. It is Mill's argument that the Principle of Causality is as certain as any empirical generalization could reasonably be. Nor, as Mill points out,[100] will failure to find a cause lead us to reject the Principle. Rather it will mean we have not searched hard enough, that we are simply ignorant as yet of the cause the Principle reliably tells us is there to be discovered. The logical form of the Principle is what renders it thus selectively immune to falsification. Like [E] and [C], the Principle of Causality is a law the statement of which involves mixed quantification. As a consequence it is neither conclusively verifiable nor conclusively falsifiable. We should, I believe, leave aside the silliness of Popper and the Popperians that what is unfalsifiable is unscientific and unempirical. So there is nothing wrong with the Principle of Causality that we have suggested, following Mill, might not unreasonably be taken to be at the top of the generic hierarchy of scientific laws. It will be the supreme guiding principle of all scientific research. Finding causal laws will tend to confirm the Principle. But, since it is not falsifiable, failure to find such laws will not imply it must be rejected. Rather, it will mean simply that the scientist must search further for the laws the Principle says are there to be discovered.

Our very success in discovering causal laws tends to confirm the principle of causality. Given its place at the top of the hierarchy of laws, induction by simple enumeration suffices to justify our adopting towards it the law-assertion attitude. Here the idea of "mere habit" or "mere association" justifying the law-assertive attitude makes some sense. We may conform ourselves to the regularity of cause and effect that we have discovered in nature, and may do so without fear of falsification by some contrary instance. Falsification is excluded by the mixed quantification of the principle of causality. But as one proceeds down to more specific levels, the straight inference from observed regularity to unconditional regularity makes no sense at all. Varietal and not merely instantial evidence is required. To the extent that it is required at that level, one cannot passively go along with the regularities nature presents to us and reasonably expect thereby to arrive at justified law assertions. Observing constant conjunctions of specific kinds of events does tend to generate asserting the generality which extends the same association to the whole population. The mind does have a generalizing tendency, in Mill's phrase. But this tendency must be curbed. The mind must step in and regulate the inferences which it tends to draw. It must so control itself that it does not assert all it merely tends to assert, but of the specific generalities only those for which it has obtained varietal evidence. The adoption of the law-assertive attitude can in any particular case be considered to be justified only if the mind has deliberately and actively judged in the light of instantial and varietal evidence. The activities of scientific research are the most sophisticated of the means by which men gather this evidence on the basis of which they adjust their law-assertive attitudes.

Hume and the Humean, we now see, provide an account of scientific explanation and of scientific theorizing in which these latter cannot be separated from the on-going process of scientific research, a process which is evolutionary, cumulative, and productive of novelty, i.e., new and better theories. Thinkers after Hume were to probe this process from other angles. Peirce in particular looked at it from a specifically evolutionary point of view. But he shares with Hume what was to become a central theme of the pragmatists' philosophy of mind, that mind, at least in the sense cf reason as the investigator of empirical nature, is a kind of behaviour, "a species of coduct which is largely subject to self-control."[101] Hume and Peirce share what Goudge has characterized as " ... the metaphysical view that there is an objective evolutionary process about which we have reliable knowledge, and that man, the cognizing subject, together with his knowledge in no way 'transcends' this process."[102] With this as the guiding principle it is evident that the process of research is itself something to be explained scientifically, and this fact has, we saw above, important consequences for an adequate account of theory-evaluation. Specific theories, such as a more or less Darwinian evolutionary theory, or theories from the social sciences, may lead to insights into the justifiability of law-assertions beyond those of Hume himself. Such insights can, as I have just indicated, be found in Peirce, who approaches research from a quasi-Darwinian perspective, and also in Kuhn,[103] who has looked at research from a perspective deriving somewhat from the social sciences.

Goudge himself has developed important views on theories and theorizing in the context of biology, in his *Ascent of Life*. We shall examine these views in detail later (Section 3.3). Biological evolution is a large-scale process which is both cumulative and involves the successive addition of new qualities and processes. A part of this process is the evolution of mind: this is a sub-process exhibiting the same features of being cumulative and generating novelty. Scientific research is part of this latter sub-process, a sub-sub-process or rather an organized group of sub-sub-processes which also examplifies the features of being cumulative and generating novelty. This emphasis that Goudge, in his discussions of explanation and of scientific theorizing, places upon the on-going process of scientific research is an emphasis no doubt deriving from Goudge's views on mind. This placing of scientific research within an evolutionary context is something Goudge shares with C. S. Peirce, about whom his first book was written.[104] Both share what is a central theme of the pragmatists' philosophy of mind, that mind is a kind of behaviour, "a species of conduct which is largely subject to self-control".[105] Goudge adopts " ... the metaphysical view that there is an objective evolutionary process about which we have reliable knowledge, and that man, the cognizing subject, together with his knowledge, in no way 'transcends' this process".[106] The process of scientific research can itself be explained scientifically. And this point has important consequences when one comes to develop an account of theory-evaluation. When Goudge discusses theories and their role in explanation and research in *The Ascent of Life* he says very little concerning the evaluation of theories. He has, however, made some important remarks on this topic in *The Thought of C. S. Peirce*.

Peirce wrote a good deal about the problem of induction, and worked out an answer to this problem along the lines of what has come to be known in the current literature as vindications of induction. This answer was in terms of the idea that induction is a self-correcting process of inference. According to Goudge, the crucial feature of Peirce's answer consists of his connecting the notion of a random sample with the notion of a sustained process of inquiry:

This self-corrective tendency of induction is ... due to the fact that induction is based on samples drawn at random from the subject matter under investigation, and that each sample is free to turn up with the same relative frequency. Consequently, the objective constitution of the subject-matter *must ultimately reveal itself*, if scientific inquiry is unchecked.[107]

The repeated use of an inductive procedure will discover the limit, or regularity, in the long run − or, at least, it will do so *if there is any ascertainable limit or regularity there to be discovered.* But if one aims at truth — as science does — and if there is no regularity in nature, then the persistent use of induction will *not* ultimately reveal the truth. The pragmatic vindication of induction can succeed only if it presupposes nature is regular, that limits exist, for which presupposition it gives no reason for accepting. This criticism vitiates all pragmatic vindications of induction, as has once again recenlty been pointed out.[108] Goudge had already made the point against Peirce in 1950:

> In the last analysis, his [Peirce's] appeal is to the criterion of inconceivability. We cannot imagine a world, he declares, in which inductive inference would be systematically misleading, i.e. a world totally devoid of order. Although this may be true in the sense that we cannot imagine the details of such a world, nevertheless Peirce has sufficiently stressed the fact that "inconceivability" is no proper guarantee of truth or falsity. It cannot serve as the ground for the doctrine in question .... It seems clear, then, that Peirce's doctrine that order is a necessary attribute of existence remains a methodological postulate which the inductive reasoner has to adopt; and is thus a material assumption about the constitution of nature.[109]

Peirce does not succeed, therefore, in establishing there is a process of inquiry that will *guarantee* our arriving at the truth; he does not provide a bridge across the Humean gap between sample and population.[109a]

But in this connection Peirce makes another point which is of considerable importance. Peirce notes that man does seem to have a natural capacity for discovering (so far as we can, within the limits of induction, tell) regularities in nature. He suggests an explanation of this natural capacity, and uses this to provide a criterion for determining whether a theory is worthy of acceptance. The terms 'abduction', 'retroduction', 'presumption', and 'hypothesis' are used fairly well interchangeably by Peirce: they denote "all operations by which theories and conceptions are engendered".[110] Peirce spent considerable time investigating these forms of inference, and connecting them to the process of induction. His mature view was that abductive inference is the only way in which we acquire ampliative knowledge. "Abduction *suggests* the theories that induction *verifies*; abduction is the *sole source* of synthetic claims which induction tests."[111] Throughout the *Collected Papers* we find Peirce formulating the criteria a good abduction must meet.[112] In the *first* place, it must be "such that definite consequences can be plentifully deduced from it of a kind which can be checked by observation".[113] *Secondly*, in making predictions based on an hypothesis we should not restrict ourselves to a set we know beforehand will be fulfilled; rather, our choice of predictable consequences should be a random one.[114] *Thirdly*, the testing procedure must be objective and unbiased; negative instances must be honestly noted.[115] *Finally*, the hypothesis should be as *simple* as possible.[116] In his early writings, Peirce took this to mean *logical* simplicity; they were the hypotheses that went least beyond the data. This view changed, however:

> It was not until long experience forced me to realize that subsequent discoveries were every time showing I had been wrong . . . that the scales fell from my eyes and my mind awoke to the broad flaming daylight that it is the simpler hypothesis in the sense of the more facile and natural, the one that instinct suggests must be preferred.[117]

'Simplicity' has become *psychological* simplicity, that which "comes to mind" easily and naturally. This is strange, because elsewhere Peirce makes the point that, since abduction must be fair and unbiased, one should not permit subjective elements to determine the choice of a hypothesis: "I myself would not adopt a hypothesis, and would not even take it on probation, simply because the idea was pleasing to me."[118] Peirce arrived at his apparently strange view about psychological simplicity being a criterion of hypothesis-selection by reflecting upon the fact that scientists had discovered in a relatively short time more true theories than the chance success of trial and error would seem to permit. This suggested there was a systematic answer to the question, how does a scientist ever come to discover a true theory? The answer Peirce suggests is a causal explanation of scientific discovery.

> You cannot say that it happened by chance, because the possible theories, if not strictly innumerable, at any rate exceed a trillion – or the third power of a million; and therefore the chances are too overwhelmingly against the single true theory in the twenty or thirty thousand years during which man has been a thinking animal, ever having come into any man's head . . . . I am quite sure that you must be brought to acknowledge that man's mind has a natural adaptation to imagining correct theories of some kind . . . . But if that be so, it must be good reasoning to say that a given hypothesis is good, as a hypothesis, because it is a natural one, or one readily embraced by the human mind.[119]

Since *homo sapiens* has adapted successfully to his environment, he must have some true theories, at first inarticulated, of course, about the nature of reality, about the world he lives in, about himself, and about how they interact. Natural selection requires this to be the case.[120] Hence, the hypotheses that come to man naturally and instinctively have something to be said for them for that very reason.[121] This causal account of discovery remains fragmentary and inadequate as it stands, but it does take the strangeness out of Peirce's idea that psychological simplicity is a normative criterion for the acceptability of hypotheses: normally, subjective elements are not relevant

THE DEDUCTIVE MODEL OF EXPLANATION 41

to the selection of hypotheses, but, by means of our scientific knowledge about evolution, we can reflect upon the process of discovery and recognize through our being able to explain it the regularity that it is more likely than not that psychologically simple hypotheses are true, which is the lawful regularity that makes it reasonable to take such simplicity as a criterion of acceptability.

There is no reason why we should rest content with some naïve version of psychological simplicity, which, in any case, hardly goes beyond our ordinary concepts of reality to the far-from-everyday theoretical concepts of science.[122] We note below the analogy between the problem-solving capacities of natural selection and those of human intelligence.[122a] We will see that the latter can be used to suggest hypotheses about adaptations. Let us now reverse the analogy. We may look upon the process of hypothesis-testing as a sort of natural selection.[123] It is a process by which a mind achieves a more adequate "intellectual adaptation"[124] to the universe. The cognizing subject is motivated by the desire for truth, by the desire to be intellectually adapted to the world, and "Not knowing one's way about is a kind of absence of adaptation."[125] Abduction, hypothesis-formation, provides a set of characteristics. Research selects among these. It eliminates the false, the maladaptations. It continues until it finds one which *is* an intellectual adaptation to the world, one which successfully ends the absence of such adaptation. We wish abduction to generate hypothesis-sets which contain an hypothesis which will enable us to intellectually adapt to the world. We wish, in other words, a set of values corresponding in the context of sophisticated research to the psychological simplicity which could be appropriate only to everyday situations. These values are provided, I would suggest, by those complex structures, consisting of generic theories, models, and research techniques, that Kuhn has called "paradigms". Paradigms determine what hypotheses "come naturally" to the mind of the researcher.[126] But *ought* these paradigm-determined critieria be accepted? At the very least, the norms for hypothesis-formation generated by these paradigms can themselves be evaluated relative to the goal of coming to know. We value them as a means by which we solve puzzles. They are therefore valued just so long as they yield hypotheses which are the solutions to scientific puzzles,[127] hypotheses which can bring about our intellectual adaptation in the world where such adaptation is recognizably absent. Paradigms are therefore also subject to a sort of natural selection: failure to solve puzzles leads to the elimination of paradigms.[128] As to the generation of new paradigms when the old are eliminated, this is dealt with by Kuhn in his account of revolutionry science;[129] but that is something we cannot go into here. There is, of course, a crucial difference or disanalogy between biological selection and the selection of paradigms: the latter is guided by conscious purpose, that of coming to know; it is a process which *really is* purposive and teleological. But that is simply another way of making the point made before that mind is "a species of conduct which is largely subject to self-control."[130] The cognizing subject or, perhaps better, the community of scientists, can reflect upon and investigate its own research methods with an aim to evaluating their worth, their capacity to improve our knowledge and fill in its gaps efficiently. In the light of this knowledge it can (if this is seen to be appropriate) modify those procedures, its own structure if you wish, so as to become *better* at problem-solving, at cognizing.[131] In this sense, the methods of science are, or can become, self-correcting, as Peirce suggested, even though such capacity for self-correction cannot, as we saw Goudge argue against Peirce,[132] provide a successful "vindication" of those methods. Now, Kuhn has often been accused of relativism. His emphasis upon the intimate connection of paradigms and criteria for the acceptability of hypotheses perhaps suggests this, though, as the above remarks make clear, relativism does not follow: more over-arching cognitive standards may provide criteria for justifying paradigms and their associated values.[132a] More important to the charge of relativism are Kuhn's unfortunate remarks about the "incommensurability" of different theories, and about the "theory-ladenness" of concepts.[133] The evolutionary view of mind and of the nature of the research process provide us with a framework that enables us to incorporate the insights of Kuhn while rejecting the charge of relativism. Kuhn shares the position of Hume, Peirce and Goudge that the process of scientific research can be investigated naturalistically and its methods and theories evaluated as more or less successful tools to be used to further the cognitive aims of science. In Kuhn, however, there is, if you wish, a sociological dimension absent from the other thinkers. In particular, Kuhn emphasizes the role of different scientific communities in generating agreement among scientists in their law-assertive attitudes. It is shared paradigms — shared law-assertive attitudes towards the same abstractive generic theories, and shared norms for law-assertion implied by those theories (and the goal of acquiring scientific knowledge) — that hold the scientific (sub-) communities together.[134] Thus, what is a law to one community is not one to another community. There is nothing wrong with this idea. Indeed, it is true: as research proceeds, what *was* law-assertible ceases to be so, and something more "true to the facts" and contrary to what was formerly asserted comes to be law-assertible. This is precisely what is to be expected, if he is — as he

THE DEDUCTIVE MODEL OF EXPLANATION 43

is — fallible. But if such communal relativity of law-assertibility is not carefully stated it can slip into a vicious — and anti-scientific — epistemological relativism.[135] There is a tendency for Kuhn to slip in that direction.[136] The view of scientific reason shared by Hume and Peirce provides us with a framework that enables us to incorporate the insights of Kunh while rejecting the charge of relativism.

1.4. THAT SOME EXPLANATIONS ARE BETTER THAN OTHERS

It is the essence of the Humean position to emphasize the activity of mind in judging in a reasonable way, and in the gathering of evidence in order so to judge. In the reasonable man the passive acquisition of law-assertive habits must come to be replaced by the active adjustment of assertions to deliberately-sought-out data. This point is often missed.[137] The reason for missing this may lie in the way the *Treatise* is written. Hume *mentions* the active nature of reasonable judgement during his discussion of causation. But he does not develop the idea in detail. For the gap at this point in Hume's discussion there is perfectly good reason. The *Treatise* is throughout a discussion of man in terms of psychological theory. Book I, "Of the understanding", deals with what we would now call the psychology of cognition and of thought. Book II, "Of the passions", deals with motivation. Book III "Of morals", deals with personality. As a matter of fact, this is a perfectly reasonable organization for a psychology text. But it means that the description of the motivation for knowledge, and the psychological types appropriate to the active search for knowledge cannot be given when Hume discusses the "Rules" such persons use. It is only in Book II, Part III, Chapter X, "Of curiosity, or the love of truth", that Hume discusses the motive which is "the first source of all our enquiries";[138] and it is only in Book III, Part III, Chapter V, "Of natural abilities", that Hume discusses[139] the qualities characteristic of persons who are able to most judiciously arrive at correct law-assertions; though, as we saw, both motive ("constancy") and ability ("sagacity") were mentioned by Hume when he discussed the use of the his "Rules".

If one's sole end were knowledge then the search for relevant data would continue until one had made but one of the possible hypotheses subjectively worthy of law-assertion. And even then in the ideal case, there would be a residual doubt possible whether the law-assertion was objectively justified. Certainly, one would have to keep one's mind open to the possibility further observations might establish the objective unjustifiability of what would at that point be a subjectively justified law-assertion. But, we should note, with Hume, who discusses the two in the same context,[140] that both sagacity and vigilance are qualities worthy of praise. On the other hand, knowledge alone is often not the only consideration. Many times one must act before all evidence can possibly be collected, or before one can afford to collect it. In these cases a law-assertion must be risked prior to when considerations of reason alone would deem it worthy of assertion. Prudence must dictate here what is reasonable. Not surprisingly, prudence is another quality that, like sagacity, is worthy of admiration.[141]

All of this, it seems to me, constitutes a fairly adequate account of causal reasoning. The difference between a *subjectively acceptable* lawlike generality and an accidental generality lies in the type of evidence available and the way it was actively sought out.[142] It is knowledge which has been *put to the test*, experimentally. In being put to the test, it has been used predictively, and the predictions were successful.[143] And it is knowledge which fits into background theories, other parts of which have themselves been put to the test, used predictively.[144] We accept it as yielding reliable predictions precisely because it has in the past been used successfully to predict,[145] or because the theory of which it is a part leads us reasonably to believe that it can be used successfully to predict.[146] For our purposes it is not necessary to explore further the nature of the scientific method, nor the logic of justifying or confirming evidence. It suffices for us to recognize that *for a lawlike generality to be worthy of acceptance the evidence for its truth must derive from the application of the experimental techniques of the scientific method, either as applied directly to the generality or indirectly, through the generality being embedded in a theory other parts of which have been put to the experimental test.*

Objectively, true lawlike generalities and true accidental generalities are both capable of yielding explanations and predictions, that is, of showing why, given one event, some other event must have occurred. But *subjectively*, we never have sufficient evidence to claim to know the whole objective truth: we know only the sample, not the population. And therefore we never know "for sure" whether an explanation we offter is in fact objectively justified. *Subjectively* we must rely upon less-than-perfect evidence. Where this evidence renders a lawlike generality subjectively worthy of acceptance, there we will be justified, so far as we can be justified, in taking that generality as providing acceptable explanations and reliable predictions. And in contrast, it we believe quite reasonably that a generality is true, but if the evidence is not of the appropriate sort, gathered via an active application of the scientific method, then we characterize that generality as an "accidental generality", and do not take it as providing acceptable explanations or reliable predictions.

For our purpose of defending the deductivist thesis, this distinction between the objective soundness of an explanation and its subjective worthiness of acceptance is an important one. The distinction arises precisely because of the Humean gap between sample and population. If the valid argument offered as an explanation is to show why one event *must* occur given the occurrence of another, then the generality among the premises will have to be true. The argument will have not only to be valid but to be sound if it is really to explain, really to show why something *must* have occurred. But the generality will always be a generality, and will therefore always suffer from the infirmities that attach to inductive generalizations. Since 'know for sure' means knowing all the instances, we never know for sure whether such generalities are true. *We therefore never know for sure whether we have an explanation or not, that is, an objectively acceptable explanation.* This is indeed so, it seems to me. I do not find it objectionable.[147] To use an analogy:[148] Ultilitarians hold an action is objectively obligatory if its consequences are in fact the most felicitous. And an action is subjectively obligatory if the consequences are, on the basis of all available evidence, believed to be the most felicitous — though in fact they may not be that, an open possibility since what are the consequences has to be judged on the basis of inductive generalizations. Similarly, it is possible to distinguish what is objectively a good explanation from what is subjectively a good explanation. A law statement is objectively worthy of acceptance for explanation and prediction just in case it is true. And a law statement is subjectively worthy (of acceptance for explanation and prediction) just in case the evidence a person has available has been acquired through an active application of the scientific method. We might take this further, and define degrees of acceptability (worthiness of acceptance) using as a basis the strength of the evidence, but this is a qualification we can for the most part ignore. We now use this to define the worth of explanations (and also predictions). An explanation (prediction) is objectively good, if it is deductive and the premises are in fact true; it is subjectively good, if it is deductive and the evidence available to the person argues scientifically the truth of the premises.[149] Just what evidence is available is a matter of sociological and psychological fact, and in that sense subjective. So, what is a subjectively good explanation varies over time, and also, one must add, between explainer and explainee very often.[150] Of course, whether the evidence one has available is good evidence for the objective worth of the purported explanation is a logical and methodological matter, and therefore perfectly objective. Similarly, whether an explanation is objectively good depends only on the deductive, i.e., syntactical connections, and on the truth of the premises, which is a matter of the way the world is; both logical form and truth (given the linguistic conventions) do not depend on the language users, both are in that sense objective. So, whether an explanation is objectively good is an objective matter. Indeed, it is precisely this fact that makes our criteria of what constitutes an explanation to be criteria of great merit: whether they are satisfied is not merely a matter of convention.

But of that, more later.[151] The present suggestion is that deduction, and true premises, an essential one of which is a statement of law, jointly are necessary and sufficient for a sequence of sentences being a scientific explanation.

The reply to this suggestion is that it does not distinguish between a scientific explanation and, say, weather forecasting ("mere forecasting").[152] Both satisfy the just-mentioned conditions claimed to be necessary and sufficient for explanations. But mere forecasting is not explanation. *Ergo*, ... etc. The counter-reply is that the distinction between "scientific explanations" and "mere forecasting" is not one between explanation and non-explanation; they both satisfy the conditions and both are explanations. The distinction, rather, is between better and worse explanations. The appropriate analogy is this: The distinction between good and evil is not a distinction between moral and not moral, but between the morally better and the morally worse.[153]

Here another objection is made. The deductive model (it is said)[154] cannot distinguish between better and worse. For deduction does not admit of degrees. The defender of the deductive model must therefore speak of explanation and non-explanation only, and not of good and bad explanations. Now, this is absurd. It is as if one could not first distinguish between explanations and non-explanations, and then go on to distinguish within the former the better from the worse explanations. A condition necessary and sufficient for an explanation is necessary for a good explanation but nothing requires it to be sufficient for the latter. To suggest that because deduction is a necessary condition for a good explanation,[155] therefore "No room is left for bad explanations",[156] is simply to misuse and misunderstand the notions of necessary and sufficient conditions. Of course, the criterion of better and worse explanation will not turn out to be a matter of either deduction or of the truth of the premises. Analogy: An action is appropriate to moral evaluation if and only if it is voluntary, and the criteria of morally better and morally worse are not defined in terms of the voluntary/involuntary distinction.

It should be evident from what we have said that there are criteria of better and worse. Consider the generic law (2.2). If mere knowledge of laws was all that was desired, the scientist would rest content with this. But he does not. He proceeds with his research until he discovers (2.3), the more specific law which (2.2) says is there to be discovered. The scientist engages in this research because he finds the specific knowledge more desirable than the generic knowledge. This is not because (2.2) is false and in that way defective as knowledge. Indeed, if (2.3) at which he aims is true, then so is (2.2), since, after all, (2.3) entails (2.2). The standard being applied compares two pieces of knowledge, and judges one as better than the other. It is to these criteria of better and worse explanations that we must now turn. These criteria are a matter of the sorts of laws involved in the explanation. Our procedure will be to try to state the ideal of scientific explanation, thereby defining the best. From that, the nature of the less perfect (worse) will follow immediately.

The ideal of scientific explanation is "process knowledge".[157] Perhaps the easiest approach to the notion of process knowledge and to the argument suggesting why it is the ideal is by way of an easy example. If we consider the law L: "Water when heated boils" we know that atmospheric pressure has been omitted. Thus, L holds only "within certain limits". If atmospheric pressure were included the resulting law would be less imperfect. For, it would enable one to explain and predict more accurately about the systems to which it applies. In addition L does not tell how long water takes to boil. Again if such information were included more accurate explanation and prediction would be possible. And such less imperfect knowledge would not only be better from the point of view of idle curiosity, simply as additional knowledge about the workings of such systems, but also from the point of view of pragmatic interests, informing one how more efficiently to plan with respect to such systems and to interfere intelligently in their operations. What the just noted move from the more to the less imperfect indicates is that in the ideal case we would want, (1) knowledge of the complete set of relevant factors within the system; (2) knowledge of the relevant factors crossing from outside into the system; and (3) knowledge of a rule (process law) describing the interaction of these factors and permitting the explanation and prediction of the values of these factors at any time given knowledge of the present values of these factors. Such knowledge is process knowledge, and such is the ideal at which science aims. In Chapter 2 we shall argue in more detail that this is the ideal. For the present, however, let us simply try to articulate more clearly just what it is to have "process knowledge".

A system is a group of entities (things, fields, etc.) in an identifiable portion of space. To have process knowledge one must know three things. *One*: One must know a complete set of relevant variables. That is, one must have a set of properties such that no other property within the system is causally relevant to the interaction of the set. That one has a complete set of relevant variables is an inductive generalization. The values of the variables which the entities in the system take on at a given time constitute the state of the system at that time. *Two*: One must know that no property outside the system affects what goes on inside the system. The generalization of this idea is that of controlled or known boundary conditions, knowledge of the influences that cross the boundary from outside to the inside of the system, and also of what leaves the system. That a system is closed, or that just such an such are the boundary conditions, again are matters of inductive generalization. *Three*: One must know a process law for the system. A process law is a law such that, for any possible state of the system, given that state (plus completeness, plus closure or given boundary conditions), then from it together with the law any future state and any past state of the system can be deduced. A process law is clearly an inductive generalization.

Newton's account of the solar system provided the first example of process knowledge, or at what it was subjectively reasonable for Newton and his contemporaries to take to be process knowledge. The masses, positions and velocities of the ten objects constitute a complete set of relevant variables. Other objects are sufficiently far away that the system is *de facto* closed. The law of gravitation enables one to formulate a process law which permits deduction of all future and past states of the system. Process knowledge has also been available in phenomenological thermodynamics. This second example makes it clear that there is nothing specifically "mechanical" about process knowledge. And, in fact, as it has been defined there has been no limitation on what might be the relevant variables: they may be mechanical, masses and velocities; or non-mechanical, temperature or field strength; or human and social.

If process knowledge is in fact the ideal of scientific explanation of individual facts and events, as I have asserted — the assertion will be justified in Chapter 2 — then we have our criterion for better and more among explanations, all of which satisfy the criterion of being deductive. The ideal is that best to be hoped for. Whatever falls short of the ideal is less than perfect. We may therefore call it imperfect knowledge.[158] To call it imperfect is not to detract from its status as knowledge. A good deal of our most useful knowledge is imperfect. The use of 'imperfect' merely locates the knowledge relative to the ideal. There are two cases worth distinguishing. (1) Here the law statement is true in all generality. Statistics often yields such imperfect knowledge. For example, we can imagine our finding it worth our while to know the probability (i.e., relative frequency in the long run) of a given number, n, of total lunar eclipses in a year. Such knowledge is imperfect. Thus, it tells us nothing about the occurrence of particular eclipses, as process knowledge would do — it tells us only about long run relative frequencies, about a certain mass event.[159] Now, in point of fact, given the laws Newton discovered we can obtain a process law for the relevant system, and from that process law deduce the probabilistic law about the numbers of lunar eclipses. Newton's process law enables us to deduce *all* future states of the system, and therefore also the long run relative frequencies. Such deduction enables us to, so to speak, fill in the gaps in our knowledge that the probabilistic law left open. But we can also have such statistical laws where no process knowledge is available, e.g., the rolling of fair dice in the standard conditions. (2) Here the law statement is false in all generality, but is true "as far as it goes". For example: Boyle's Law, Kepler's Laws, Galileo's Law of Falling Bodies. We know these to be literally false: pressure does not vary inversely as the volume, the planets do not move in exactly elliptical orbits, objects do not fall towards the earth at an acceleration which is constant for all distances from the earth. On the other hand, they do hold "within certain limits", as one says. Within those limits, what those law statements say is true; within those limits the law statements are true generalities. That is why the discoveries of the mentioned laws constituted increase in knowledge. They are false in all generality because they do not hold beyond those limits. They are imperfect knowledge because they omit in their statement those limiting conditions.[160] That is, the knowledge they embody, as in the case of statistical laws, has gaps in it. Such gaps can be filled by process knowledge. The limits, and the exact extent to which such imperfect knowledge in fact holds true, is provided by process knowledge — or at least, if process knowledge is available in the relevant area. In the case of our examples, this is now the case, though at one time it was not so, and certainly there are many examples of imperfect knowledge where no process knowledge is available, e.g., in the social sciences. When the more perfect knowledge is *not* available, then the limits within which the imperfect knowledge holds will not be known. Not knowing them, one cannot identify them. That means there is always (over and above the risk always implicit in the use of inductive generalizations) a risk involved when such knowledge is used to explain or predict: there is always the risk that one is applying it outside the limits under which it holds, applying it where it is *not* knowledge. In such a case, one does not have an explanation, at least not objectively. For, outside the relevant limits, the generality is false, and explanations require true — objectively true — premisses. Of course, an explanation based on applying imperfect knowledge outside its unknown limiting conditions, while not objectively an explanation may subjectively be one, as was the case with Kepler and Galileo. This will depend on what evidence one has available, and so on. In any case, the acquisition of more perfect knowledge in the relevant area will give one knowledge about the limits of the imperfect knowledge, and therefore about the extent to which one's subjectively worthy explanation failed to be objectively worthy, basing one's judgement of that failure on what one has come to accept as a subjectively worthy explanation.

We have discussed some of these points previously when we looked at (1.1), (2.2), and (2.3), in Sections 1.1 and 1.2 above, and we shall return to them directly. But we must further illustrate the role which the distinction between better and worse explanations plays in our system of knowledge.

Our distinction between the more and the less imperfect is one that is in fact common. For example, it is what underlies a good deal of causal thinking and of our criticism of others for having committed certain causal fallacies. Thus, consider the fallacy, so-called, of common cause.[161] Suppose someone explains violence in the streets (G) as caused by (F) violence on television. The criticism would be that it is not so much that F causes G but rather that F and G are both effects of a common cause, for example, certain tensions (H) implicit in the social structure of our society. In asserting H is the common cause of the two effects F and G we are committed to such subjunctive conditionals as

> If H were absent, then so would F and G be absent.
> If F were absent, then G would still be present if H were present.

These are of particular importance when pragmatic interests are involved. They indicate that even if we were to interfere in the situation, and control F, so long as H was left alone, we would still have G. The assertion of these subjunctive conditionals presupposes certain complex assumptions about the lawful relations among F, G and H. The person being criticized holds F causes G, and is prepared to assert that

> If F were absent then G would be absent.

In our criticism we are accusing him of omitting at least one relevant factor, H, and in citing H as the common cause we are indicating in a general sort of way the specific way in which it is relevant. So we are criticizing him from the standpoint of less imperfect knowledge.

Our charges that a person has committed the *post hoc, ergo propter hoc* fallacy are also usually based on a contrast of more and less imperfect knowledge. Consider the aged uncle who takes patent medicine M, whose cold always disappears some time thereafter (C), and who insists that M causes C. We would probably insist that the relation was not causal, and that the uncle had illegitimately inferred *propter hoc* from *post hoc*. In asserting that M does not cause C we are not, of course, denying that

M is regularly followed by C.

but rather are holding that

C would occur even if M were absent.

We assert this subjunctive conditional on the basis of lawful knowledge that we have, but the uncle does not. This knowledge precludes M being the cause of C. We would also probably be able to explain why M is regularly followed by C; for example, colds disappear, C occurs, naturally, in just the same length of time, whether M is present or not, but for this reason it turns out that whenever a cold appears and the medicine is therupon taken, M will be regularly followed by C. From the vantage point of our greater knowledge, we can judge the regularity that M follows C is *not* worthy of the law-assertive attitude. Which is not to say that it is not true. In criticizing the inference as illegitimate we are saying, not necessarily that the uncle ought to have the knowledge we have, but at least that he ought to recognize the situation as one in which one might reasonably expect there to be other factors possibly relevant to C. Such a proposition, while more imperfect than our knowledge, is nonetheless sufficient to establish that M causes C should be recognized as competing with a number of other possible hypotheses. For that reason it will remain unworthy of the law-assertive attitude until further research is done. The knowledge that there are things other than M relevant to C is a piece of knowledge we expect the uncle to share with us. Insofar as he does, then he has wrongly adopted the law-assertive attitude in this context. Given the background knowledge we may reasonably expect the uncle to share with us, it turns out that the regularity is, for him, *subjectively unworthy* for the law-assertive attitude. It is for this reason that we can charge him with a *fallacy*, that fallacy which we call *post hoc, ergo propter hoc*.

One must make similar comments about another standard example of the *post hoc* fallacy. One version [161a] goes in terms of the ancient Chinese belief that an eclipse of the moon consisted of a dragon devouring that object. Acting upon this belief they then exploded fireworks so as to scare the dragon away, leaving the moon behind. Their attempts were always successful for the moon always reappeared. It was their belief that there was a causal relationship between exploding fireworks and the reappearance of the moon, and for evidence they could cite many instances. However, in spite of the regular connection, no causal connection exists, and their inference from the regularity of the connection to the causal efficacy of their act is an example of the *post hoc, ergo propter hoc* fallacy. From our vantage point of superior knowledge we can see indeed that those who practised the custom of exploding fireworks were objectively unjustified. We can also see, however, that they were subjectively unjustified. In their explanations they mentioned not merely the fireworks but also the dragon. In their explanation they introduced dragons and the anxieties of dragons as relevant variables. But these factors remained purely hypothetical. Their existence was never confirmed independently of the effects they were supposed to produce. Without such confirmation, the explanations should have remained as hypotheses, requiring further research before any law-assertive attitude was adopted towards them. To the extent that the law-assertive attitude was adopted prior to such research the explanations were subjectively unworthy.[161b] So we accuse the ancient Chinese of the story of committing the *post hoc* fallacy.

The point we are making here is, of course, exactly the same point Mill made when he discussed Reid's argument against Hume. Reid argued [162] that upon Hume's view day must cause night and night must cause day, since these are invariably connected. Mill replied [163] that it would indeed be a mistake to call these regularities causal since we know they are not unconditional. The more adequate explanation we are in a position to give of these sequences is in terms of a luminous body (the sun), the absence of any opaque body between the sun and the earth, and the rotation of the earth. Anyone who held day caused night would be making a subjectively unworthy law-assertion.

No one, probably, ever called night the cause of day; mankind must so soon have arrived at the very obvious generalization, that the state of general illumination that we call day would follow from the presence of a sufficiently luminous body, whether darkness had preceded or not.[164]

This sort of response of Mill to Reid is also entirely appropriate to an example given by Broad intended to refute the "Humean" claim that causation is constant conjunction:

> I am quite sure that the hooter of a factory in Manchester does not cause the workmen of a factory in London to go the their work, even though the Manchester hooter does always blow just before the London workmen start to wend their way to the London factory.[165]

We can see that we do not adequately explain the entry into the London factories by citing the blowing of the Manchester hooter because we know a good deal more than is mentioned about factors relevant to the behaviour of the workmen. We know, for example, that the workmen enter their factories upon hearing a hooter, that sound beyond a certain distance cannot be heard, that the Manchester hooter is while the London hooter is not beyond hearing distance, and that hours of work are generally the same throughout England from London to Manchester. On the basis of this *lawful* knowledge we can assert that it is the London hooter that causes London workmen to return to work, and that even if the Manchester hooter were not to sound, the London workmen would still return to work at the same time as always. The regularities Broad cites are pieces of very imperfect knowledge. We all in fact can give much more adequate, much less imperfect, explanations of these regularly connected events. And because any reasonable man might be expected to offer an equally adequate explanation, we are quite prepared to judge that for anyone the regular connection between the Manchester hooter and the return to work of men in London is a regularity for which the law-assertive attitude would be subjectively unworthy.

As all these examples make clear, less imperfect knowledge can explain the more imperfect. Or at least, can explain the more imperfect to the extent that it is worthy of explanation, that is, to the extent to which factually true regularities are cited in the more imperfect explanations. Thus, the explanation by the ancient Chinese of the end of lunar eclipses cites two things, the exploding of the fireworks and the dragon and its anxieties. The former exist, and the regularities involving them are explained by our less imperfect knowledge. But the dragons do not exist, and to that extent the regularities cited by the ancient Chinese are unworthy of explanation. We shall make more of this point shortly. For the present, however, we need only notice how the acquisition of greater knowledge, the transition from the more to the less imperfect, yields explanations of the more imperfect knowledge. We have at this point started to move from a discussion of the explanation of individual facts and events to the explanation of laws. Before turning to that subject, however, let me make a few remarks about "explanation sketches".

This notion was originally introduced by Hempel.[166] The basic idea is that where a great deal can be taken as understood, there much need not, and therefore will not be stated, i.e., in such circumstances one can "sketch" the explanation. The explanation sketch will in general be an enthymematic statement of the explanation being advanced. In this sense, what is offered in the explaining transaction when an explanation sketch is advanced is "incomplete". The idea is not problematic. In many cases the explanation sketched will be one involving imperfect laws. This provides us with another sense of 'incomplete': the explanation sketched will be incomplete just to the extent that the laws involved are imperfect. Hempel, unfortunately, was not careful to distinguish these two sense of 'incomplete',[167] though one must add, to be fair, that it is not obvious the extent to which it was necessary to be clear on this point for the paper to achieve its intention. Nonetheless, the failure to be perfectly clear on the matter was unfortunate. For, it permitted a great deal of discussion to arise which traded on just such an ambiguity. In the *first* place, the use of 'explanation sketch' suggested that, e.g., explanations given by historians were not as good as those of, say, physicists. This was certainly not in Hempel's mind (he uses physical as well as historical examples). And in any case it depends upon an illegitimate blurring of the distinction between the transaction of explaining (where the law is omitted – only the sketched is offered) and the explanation (where the law is included – it is part of the context): even where an explanation sketch is offered in the transaction of explaining, the explanation which is involved *does* include a law. So the fact that historians offer explanation sketches more often than physicists does not mean there is any difference so far as the inclusion of a law in their explanations is concerned. All that follows is that historians more often than physicists can presuppose the law portion of their explanations is available in the relevant context. In the *second* place, the fact the laws to which historians appeal are imperfect again gave rise to the suggestion that historical explanations were not as good as those in physics. Nor would this be a *suggestio falsi*, insofar as it is simply a recognition of the imperfection of the relevant laws. But it is wrong insofar as it involves the idea that historical explanations are *in principle* inferior to those in physics. The incompleteness of the laws currently available in no way argues for the latter idea. But this second point was compounded by a *third*. The fact that a law was not explicitly present in the transaction was understood to mean that no law was present in the explanation. This incompleteness in the first sense was then confused with incompleteness in the second sense. The consequence was that wherever an explanation involves imperfect knowledge (is incomplete in the second sense) it comes to be understood as involving no laws (incomplete in the first sense). This move is, of course, clearly illegitimate, but equally clearly it has been made. The consequence is that critics of the deductive model have come to believe that according Hempel explanations based on imperfect laws are not deductive (it being supposed no law is available for the deduction to go through), nor *ipso facto* explanations. Indeed, this is the main thrust of much of the criticism of Hempel, of the deductive model, and of the notion of "explanation sketches". The criticism resulting from these sorts of confusions we shall have to look at in detail as we proceed. But because the idea of an "explanation sketch" does tend to generate confusion, we shall now, having indicated the essential points, not mention it again.

The philosopher who has recently most closely approximated to the approach and distinctions here adopted is J. L. Mackie.[168] He begins from the side of imperfect knowledge and works towards the ideal. The starting point is rather simple causal statements such as "striking matches causes them to light", in which the presence of one property is indicated to be sufficient for the presence of a second property. Several other properties may also be sufficient for the presence of the second. Thus, being put in a flame is also sufficient for a match to light. So the original property cited as cause is really an unnecessary sufficient condition. But further, in general what is cited as a cause is usually just one necessary condition among several which only jointly, and not individually, are sufficient for the effect. Thus, it is only striking when the head is dry and in the presence of oxygen that effects the lighting of the match. So what is cited as the cause is an insufficient necessary condition which is part of an unnecessary sufficient condition. For short, causes as we ordinarily speak are "INUS-conditions".[169] A full law will state all sufficient conditions and within each sufficient condition it will state all the necessary conditions. Then, practical considerations will determine our selecting one of the many INUS-conditions as "the *cause*".[170] We shall ourselves argue in very much this way.

Mackie indicates, quite correctly, that in most everyday situations the full law is not available. In our terms, the knowledge is imperfect. Mackie suggests two forms of imperfection. One is that the law will be applicable only under certain conditions not mentioned in the law statement actually used. These various conditions sufficient for correctly using the law (each of which will have necessary conditions within it) form what Mackie, following John Anderson,[171] refers to as the "field" of the law used.[172] The field thus constitutes what we referred to above as the "limiting conditions" for the applicability of the law. The conditions defining the field may be known. Most often in everyday situations they are not. We may know *there are* such conditions but not just what those conditions are.[173] Use of the law in giving INUS-explanations will therefore depend on the trained judgement of the user.[174] We shall argue below in very similar fashion.[175] The other sort of imperfection Mackie mentions is that, within a known sufficient condition, not all the necessary conditions may be known. We may well know that *there are* these further necessary conditions but not known what they are. Such a law, in terms we used above, has gaps in it, and in fact Mackie himself calls such laws "gappy". In our terms, Mackie does not indicate all possible forms of imperfection. But he has certainly indicated two important sorts. And he argues, quite correctly, as we shall argue, that such imperfection does not prevent the use of such laws in explanations.[176]

Mackie does not spell out what he considers to be the ideal of scientific explanation. Presumably one must eliminate the imperfections. In his terms this amounts to eliminating the gappiness and to fully articulating the field. That the former has limits is clear. That the latter does is not. As Russell once argued,[177] is not every event connected to every other? and does not this indicate the field will always be infinitely complex? In fact, Mackie accepts this argument.[178] But that makes the separation of field from non-field a matter of convention. One would then have to argue for one convention as preferable to another, though Mackie provides no such argument. However, I do not think Russell's argument is valid. From the fact that every particular event is connected to every other, it does not follow that every *type* of event, every property, is lawfully related to each other.[179] Thus, what is relevant to the positions and velocities of the objects in the solar system relative to each other are just the masses, positions and velocities of those objects. We have every reason (inductive reason) to suppose the colours of the planets are not relevant. Was not Newton able to achieve his success in explaining the motions of the solar system without any reference to planetary colours? What I am suggesting is that the notions, presented above, of completeness and closure provide a criterion for when the field has been fully articulated. To be sure, the criterion is one such that the evidence it is or is not fulfilled has to be inductive evidence; but that is no difficulty.

Whether Mackie would accept this, I do not know. But certainly, as far as his position goes it is quite compatible with the one we are defending.

THE DEDUCTIVE MODEL OF EXPLANATION   57

The latter goes beyond Mackie's position, of course, in actually defining an ideal and defending the thesis it is in fact the ideal. I believe Mackie must ultimately do something of this sort also. If I am right, the ideal he should end up defending is just that of process knowledge. In any case, there is one virtue to our approach, which begins straight off by defining the ideal of scientific explanation, namely, the continuity between everyday and scientific explanations, their essential identity of form, is obvious right from the beginning. One of the ways of proceeding of some who have recenlty attacked the deductivist position is to attempt to drive a wedge between ordinary and scientific explanations. Thus, as we shall see, Scriven distinguishes explanations based on (what he calls) laws. Our approach yields an effective strategy for meeting such attacks on the deductivist position. But there is no doubt Mackie could in his own way also meet these attacks, without defining some ideal. So what Mackie and ourselves are about remains, I believe, effectively the same. It is more, perhaps, a matter of emphasis than anything else.

Let us now return to the explaining of laws. Explanation of laws can be divided into two cases. One is the explanation of the laws involved in process knowledge. The other is the explanation of imperfect knowledge.

The explanation of the laws involved in process knowledge occurs by locating them within the context of a theory. A theory involves at least a body of laws so logically interconnected that they can be arranged into an axiomatic system.[180] The use of a theory does not require that they in fact be so arranged. A scientist can trace deductive connections among laws without an explicit axiomatization at hand, and certainly without a full axiomatization.[181] All that is required is that the deductive connections be such that they *can* be so arranged.

A law can be fit into such a theory in two ways. It can, in the first place, be a deductive consequence of the axioms. Thus, the law of conservation of linear momentum follows from the axioms of classical mechanics. Or, the law can, in the second place, be an instantiation of the axioms. Thus, the axioms of classical mechanics say that there are force functions, and lay certain conditions on such functions. But such restrictions leave open whether the function, say, involves the inverse square of the distance, the inverse cube, or whatever. Each is an instantiation of the general claim that there are force functions. Now, Newton's law of gravitation is such an instantiation.[182] The law of gravitation, and so the process law for the solar system, is explained by the fact that it is an instantiation of a more general theory, namely, the theory which we call classical mechanics. The general pattern of this sort of theoretical explanation is exemplified by the relation between (2.2) and (2.3) discussed in Section 1.2, above.

What makes a *theory* and therefore the explanation of laws in terms of it, better or worse is the *scope* of the theory. (Note that we are not here talking of laws being better or worse at explaining individual facts — where the idea of process, of specificity and predictive power, gives the criterion of better or worse — but rather of better and worse among the explanations of laws. As we shall see, the two criteria are in a way opposite, for scope leads one away from the specific towards the generic.) The more laws a theory is capable of explaining, the more laws it is capable of tying together in its deductive net, then the greater scope it has, the better it is as a theory, and the better are the explanations in terms of it. In order to group together several specific laws theories generally have to consist of generic laws. They must be abstractive. But such theories tend to become of real scientific interest only when there is in addition among their axioms what has been called a composition law. This is a law which, functioning as a premiss, enables one to derive the laws for complex systems from those for simpler systems.[183] Another feature of theories, parasitic upon the former, by which scope is achieved, is reduction.[184] We may, however, for our present purposes leave aside the question of the more precise delineation of these three features which give scope to a theory. For present purposes, what is important is the connection between explanation of laws by theories, on the one hand, and deduction on the other. To put it briefly, explanation of laws by theories is deductive.

The deductive model (taken now to include the case of instantiation) thus applies not only to the explanation of individual facts, but also of laws. And, that this is so has been fully recognized by those who have defended the deductivist tradition in the philosophy of science. This shows how radically misleading and, one must say, unfair if not ignorant, the following remark of Collins:

Since subjects for characteristically scientific explanations are relationships expressed in generalizations and not particular occurrences, we should find it natural that the explanations given advert to other generalizations or laws. This is not to say that the law-deduction theory is satisfied by many explanations actually given in the sciences. The law deduction theory is entirely organized to deal with the explanation of particular occurrences, so that on the surface it is not even illustrated, much less confirmed, by typical scientific explanations.[185]

In the first place, the deductive model *does* apply to explaining laws by theories (i.e., other laws or systems of laws). In the second place, it is not true that scientific explanations do not apply to particular occurrences. One has merely to think of Newton's explanation of the operations of the solar system. Further, that explanation was deductive. In the third place, even if explanations of laws by theories are "typical" it does not follow either that there are *no* explanations of particular occurrences or that such explanations of particular occurrences are not deductive. In the fourth place, one wants to ask why (if it is true) explanations of laws is more "typical" than explanations of particular occurrences. It may turn out, for example, that it is a more fruitful research policy to give more time and journal space to theoretical elaborations, and the extensions of theoretical explanations of laws: that would explain what was "typical", without impugning the scientific status of explanations of particular occurrences, of individual facts and events.

The other case of explanation of law is the explanation of imperfect knowledge by perfect knowledge. This divides into two cases. It may in the first place, be a deduction of a law, which is imperfect, but which holds in all generality, from a process law (which also holds in all generality). For example, we have the deduction of the statistical law about eclipses from the process law for the solar system. Clearly, this is just a special case of the explanation of law by theories. In the second place, the explanation may be of a law which does not hold in all generality. Here the explanation consists of a deduction of a law such that, it holds in all generality, and also such that, it shows the limits under which the law to be explained does hold. If the imperfection of the law had not been recognized, this explanation will, naturally, be accompanied by an account of the evidence available before and after, which led to the previous failure to recognize the limits, and to the subsequent success in correcting the error.

Let us develop this second sort of explanation of imperfect knowledge by means of an example. We suppose there is the imperfect law

(4.1) $\quad (x) \, [Kx \supset (Fx \equiv Gx)]$

which says that systems or objects of kind K are F just in case they are G.[186] Since (4.1) is imperfect, we are supposing it is false in all generality that Ks are F just in case they are G. Rather, let us say, such holds only in case the Ks are also Cs, i.e., that it is true in all generality that

(4.3) $\quad (x) \, [Kx \supset (Cx \equiv (Fx \equiv Gx))]$

(4.3) is more perfect than (4.1), and we are supposing it will be used to explaine (4.1). If it were true that

(4.4) $\quad (x) \, [Kx \equiv Cx]$

then we could deduce (4.1) from (4.3). From (4.3), by universal instantiation, we obtain

$$Kx \supset (Cx \equiv (Fx \equiv Gx)]$$

which, since '$p \supset (q \equiv r)$' entails '$p \supset (q \supset r)$', entails

$$Kx \supset (Cx \equiv (Fx \equiv Gx)$$

or

$$(Kx \;\&\; Cx) \supset (Fx \equiv Gx)$$

from which, together with the universal instantiation of (4.4)

$$Kx \equiv Cx$$

we obtain

$$(Kx \;\&\; Kx) \supset (Fx \equiv Gx)$$

or

$$Kx \supset (Fx \equiv Gx)$$

which universal generalization transforms into (4.1). Thus, under assumption (4.4), (4.3) entails (4.1). It is also possible to prove that under the same assumption, (4.1) entails (4.3). That is, under the assumption, the two are logically equivalent.

It is clear why we can reasonably say that (4.3) explains (4.1) by showing the limits under which (4.1) holds. The condition which limits the applicability of (4.1) to Ks is given by C: (4.1) yields true predictions and correct explanations of the behaviour of Ks if the Ks are also Cs, and yields inadequate explanations and predictions in the absence of C in Ks. We do not deduce (4.1) from (4.3). Rather, we deduce the limits under which it holds. In fact, we see immediately that there is a real sense in which we do *not* explain (4.2). Upon discovering (4.3), we discover the falsity of (4.1). *Objectively*, then (4.1) has never functioned in an explanation (as we have defined the term) or in a reasoned prediction, since both these require the premisses to be true. That means, supposing the aim of science to be explanation and prediction, that *objectively* (4.1) is of *no* interest to the scientist.[186a] *Objectively*, (4.1) is not among the propositions of science. Whence, *objectively*, the scientist need not, and does not, explain it. Subjectively, (4.1) was worthy of explanation. But as evidence was acquired which falsified (4.1) and confirmed and did not falsify (4.3) it was discovered that (4.1) was objectively unworthy for explanation. What (4.3) does is not explain (4.1) so much as show why, under certain conditions, namely, when Ks are Cs, it is possible to *get by* with (4.1). (4.3) does not explain (4.1) but rather shows the extent to which the subjectively worthy appeals to (4.1) were in fact objectively justified. It is clear, of course, why the defender of the deductive model must deny that (4.3) literally explains (4.1). For, that would require him to hold (4.3) entails (4.1) when (4.3) is (we may suppose) true and (4.1) is false, something which violates the very idea of deductive logic. But we see that it is not paradoxical to say that, speaking literally, (4.1) is not explained.

What (knowledge of) (4.3) does is make clear why, when the fact that C is relevant is not known (as previously it was not), (4.1) could still be used in certain contexts to successfully explain and predict. Whenever a K is a C, (4.3) yields adequate explanations and predictions. This holds objectively, even when, subjectively, C is not known to be a causally (lawfully) relevant factor (variable). What makes knowledge of (4.1) without knowledge of (4.3) imperfect is just this failure to know the relevance of C. In the absence of knowledge of C it is easy to understand why one might mistake a limited generality (4.1) for a law like (4.3) which holds in all generality. The reason is that when the relevance of C is not known, what one is in effect assuming (if only implicitly) is the truth of (4.4). But under that assumption (4.1) and (4.3) are logically equivalent. To put the point a bit differently, at first, the evidence available makes (4.1) subjectively worthy for explanation and prediction. So that evidence is such that we are ignorant of the relevance of C. To ignore C is to assume (4.3). Thus, the evidence making (4.1) subjectively worthy makes (4.1) equivalent to (4.3). In other words, the evidence is such as to make it subjectively correct to assume (4.1) holds in all generality, when in fact, objectively, (4.1) is limited and only (4.3) holds in all generality. Only at a later stage does the role of C become evident, and (4.3) then surplants (4.1) as the law statement yielding subjectively worthy explanations and predictions. When it does, we see the extent to which (4.1) failed to be objectively worthy (at least to the extent that what are now subjectively worthy explanations are also objectively worthy), i.e., we see explicitly the imperfection of (4.1), and we also see why the evidence we did have made (4.1) subjectively worthy [when not objectively worthy] and why the evidence we now have makes (4.3) subjectively worthy [and objectively more worthy than (4.1) − since this evidence falsifies (4.1) and confirms and does not falsify (4.3), and since the worth of the evidence, being a matter of logic, is something which is objective].

This discussion enables us to look at Galileo's law of falling bodies. This law has been appealed to as falsifying the claim that explanation of laws by laws is deductive. Both Scriven[187] and Feyerabend[188] make this appeal. In fact it is spurious. The above discussion shows it to be so. Galileo's law states that massy objects move towards the surface of the earth on a line perpendicular to it according to the formula

(4.7) $\quad s = \frac{1}{2} k t^2$

where s is the distance fallen, t the time of fall, and k the rate of acceleration, where K is a constant:

(4.8) $\quad k = g$

But (4.8) is in fact false. k is not a constant. Rather,

(4.9) $\quad k = G m/r^2$

where G is a constant, m is the mass of the earth, and r the distance of the object from the centre of the earth. Since r varies continuously with the fall, so does k. On the other hand, if r_1 and r_2 are two distances and k_1 and k_2 the corresponding accelerations, then if the difference between r_1 and r_2 is very small, so is the difference between k_1 and k_2. Thus, for small differenes of r, k is approximately constant. That is, for small differences in r, (4.8) approximates to the truth. Thus, (4.7) understood via (4.8) is imperfect knowledge. It is literally false if taken to hold in all generality, as one might argue Galileo understood it. On the other hand, (4.7) understood via (4.9) is true, and, in fact, process knowledge. This process knowledge for earth-falling object systems was discovered by Newton. This process knowledge shows the limits within which (4.7) understood via (4.8) holds true, namely, in circustances where the distances fallen are small. In fact, if G, m and r are known — as they are — then the exact changes can be calculated. This is simply part of its being process knowledge. Now, Galileo examined objects falling only a short distance. Within that distance, and given the limits of measurement, it is not possible to obtain any evidence but that which confirms (4.8). Thus, even though (4.8) is false, and objectively unworthy for explanation and prediction, it was, given the evidence available, subjectively worthy so far as Galileo was concerned. Newton went further and collected more evidence, e.g., about the fall of the moon. This evidence showed (4.8) to be false, objectively unworthy for explanation and prediction; and confirmed (4.9). And (4.9) in turn showed exactly the limits which render (4.8) and (4.7) understood via (4.8) imperfect. (4.7) understood via (4.8)

THE DEDUCTIVE MODEL OF EXPLANATION 63

corresponds to (4.1); (4.7) understood via (4.9) corresponds to (4.3); and the assumption that differences of r are small corresponds to (4.4).

Scriven and Feyerabend object to the deductive model because they hold that, on the one hand, (4.9) *does* explain (4.8), i.e., Newton *did* explain Galileo, while, on the other hand, (4.9) does not entail (4.8) and, indeed, could not entail (4.8) since (4.9) is true and (4.8) is false. So much the worse for the deductive model, they conclude. Hardly so quickly, the defender retorts. There is, after all, another alternative. They can conclude that explanation and deduction do not amount to the same thing because they affirm both the failure of deduction and the success of the explanation. One can retain the equivalence of explanation and deduction (from true premisses with a law) while affirming (what is patent) the failure of deduction by simply denying that one has a successful explanation. Which is what we shall do. (4.8) is false. It is therefore of *no* interest to science, and *ipso facto* not something to be explained. For this reason, the appeal by Scriven and Feyerabend to this example is completely irrelevant. On the other hand, we must do them the courtesy of accounting for the plausibility of their assumption that (4.9) explains (4.8). This, by the way, is a courtesy they fail to grant their opponents. If one is charging an opponent with intellectual errors, one is under an intellectual obligation to account plausibly for why the errors should come to be committed in the first place. That is the essential diagnostic task. Instead of providing diagnoses in terms of which they can account for the plausibility of the deductive model and the arguments for it (when they direct themselves to the latter, which is rare), these authors are content to simply refute on the one hand, which is the easy half of the task, and to ignore the more difficult diagnostic task, being content, on the other hand, to resort to name calling, suggesting simple perversity and resorting to epithets like "modern schoolmen". Be that as it may, however, the plausibility of the claim that (4.9) explains (4.8) must be accounted for. Fortunately, given our discussion of (4.3) and (4.1) such an account is easily provided. In the first place, (4.9) shows us the conditions under which we can successfully *get by* using (4.8) in explanations and predictions. In the second place, the nature of the evidence for (4.8) and (4.9) makes it clear why (4.8) should be for Galileo subjectively worthy for explanation and prediction even when it is objectively unworthy (save within the limits (4.9) specifies). Thus, we explain using (4.9), not so much (4.8) itself, but the successes and the failures of (4.8). And we account for why Galileo should notice only the successes (making (4.8) subjectively worthy to him) and fail to notice the failures (which make (4.8) objectively unworthy).[188a]

Thus far in our discussion of imperfect knowledge we have more or less taken (4.1) to be representative. As an example of imperfect knowledge (4.1) was understood as claiming to hold in all generality of Ks when it was in fact objectively false. The contrast is to (4.3) which does (we are supposing) hold for all Ks. What makes it reasonable to say (4.1) is knowledge even though it is false is the fact that since it does hold within the limits (4.3) indicates it is indeed an "approximation to the truth". There is in fact an intermediate case, between those of (4.1) and (4.3). And in fact I believe this to be far more common than the (4.1) case – though, for reasons I shall try to indicate, this intermediate case might often be mistaken for cases of the (4.1) sort.

(4.1) is originally asserted (as in the case of Galileo's Law of Falling Bodies) because that there are limits is not known. In the intermediate case, something like (4.1) is known *and* it is known that *there are limits*, but it is not known precisely what these limits are. Indeed, it is often known that these limits are of some certain sort. This contrasts to (4.3). As we saw, involved in (4.3) is knowledge of something like (4.1). Moreover, (4.3) shows that there are limits to this knowledge. But – this is the contrast – (4.3) not only says that *there are* limits; it also says *what* they are, specifically. The intermediate case we are now looking at says only that there are limits, at best determining these limits only generically, as being of a certain sort.

This intermediate case can be exemplified in a way with which we are already familiar, in a schema similar to that of (4.1) and (4.3). In effect this schema is simply the existential generalization of (4.3) with respect to C. If we suppose that the limits are of kind $\mathscr{C}$, then this intermediate case can be represented by the schema

(4.2) $\quad (\exists f) \{ \mathscr{C} f \& (x) [Kx \supset (fx \equiv (Fx \equiv Gx))] \}$

(4.2) is clearly imperfect compared to (4.3). (4.3) yields fully determinate predictions: if we find a K which is F and G we can deduce it is specifically a C. (4.2) does not yield determinate predictions: for a K which is F and G we can deduce only that it is generically an object which is $\mathscr{C}$-ed; that is, that there is a property f which is exemplified by it and which is a property of sort $\mathscr{C}$. In this sense, as in the case of (4.1), and as in the case of statistical generalizations, our knowledge still has "gaps" in it. It is (4.3) that fills in those "gaps".

(4.2) is a law. It is, moreover, (we are assuming) true. Assuming C is $\mathscr{C}$ then (4.2) follows from (4.3), and if (as we are supposing) (4.3) is true then so is (4.2). That (4.2) is true in all generality where (4.1) is not is the contrast between these two cases. This has as a consequence that any explanation which uses (4.2) to explain individual facts is going to be objectively worthy of acceptance. An explanation involving (4.1) was never, strictly speaking, a scientific explanation. In contrast any explanation involving (4.2) is always, objectively, worthy of acceptance as a scientific explanation.

Of course, the knowledge embodied in (4.2) is always imperfect relative to that embodied in (4.3). An explanation using (4.2) will therefore never be as good an explanation as one involving (4.3). But, as we know, that does not make it any the less an explanation.

Since (assuming C is $\mathscr{C}$) (4.3) entails (4.2) by existential generalization, we see immediately how the less imperfect can explain the more imperfect: by simple deduction, existential generalization. To turn this about, we see that explanation of the more by the less imperfect is obtained when one discovers a less imperfect law which is a true instantiation of the more imperfect. This instantiation is what fills in the "gaps" in the more imperfect knowledge, and in doing that it explains the imperfect knowledge. This explanation is by deduction, and no qualifications are needed, as in the case of explaining (4.1) by (4.3).

A good deal of imperfect knowledge is, I believe, of the sort (4.2).[189] Boyle's Law for gases is usually stated in terms of pressure varying inversely as the volume: for gaseous systems

(4.10) $\quad pv = c$

This makes it seem like the (4.1) case, since, as we all know, temperature is a relevant factor. The less imperfect generalization that explicitly introduces temperature is: for gaseous systems

(4.11) $\quad pv = kT$

The relation of (4.11) to (4.10) is that of (4.2) to (4.1). (That they are gaseous systems replaces K.) However, this is not true historically. Boyle actually *knew* temperature was a relevant variable. The trouble was he did not know in what way it was relevant.[190] He knew that *there is* a functional relation connecting p, v and T. He knew, moreover, that this functional relationship had to satisfy the condition that (4.10) be a good approximation within a certain range. This places generic limits on the specific sort of functional relationship that Boyle knew (had reason to believe) was there. So Boyle knew this functional relationship could not be of the form

(4.12) $\quad p = k' T^v$

since there is no way one can obtain from (4.12) that (4.10) is an "approximation to the truth". This formula might even be written on a separate line (as we have done). The formula is not the whole law. Indeed, it is just the top of the iceberg. The statement of the whole law would spread over onto other lines (as we let 'gaseous system (= K)' spread over onto the preceding line). It is what is on the adjoining lines that will determine whether the formula is meant in such a way as to have a (4.1) case or a (4.2) case. If one concentrates on the formula, ignores the adjoining lines,[191] one will easily mistake cases like (4.10) for cases like (4.1).

Cases like (4.2) are, I think, the rule. Certainly, it is clear that all imperfect knowledge can be more or less represented after the fashion of (4.2). If our knowledge is imperfect there either we do not know a relevant variable or we do not know the boundary conditions or we do not know the functional relations which are the process law. If we recognize the presence of imperfection (so the case will not be like (4.1)), then if our aim is to replace this imperfect knowledge with process knowledge, then that aim (if it is reasonable) must be guided by the belief that *there is* a relevant variable to be discovered, that *there is* some boundary condition to be discovered, that *there is* a set of functional relations constituting a process law. If our knowledge is imperfect and that imperfection is recognized then the most appropriate way to express it is as a generality which also involves our existential quantification over properties of the systems to which the generality applies, that is, the systems which are in the range of the universal quantifier which makes the proposition a generality. The existential quantification ensures that the statement about the systems the law describes will be of a generic sort, rather than specific. It ensures the generic nature, and thereby the "gappiness", of imperfect knowledge.

There are, however, peculiarities about laws of the sort (4.2), particularly when it comes to predicting. These peculiarities have been exploited by Scriven and others in order to attack the thesis, entailed by the deductive model, that explanation and prediction are symmetric. We shall have to return to this matter in both Chapters 2 and 3.

1.5. THAT TECHNICAL RULES OF COMPUTATION ARE LAWS

The position we are defending is that of the deductivist. And we are interpreting that position in a traditional empiricist manner: the concepts appearing in the laws must satisfy the empiricst meaning criterion: they must either designate directly observable properties and relations or be explicitly defined in terms of concepts which do.[192] Since many of the most important concepts of science are, in a straight-forward sense, operational, the position has been called operationist. Now, Popper identifies such operationism as a version of what he calls "instrumentalism".[193] And he offers a lengthy argument against instrumentalism.[194] We should investigate the extent to which his arguments tell against the position we are here defending.

We insist that the concepts of science *ought* to satisfy the empiricst meaning criterion. It is to be granted that a good deal of what is science was once embedded in various sorts of metaphysical frameworks, Aristotelian, atomistic, and so on, the basic concepts of which did not satisfy the empiricist meaning criterion. This means those metaphysics must be labelled non-scientific. This must be done even though it is also granted, what is evidently the case, that good science eventually grew out of this non-scientific or pre-scientific metaphysics — metaphysics which must, from the empiricst standpoint, be labelled non-sense. Popper has remarked[195] that myth, e.g., atomism, has often developed testable components, and commented against the empiricist that "It would hardly contribute to clarity if we were to say that these theories are nonsensical gibberish in one stage of their development, and then suddenly became good sense in another."[196] One can only respond to such as this, surely, by insisting that it is clarity alone that demands one distinguish the myths with no empirical meaning from their linear descendents which do have empirical meaning. Rhetoric, and even rhetoric in the name of clarity, cannot confuse distinctions which clarity demands can and must be made. It is, of course, to secure that empirical testability upon which Popper so correctly insists that we require the concepts of science to conform to the empiricst meaning criterion. But the same insistence upon complete hook-up to the empirical world means that laws of science are straight matter-of-fact generalities. This means that they automatically satisfy the demands which a technologist places upon knowledge, that it be *useful*, capable of application. The technologist has certain pragmatic aims, and demands knowledge as a means towards satisfying those aims. The desire for knowledge as a means lays down the condition that any knowledge of interest to the technologist must be capable of application. It must therefore apply to the empirical world in which the technologist works, and any knowledge which goes beyond that world is of no interest to him. Thus, the knowledge in which the technologist is interested is of the same general sort as the knowledge in which the empiricist scientist is interested. (The 'empiricist' in 'empiricist scientist' is literally redundant, but is needed here to distinguish the view from other, false, views of what science is, e.g., Popper's.) It is this identification of the scientist's interests with that of the technologist to which Popper objects. Those interests are, he argues, quite divergent. It is this argument against the position we are defending — operationism, or instrumentalism, or empiricism, or whatever you may wish to call it — that I propose here to examine.

Popper argues[197] that there is a profound difference between "pure" theories and technological computation rules. The ways in which computation rules are *tried out* are different from the ways in which theories are *tested*. This difference stems from a logical difference between theories and computation rules. Theories are tested by attempting to refute them. And if refuted, they are discarded. We select for our tests those crucial cases where we should expect the theory to fail if it is not true. In contrast, computation rules are tried out in various areas in order to discover the limits of their applicability. The aim is not to see where they fail, but to see where they work. And here, "work" does not presuppose truth. For instrumentalist purposes of application a theory may continue to be used even after its refutation, that is, used within the limits of its applicability. Thus, an astronomer who believes that Newton's theory has turned out to be false will not hesitate to apply its formalism within the limits of its applicability.

> Summing up we may say that instrumentalism is unable to account for the importance to pure science of testing severely even the most remote implications of its theories, since it is unable to account for the pure scientist's interest in truth and falsity. In contrast to the highly critical attitude requisite in the pure scientist, the attitude of instrumentalism (like that of applied science) is one of complacency at the success of applications.[198]

In this argument Popper is not as clear as one might reasonably expect of one who so strongly proclaims the value of clarity. There is no careful distinction between knowledge or what is known, the reason why one finds it desirable to know what is known, and the means by which one comes to know what is known. What is known is a body of general facts, laws of nature. One may find this knowledge valuable for its own sake. In that case one is simply idly curious, and just wants to know for the sake of knowing. Or one may find knowledge valuable as a means. One wishes to know laws of nature because such knowledge is knowledge of causes by which valued ends may be achieved. In this case it is those further ends, one's pragmatic interests, that render the knowledge valuable. The difference is attitude, aiming at knowledge out of idle curiosity and aiming at knowledge out of some pragmatic interest, that distinguishes the pure scientist from the technologist. The means to acquiring knowledge is the experimental method, the rules of scientific research. So long as both the technologist and the "pure" scientist aim at the same thing, knowledge of lawful regularities, the methods of research will be the same. Popper distinguishes the technologist from the "pure" scientist by ascribing to the former the aim of *applying* knowledge and to the latter the aim of *testing* it. But there is no legitimate contrast here. Knowledge is applied *after* it is acquired. Testing occurs in order to acquire it. Testing is *prior* to the acquisition of knowledge. Since the technologist aims at applying knowledge, he must first acquire it. So he has an interest in testing as much as any "pure" scientist. Testing and applying do not excluded each other.

But Popper has another contrast which he uses. This is one of research methods. I have in mind here his distinction between "testing" and "trying out". The scientist tests in order to acquire knowledge. The technologist tries out in order to acquire an applicable computation rule. Testing aims at refuting, at eliminating falsehoods. Trying out does not aim at eliminating falsehoods. The latter is evident from the fact that the technologist applies, and successfully applies, rules that are known strictly to be false. The question we must ask, then, is whether the technologist actually does use a known-to-be-false rule. To this question we can, I believe, give a negative answer: the technologist does not knowingly apply a rule he knows to be false. (Unknowingly he may do so, but that is a quite different matter. It happens regularly, of course, and for the simple reason that technologists are fallible. But so are scientists, which means the point is irrelevant to the distinction Popper is trying to get us to accept.)

The technologist is motivated by some pragmatic interest. He is interested, for example, in K's being G. As usual, we may suppose K the class of persons, and G some disease. The technologist (physician) is interested in the germs which are the cause, F, of this disease. For if he knows that cause, he can prevent the disease by removing that cause. If he locates a K which is F he immediately acts to change it to not-F. For, he knows F is necessary and sufficient to G, and aiming to remove G, he acts to bring the K to be not-F. The knowledge he is applying in his determination to make the K not-F is the knowledge embodied in the rule that

(5.1) $F \equiv G$

However, he knows that there are Ks for which this rule does not apply. He knows that the rule works only for Ks which are C. So he applies (5.1) to an individual if and only if the individual is a K which is a C. If (5.1) is thought of as universally quantified

(5.1*) $(x)(Fx \equiv Gx)$

then the technologist's applying (5.1) if and only if the K is also a C is equivalent to saying the variable in (5.1*) ranges over all *and only* those things which are both K and C. His action in restricting his manipulations of F to Cs shows the scientist treats (5.1*) as this sort of limited generality. The same point can be made in another way. To say the technologist applies (5.1) is to say he manipulates F in order to bring about the desired state of G. But he applies (5.1) only in the presence of C. So the *knowledge upon which he acts* is more correctly represented by

(5.1**) $C \equiv (F \equiv G)$

than by (5.1). Both ways of putting the point make clear that if one makes explicit everything the technologist knows as he acts regularly to achieve his desired end then that knowledge would be represented by (4.3)

$(x) [Kx \supset (Cx \equiv (Fx \equiv Gx))]$

The technologist's use of (5.1) makes it appear the knowledge upon which he is acting is (4.1)

$(x)(Kx \supset (Fx \equiv Gx))$

And the technologists' use of (5.1) even after the recognition of its limited validity makes it seem he is relying upon (4.1) even after the latter has been falsified. (Compare our remarks above on (5.1).) But the appearance that he is using the false generality (4.1) is just appearance. What he is in fact using and acting upon is (4.3). And (4.3) is true. Popper notwithstanding, and notwithstanding appearances to the contrary, the technologist does not (knowingly) act upon known-to-be-false laws. And these comments hold equally of Popper's astronomer who uses Newtonian mechanics. The astronomer is not assuming what is false, that Newtonian mechanics holds in all generality, and using it anyway; rather, he is assuming, what is true, the limited validity of Newtonian mechanics within the realm in which he is interested, and is acting upon that truth.

The only question that now arises is whether the discovery of C which limits the validity of technologist's rule requires a methodology other than that used by the "pure" scientist. Does that discovery require "trying out" rather than "testing"? The answer is surely negative. C is discovered by the experimental method. To be sure, the technologist aims at knowing C in order to know the limits of applicability of his rule. But this is also to aim at knowing a general fact, namely, the fact (4.3). And this sort of general fact is in no way essentially different from the general facts the "pure" scientist aims at knowing. From which it follows that the method justifying the law-assertive attitude is the same for both persons. In both cases one puts the generality to the experimental *test*.

Popper identified operationism with instrumentalism, and instrumentalism with technology. He then argued the technologist was to be distinguished from the "pure" scientist and concluded that operationism was inadequate as a philosophy of science. We now see that his distinction between technology and "pure" science, at least as made in the way Popper needs to make it for his argument to go through, is not tenable. So Popper has not established the views we are defending are an inadequate philosophy of science.

CHAPTER 2

THE REASONABILITY OF THE DEDUCTIVE MODEL

Thus far all we have done is define 'explanation'. To be sure, we have shown that some purported counter-examples do not show this to be an unreasonable definition. And much of the present chapter will also be devoted to an examination of such proposed counter-examples. In particular, we will try to show that, insofar as ordinary causal discourse does provide explanations, there is nothing in it which runs contrary to the deductivist thesis. But to refute one's opponent is hardly to make a positive case. Now, in one sense at least, I don't think a case is needed, so far as the descriptive utility of our definition is concerned. It seems to me that scientific explanations are as a matter of *obvious* fact (cf. Newton) deductive. But we must not be content with mere description. We wish also to be normative. It is therefore necessary to show that it is *reasonable* so to define our terms that explanations are constituted by deduction from true premises, among which at least one generality which is essential to the deduction. It is to this task that we now turn. It shall be argued in Section 2.1 that it is *reasonable* that scientific explanations be deductive; and then, second, that it is *reasonable* that process knowledge is the ideal of scientific explanation. Subsequent sections will challenge criticisms of these two theses.

These criticisms turn largely on trying to draw a wedge between explanation and prediction, or between causation and correlation, where the former explains and the latter does not. Our discussion in the previous chapter showed that, contrary to what Popper has suggested, the defender of the deductive model has no easy way to exclude "mere" correlations from the category of laws that yield "real" explanations. There is a continuum that leads from "mere" predictive devices, like Ptolemy's, to laws, like Newton's for the solar system, that are acknowledged by all to be fully explanatory. Popper notwithstanding, the logical form of the one is that of the other and the evidence supporting the one is not different in kind from the evidence supporting the other. Historically, of course, such a distinction was drawn. But, as we saw, it was drawn by reference to non-scientific, metaphysical, criteria for what it is to be a successful explanation. The point is that, for the deductivist, or at least for one of the Humean sort, the distinction makes, in the last analysis, no sense: causal laws and correlations are *not* different in *logical kind*. It is this last that is used to challenge the deductivist: causation and correlation *are* different in kind, it is suggested, and examples and arguments are advanced in support of that position. It is to these that we must address ourselves.

Of course, if the normative case of Section 2.1 of this chapter is successful, then no reply is needed, so far as the principle is concerned. Still, one's case is always stronger if one can show exactly where one's opponent's case has gone wrong. But in general the point we will be after is not merely that of taking the anti-deductivist's case apart. For, the arguments and the examples of the anti-deductivists do not come from nowhere: they do in general have a valid point. It is just that the anti-deductivist conclusion does not follow from those valid points. Our discussion will therefore for the most part consist in finding what is the real point that is being made by the various critics of the deductive-nomological model, and in showing that the anti-deductivist conclusion that is claimed to follow does not, or, in other words, showing that the deductivist can agree with the essentials of the point being made while yet continuing to defend the deductive model of explanation.

Central to the strategy we will employ in discussing the critics of the deductive model will be the distinction, to be defended in Section 2.1, between explanations that are better or worse than others by being based on knowledge of laws that are less or more imperfect. For example, we will argue that while correlation *is* explanatory — because it is deductive and yields predictions — nonetheless, the explanations are not as good as, since more imperfect than, explanations in terms of causal laws: critics who use the distinction have hit upon a distinction that is indeed valid, but it is a distinction between a weak explanation and one that is less weak, rather than, as they contend, one between a non-explanation and an explanation. Once this is seen it is clear that the deductivist can accept the distinction while yet defending the deductive model. We shall elaborate upon this particular point in detail below. And the strategy here will be similar to that deployed throughout the present chapter: the critics of the deductive model have a point, but mistake it for a criticism of that model because they overlook the relevance of the distinction between explanations that are better and worse, that is, the less and the more imperfect.

Before we begin, however, there is one point that is worth mentioning, in order to dispell right from the beginning one illusion about the deductivist position: *None of the arguments that we shall be advancing in defence of the deductive model will make any essential reference to physics*. We will not, therefore, be appealing to the prestige of this science in order to bolster

THE REASONABILITY OF THE DEDUCTIVE MODEL   75

irrationally an attempt to transfer deductive explanations from it to the other sciences. What we will be doing will not be subject to the following criticism:

> It turns out that explanations given in everyday life or historians are denied the title "scientific explanation" not because they are not given in a science or by scientists, but because they fail to deduce what is explained from premisses including at least one law. Science is supposed to be a source of examples of *good* explanations and not the locus of a special concept of explanation. "Scientific" has chiefly an honorific function in "scientific explanation". The impression that there is a special or better scientific way of explaining things, as compared with everyday or other "pre-scientific" ways, is an illusion created by an inappropriate appeal to the prestige of physical science.[1]

To comment: In the first place, everyday and historians' explanations *are* deductive (as we shall argue below), and therefore satisfy the definition given in the present essay. In the second place, "scientific explanations", i.e., those (according to Collins) of physics, do not yield a special concept of explanation. Those of physics *and* those of everyday life are all deductive, including one law among the premisses. In the third place, "scientific explanations", those of physics, *are* good, and better than those of everyday life. This is not because the one is and the other is not deductive (since both are deductive), but because only in physics is there process knowledge. The explanations of everyday life are inferior because they involve only imperfect knowledge. That means, in particular, that phsyics *deserves* the prestige. But, of course, not simply because its explanations are deductive. In the fourth place, there *is* a "pre-scientific" way of thinking, that of Aristotle, which does in fact infect a good deal of our everyday attempts at explanation, and which leads us to overlook the imperfection of those explanations. It is this "pre-scientific" mode of thought which leads us to think our imperfect explanations of everyday life are less imperfect and therefore better than they are in fact are. (Of this, more in Chapter Three.) And, viewing our everyday explanations as better than they are we are tempted to sneer at the justly deserved prestige of the explanations in physics.

2.1. WHY OUGHT THE DEDUCTIVE MODEL BE ACCEPTED?

We have said explanation is by deduction from true premisses which include at least one law statement. Is it reasonable for explanations to be of that sort?

Begin with explanations of individual facts and events. Here we have an event of a certain kind or type to be explained (the explanandum) and an event of a certain kind or type which is to explain it (the initial conditions). As I have said, I take it to be part of the idea of a scientific explanation that if the explaining events occur then the event which is the explanandum must occur: the occurrence of the initial conditions somehow *necessitates* the occurrence of the explanandum.[2] Assume that one wants such explanations. The question then to be raised is precisely what is the sense of 'must' and of 'necessitate' which were used to state the idea of what is wanted. These are problematic terms, and must be explicated.[3] The deductive model provides such an explication. The law statement (if it is true) states a connection among kinds of events, relating specifically a kind instanced by the initial conditions to a kind instanced by the explanandum. The deductive model holds that the explanation consists of the valid deduction of the explanandum from the true law statement and the initial conditions. Now, the notion of deductive validity is such that, if the premisses are true and the argument is valid then the conclusion *must* be true, i.e., given the initial conditions and the law the explanandum must occur; or, at least, the former so explicates the latter. This explication is in terms of *objective* criteria. Whether or not the premisses are true is objective, depends only on what the sentences are about. Whether or not the argument, i.e., the explanation, is deductively valid is objective, depends on the logical form of the argument, on the meaning of sentences involved. Neither truth nor validity depends upon either the explainer or the explainee in any way, e.g., they don't depend on whether the explainee feels satisfied by the explanation. (Indeed, that is precisely why we say truth and validity are objective.)

So far as I know there is no other explication of the problematic 'must'. Of course, as we noted previously in Chapter 1, some have thought that there are non-lawful "nomic" connections relating initial conditions to explanandum, these "nomic" connections somehow meaning that the former necessitates the latter. This view requires one to introduce non-truth-functional connectives, "nomic ties". The standard explication of the analytic/synthetic distinction requires us to have none but truth-functional connectives.[4] We should, therefore, exclude these "nomic ties". If we do this then we must provide an alternative account of why one event, *qua* being of a certain type, should "necessitate" some other, logically independent, event, where the second event is an event of some different type. One wants a relation of some sort between the events, but between them *qua* being of certain types. A relation simply among the types will not suffice. For even if F bears R to G the fact that F is exemplified in an individual in no way establishes G will also be exemplified. So this won't do to explicate the idea that the exemplification of *F* "necessitates" the exemplification of *G*.[5] Nor will a relation simply among the events suffice. For then that the events are of any type at all becomes quite irrelevant to the connection obtaining, and the "necessitation" of the one by the other to occur. The only "relation" available for "tying together" events *qua* of certain types in such a way as can make reasonable sense of the "necessitation" of the one by the other is the "relation" represented by "$(x)(Fx \supset Gx)$". At least so far as I know there is no other explication of the notion of one individual event "necessitating" another individual event, no other explication of the problematic 'must'.

Anticipating that philosophers will find no other explication of the 'must' besides that provided by the deductive model, we may conclude that the latter's explication is successful.

Now the important point to be emphasized is that the deductive model yields *objective* criteria of what is to count as an explanation. Truth of premisses and deductive validity of the argument are jointly necessary and sufficient for a sequence of sentences to constitute an explanation. Once the linguistic conventions have been laid down whether these two conditions obtain in no way depends on either explainer or explainee. Suppose the explainer then presents a set of sentences as an explanation, or the explainee accepts a set as an explanation. Since the criteria are objective we can say whether the explainer *ought* to have presented them or whether the explainee *ought* to have accepted them as constituting an explanation. Because the criteria are objective we are able to *judge and evaluate* the behaviour of explainor and explainee in the transaction of explaining.

Of course, if the explainee does not want a scientific explanation, that is, one which shows how one event necessitates another, in the sense of 'necessitates' that the deductive model explicates – if such an explanation is not wanted, then he is not to be judged adversely for rejecting an explanation satisfying our objective criteria or for accepting an explanation that fails to satisfy them. Our evaluations about whether or not one ought to accept an explanation depend, of course, upon what one wants. But, *if* one wants such an explanation, then we have such objective criteria.

As for wanting such an explanation, can one argue that one *ought* to want such explanations? Well, to a certain extent one can. If such explanations are available, then one does thereby have knowledge of a sort that makes intelligent interference and control possible (supposing such interference to be technologically feasible): *knowledge of causes is knowledge of means*. Thus, pragmatic interests have such explanations as their aim or object. In this case one ought to want such explanations because the knowledge they embody is a knowledge of means to ends one has. But if the question is why or whether one ought to want such explanations as ends in themselves, why or whether one ought to be idly curious about lawful connections, then no argument is available. Except that the acquisition of such knowledge is, for some people, an end in itself, an activity that is a joy to them, and, other things being equal, people ought to be allowed to exercise and develop their faculties and capacities in those ways that give them pleasure.

We have, then, certain *cognitive interests* – idle curiosity and pragmatic interests – in empirical knowledge of natural (including social) processes. These cognitive interests are not only in what *is* happening but also in what *will* happen in the process. Now, to know what *will* happen is to *predict*. But our pragmatic interests do *not* want this knowledge to be *mere guessing*: rather, it must be *reasoned prediction*. Our pragmatic interests aim at knowing the *future by reasoned inference from the present*. To put the same point sightly differently, the knowledge of the future at which our pragmatic interests aim is *knowledge of the future based on knowledge of how the present determines the future*. Pragmatic interests must aim at knowledge of how the present determines the future because otherwise it would not be knowledge that was relevant to *control*, and what is control other than determining the future by manipulating the present? Our pragmatic interests, then, aim at knowledge of how the present determines the future. But the latter is knowledge of how the future *must* be, given the way the process presently is. However, as we have said, this problematic 'must' receives its only reasonable explication in terms of *deduction from laws*. Again, for purposes of intelligent interference, our pragmatic interests aim at knowledge of what *would* happen *were* the present state of the the process to be different, and this again must be a *reasoned inference* from a contrary-to-fact assumption about the process to what would go on to happen in the process if the assumption were true. And again, such a reasoned inference can only be in terms of *deduction from laws*. Our pragmatic interests therefore aim at knowledge of *matter-of-fact generalities* that permit prediction and contrary-to-fact inference. But we can also be simply idly curious in *this sort of knowledge*. This idle curiosity will cast a wider net than the pragmatic interests, including not only the future but also the past, and not only what would happen were certain things to be different but also what must have had to have happened in the past if things were now to be other than they are. But the *sort* of knowledge at which this idle curiosity aims is the *same sort* at which our pragmatic interests aim, viz, *knowledge of matter-of-fact generalities capable of being used to predict and to support counterfactual inferences.*

Now, when the cook *understands* the recipe and how it will and ought to work, then he knows what will happen when various things occur, and what will happen when he interferes in certain ways, and what would happen if he were to do certain things. Similarly, the mechanic understands the engine when he knows what will happen when certain things are done and what would happen if certain other things were to be done. Thus, *when one has the knowledge of empirical generalities that describe the process and at which our pragmatic interests aim, then one understands the process*. But to understand is to explain: *to have knowledge that provides understanding is to have knowledge that explains*. That is simply how the words 'understand' and 'explanation' are used. The sort of understanding at which our pragmatic interests aim is therefore *knowledge that explains*. But this knowledge is *causal knowledge*, or, what is the same thing, we are suggesting, *knowledge of true matter-of-fact generalities*. We thus see that our pragmatic interests aim at knowledge of laws that permits prediction and that such knowledge provides explanations. Thus, *our pragmatic interests aim at knowledge and understanding in which explanation and prediction are symmetrical*. Idle curiosity aims to acquire this same sort of knowledge about all aspects of a process, and not just those aspects that our pragmatic interests lead us to be concerned with. But it is the *same* sort of knowledge. *We may therefore conclude that the cognitive interests, that lead us to be concerned with the sort of knowledge that empirical science provides, define a concept of understanding and of explanation in which those notions are completely exhausted by the notion of predictability on the basis of laws* (*true matter-of-fact generalities*).

This principle, that the notion of lawful prediction exhausts the concept of scientific explanation we may call the "principle of predictability". We shall make use of this principle throughout our discussions, particularly in Section 3.2. In the present chapter we shall be concerned to dispute those who dispute the thesis of a symmetry between explanation and prediction, and who claim that there are explanations that are not predictions and that there are law-based predictions that are not explanations. The important point to be made is that *by appeal to our cognitive interests in empirical knowledge of natural* (*including social*) *processes*, we have shown that the concept of explanation that is defined by the principle of predictability and the symmetry of explanation and prediction thesis is a *justified concept* , that is, justified relative to those cognitive interests: *explanations of the sort picked out by this concept satisfy those cognitive interests*. We have thus shown that, given those cognitive interests, this concept of scientific explanation *ought* to be accepted. In other words, our thesis is a *normative one*. If, therefore, many ordinary explanations do not conform to this standard, then that does not mean that they are counter-examples to our point of view. All that follows is that some have accepted explanations that they ought not to have accepted. On the other hand, the cognitive interests to which we have appealed are common enough, at least the pragmatic interests, and we should therefore expect that many ordinary everyday explanations do meet these standards. This, indeed, is just what we shall argue.

We have now, I believe, justified, so far as is possible, the choice of the deductive-nomological model as definitory of the explanation of individual facts and events. Two comments will clarify some matters. (1) The laws that occur in the explanation may be either those of process knowledge or those of imperfect knowledge. (2) We may have two or more incompatible law statements which are taken hypothetically (that is, which are such that the evidence available is not sufficient to exclude one, make the other subjectively worthy), but which are also such that it is reasonable, given background theory, etc., to treat both as possibly satisfying the criteria which define an explanation via laws as objectively worthy. Here we have two tentative *candidates* for justified presentation and acceptance in an explaining transaction. Candidacy for explanation, in this sense, is a matter of the subjective worthiness of the explanations. When we have such candidates for explanations, we are not certain about the truth of the premisses – the lack of evidence is such that the law premiss is not subjectively worthy. It is quite compatible with this that the argument be deductive. That is, the fact that premisses are uncertain at times, meaning that at times we subjectively have but candidate explanations – this fact in no way confutes the deductive model. All it indicates is that further research, the active application of the scientific method, is needed to transform candidacy into acceptability.[6]

In the case of explaining laws by theories, the case for deduction as a necessary condition is much the same. We suppose that theories show why laws must be this way rather than some other way. We then explicate this 'must' via deduction from (or instantiation of) true premisses. Again the criteria are objective. They therefore yield objective standards for judging whether an explainee ought to have accepted a set of sentences as explaining a certain law (supposing he wanted an explanation of the sort the deductive model explicates).

We shall not in the present essay, in detail enter into the question of the criteria for the worth of theories, however.[7] In this section we shall attempt no more than to justify the claim that process knowledge is the ideal of scientific explanation of individual facts.

Scientists, *qua* scientists, aim at knowledge of laws in order to explain and predict. The motive behind this aim may be of two sorts. It may, in the first place, be the motive of idle curiosity. Or it may, in the second place, be some pragmatic interest. These two motives serve to distinguish "pure science", the motive of which is idle curiosity, from "practical research", which has some "application" in mind. The distinction between "pure" science and science which is not pure is not one of subject-matter: in both cases, the aim is knowledge of laws. The difference is simply one of motive. One thing is clear, however: the motive of idle curiosity casts a wider net than that of any pragmatic interest. If one's interest is pragmatic one will stop the search for knowledge as soon as one's knowledge of laws is sufficient to enable one to achieve the end determining one's pragmatic interest. In contrast, if one's interest is simply idle curiosity one will continue the search for knowledge even if there are no (immediately obvious) "practical" consequences. Clearly one can have pragmatic interests which motivate investigation of laws when the motive of idle curiosity is totally absent. And, in fact, only since Galileo has idle curiosity ever been a motive in the search for lawful knowledge. In the Middle Ages it was theology, not science, that was studied for its own sake. (Or was knowledge of it, too, motivated by a pragmatic interest, viz, *ad maioram gloriam Dei*?) In the Renaissance it was humanistic or literary learning that was studied for its own sake. Only with such figures as Galileo, Bacon and Descartes did science acquire sufficient status to be considered worthy of one's idle curiosity.

We are asking, what is the ideal of scientific explanation? Scientific explanation is in terms of deduction from laws, so this question amounts to asking what science might reasonably aim at by way of knowledge of laws. The 'reasonable' here is meant to directly exclude two things. First, it is not reasonable to attempt to overcome the limitations of induction. So the ideal cannot be one in which such limitations are overcome. Second, it is not reasonable to attempt to achieve absolute precision of measurement. So that, too, is excluded from the ideal. In general, we may say, one excludes from the ideal anything it is not possible to attain.

It is important to recognize this. Some, unfortunately, do not. Thus, some have defined the ideal as "complete truth". Ackoff, for example, states that the "scientific ideal of perfect knowledge" is the "complete attainment of truth".[8] If this is the end of science, then *all* knowledge is imperfect relative to this ideal. The imperfection-of-*all*-knowledge thesis is re-inforced by the suggestion that *all* knowledge is approximate.

All models of problem situations are approximate representations of these situations. They are generally simpler than the situations they represent, which are usually so complex — although in many cases, fortunately, only in unimportant detail — that an 'exact' representation (even if possible) would lead to hopeless mathematical complexity. Therefore, the problem confronting the model builder is to attain a best or good balance between accurate representation and mathematical manageability. Practical considerations (e.g., limited time, money, personnel, or computing facilities) almost always require some compromise of accuracy.[9]

However, if a factor is not relevant to the interactions one is interested in then to omit it is not to omit an important factor, even though it may be a truth. Thus, one can omit the colours of the planets when explaining positions and velocities; colour is in fact not a relevant variable. The omission of factors irrelevant to the processes one is interested in explaining is not in any reasonable sense a defect. So process knowledge, though it may not be the "whole truth", and may in that sense be an approximation, need not be considered defective or any the less perfect. What must be contrasted to this is imperfect knowledge which omits not just part of the "whole truth" but also part of the *important* truth, a relevant variable, or closure conditions, or details of the process law. Furthermore, all knowledge of laws is limited in two respects. All such knowledge is inductive and cannot overcome the Humean limitations of inductive inference. In addition there are in fact limits on the accuracy of measurement,[10] and no knowledge can overcome these. Both these limitations infect both perfect and imperfect knowledge, and to this extent all knowledge is an approximation. But these two limits cannot be overcome. The world being as it is, it is unreasonable to so define the ideal as to overcome these limits. On the other hand, within the framework these limitations impose on all knowledge of laws, it is reasonable to try to overcome the limitations of imperfect knowledge, and therefore to define process knowledge as the ideal of scientific explanation. Of course, in any concrete case, practical considerations of time, money, etc., may lead one to compromise and settle for imperfect knowledge, and an approximate representation of the process that one is interested in. But in principle one could eliminate compromises of this sort that lead one to rest content with knowledge known to be imperfect. In contrast, one could not in principle overcome the limits of measurement and induction: accepting these limits is not a matter of compromise, but simply coming to terms with the way the world is.

THE REASONABILITY OF THE DEDUCTIVE MODEL

In short, by defining the scientific ideal as the "complete attainment of truth" Ackoff proposes an unreasonable ideal, and therefore one that in no way compromises either the distinction of this essay between process knowledge and imperfect knowledge or the argument that the former is the ideal of scientific explanation.

With this qualification in mind, that we must exclude the unreasonable, we may sketch an answer to our question, what science might aim at by way of knowledge of laws. But it will pay to once again transform it slightly.

We have suggested that science aims at knowledge of laws out of two motives, idle curiosity and pragmatic interests. Our question therefore becomes this: what might science (reasonably) want to know by way of laws out of either idle curiosity or pragmatic interest, so far as the explanation of individual facts and events is concerned? But even this question should be made more precise.

To do this, we must recall how we above justified the deductive-nomological model of explanation. What we did was appeal to the cognitive interests of idle curiosity and our pragmatic interests in empirical knowledge of natural (including social) processes. Relative to these we argued that we *ought* to accept the deductive-nomological model, that concept of explanation in which explanation and prediction are symmetrical and in which the principle of predictability exhausts the concept of explanation. We are now raising the issue whether such interests might not also justify our claim that process knowledge is the *ideal* of scientific explanation. If such an argument is to be made, then, given that the principle of predictability defines the concept of scientific explanation, it would seem that one explanation is better than another provided that it is *better at predicting: Two arguments are of equal explanatory power just in case that they yield just the same predictions, and one argument will have greater explanatory power just to the extent that it is better at predicting*. This definition of 'equal explanatory power' will be of importance later, in Section 3.2 in particular, but the immediate task is that of getting hold of the idea of one argument being "better at predicting" than another. The difference here cannot be between arguments with a false generality as their major premiss and those with a true generality. For, the former do not yield reasoned predictions and therefore cannot be counted as explanations. So, the difference between better and worse must be one between the laws involved and the sorts of facts that the logical forms of those laws permit one to predict. The relevant point seems to be that some laws will yield more detailed and more specific predictions about processes than will other laws. "Water when heated boils" leaves out many details of the thermodynamic process that it describes and that can be supplied by the more detailed laws of the science of thermodynamics: how much heat is needed, how long it will take to boil, just in what way the omitted variable of air pressure is relevant, and so on. Take the case of length of time required to boil. The statement that "Water when heated boils" asserts that *there is* a time at which the water will boil but does say *specifically* what that time is. Thermodynamics, in contrast, can make more specific predictions. The prediction deriving from "Water when heated boils" is a *determinable prediction*, whereas from the laws of thermodynamics we can derive *determinate* predictions. It is the difference between a law like

$$(x)(Fx \supset Gx)$$

which yields determinate predictions, and a law like

$$(x)[Fx \supset (\exists g)(\mathcal{G}g \;\&\; gx)]$$

which yields only a determinable prediction. The second is, to use Mackie's useful phrase, "gappy" and this gappiness, this indeterminacy in its predictions, makes it less adequate as a predictor than the first. This makes clear the line of argument that we must take if we are to establish that process knowledge is, relative to our cognitive interests, the ideal of scientific explanation: if we can argue that, relative to the cognitive interests that justify and define the deductive-nomological model, is better at telling us what we could want to know than any other sort of knowledge, then we shall have established that it is indeed the ideal. And the knowledge that process knowledge must yield is with respect to the making of predictions (including postdictions) and counterfactual inferences.

In this respect, then, does process knowledge yield what one might (reasonably) want to know out of either idle curiosity or pragmatic interest? Well: with respect to the relevant variables, it tells you what will happen in the system, what did happen, what would happen if certain variables were to have different values, what would have had to have been the case if a variable were now to be of a different value, whether a variable can take on a certain value in the system, and if it can, what change must occur (be made to occur) if that variable is to take on that value. That, I suggest, is everything that one could reasonably want to know about the system in respect of the relevant variables out of either idle curiosity or pragmatic interest. What more could one want to know? Nothing, so far as I can see. Certainly, it is knowledge sufficient for explanation. A person understands and can therefore explain a system, e.g. a machine or a production process or a system consisting of a wireless transmitter at one place and a receiver at another — he can explain any such system just in case he call tell us what will happen to the machine or process or system, how the action of the parts is functionally related to the action of the other parts, what would happen if the parts were changed in certain ways, and so on. Process knowledge tells one everything one needs to know in order to intelligently interfere in the operations of the system and bring about (if one can) what one wants to bring about. In short, both our idle curiosity and our pragmatic interests will be fully satisfied by process knowledge.[11]

I therefore conclude that if we have process knowledge (whether or not we have it is an objective matter) — *if* we have such knowledge, then we know all we could reasonably want to know about a system in respect of the relevant variables. That means process knowledge is indeed the ideal of scientific explanation.

In point of fact, of course, a scientist may be satisfied with less than the ideal. Three reasons for this are worth mentioning. *First*: It may turn that in a certain area it is unreasonable to aspire to the ideal because we come to know that in fact the ideal does not hold in that area. Nature has no obligation to conform to our ideals. The world guarantees no story a happy ending, not even that of science. To suppose otherwise is infantile. To take an example, consider the sequence of events which consists of a long sequence of losses of an unbiassed coin. With respect to such a sequence it is known that the best we can do is acquire knowledge of statistical laws (the probability of the state heads is one-half, of tails one-half). It is known that process knowledge cannot be attained for such a system. It is therefore unreasonable to try to attain such knowledge in this case. *Second*: It may be that there is no knowledge which makes process knowledge unreasonable as an ideal for an area, while there is sufficient reason, owing, e.g., to the complexity of the subject, to rule it out for any immediate purposes, or, indeed, for purposes in the foreseable future. This is the state of all social science. *Third*: Both the first two reasons for rejecting process knowledge as the ideal applied whatever the motive was for science. This third reason occurs only where the motive is a pragmatic interest. Idle curiosity always wants to know as much as is reasonably possible. Not so pragmatic interest. Medicine would be satisfied with a cure for cancer. Doctors will leave it to physiologists to work out the complete physiological process by which the cure is effected. Thus, if one's interest is pragmatic, knowledge far short of the ideal may suffice to satisfy.

We noticed earlier that idle curiosity so to speak casts a wider net than any pragmatic interest. It follows that in general only idle curiosity, not pragmatic interest, will lead one to aim at the ideal, to aim at process knowledge. Pragmatic interest will lead one to be satisfied often with much less, with knowledge which is imperfect relative to the ideal. Thus, recognition of the ideal, that is, recognition of process knowledge as the ideal of explanation and prediction, presupposes the recognition that it is possible to be merely idly curious about matter-of-fact generalities. If one does not recognize the possibility of such an interest, if one recognizes only pragmatic interests as the motive for acquiring knowledge of natural laws, then one will continually rest satisfied with imperfect knowledge and simultaneously will not recognize there is a sort of knowledge in which imperfections are overcome, with the consequence that one will not recognize that one's knowledge is in fact imperfect. Imperfection is defined only relative to the ideal: if the latter is not recognized then neither will the former he recognized. Until Galileo, matter-of-fact knowledge was not thought *worthy* of idle curiosity. For Aristotle, one had to go beyond such knowledge, to the rational intuition of Natures. For the mediaevals only religious truths were worthy of idle curiosity. For the Renaissance humanists only the classics of literature were worthy. This accounts, in part at least, for their never recognizing the ideal of scientific explanation, and, failing that, for their failing to recognize that the knowledge of matter-of-fact generalities with which they rested content was in fact imperfect knowledge.

2.2. ARE THERE REASONED PREDICTIONS WHICH ARE NOT EXPLANATIONS?

We have seen the deductive model of explanation permits the use of imperfect knowledge for the purposes of explanation. A number of objections to the deductive model fail by virtue of their not taking this fact into account. In particular, I have in mind several objections the point of which is to attempt to show there is an asymmetry between explanation and prediction, contrary to what the deductive model entails. The idea of this sort of objection is to attempt to present examples which are intuitively explanatory but without predictive power, or which have predictive power, but which are intuitively not explanatory. An immediate difficulty arises with such a mode of argumentation: it is not conclusive. It could not be conclusive until criteria are laid in terms of which we can evaluate the extent to which our intuitions are justified. It is always open to those such examples are intended to refute to question the intuitions, to reject an example as one which explains or fails to explain as the case may be. And, in fact, part of our defence of the deductive model will consist of just this sort of move. It will be suggested that the examples of predictions which are supposed to fail to also be explanations are in fact explanations after all. That is, it will be suggested we are given no good reason for supposing, what is claimed, that the examples are not explanations. The other move which can be made against the attack by examples is to accept them, and then argue they don't refute one's view. Thus, we shall accept certain examples as being explanations and as not being predictions, and argue that nonetheless they don't refute the deductive model and its corollary, the symmetry thesis with respect to explanation and prediction; more accurately, we shall argue that these explanations only in a certain way fail to yield predictions, and that failing in this way to yield predictions does not violate the symmetry of explanation and prediction which is entailed by the deductive model.

First, we must clear up some minor matters concerning 'prediction'. By 'prediction' we mean *reasoned prediction*. It is not a mere guess. Nor is it a matter of what one might wish to be the case. Rather, it is an *inference*, based on factual evidence, to other facts. What is claimed is that *scientific explanation* is symmetrical with *reasoned prediction*. In explaining the rules of chess, we make no predictions. But it is not claimed that such explanations are symmetrical with predictions. Only scientific explanations are said to be symmetrical. Objections to the symmetry thesis by Scriven and Collins[12] based on the chess example therefore miss their target completely. Again, 'prediction' means an inference from the known to the unknown: on the basis of what is antecedently known one predicts *laws*, even though laws so to speak hold at all times, and not just in the future. And one may predict the past, or, as is sometimes said, retrodict. And one may predict the future.[12a] It is therefore false to claim the deductive model with its symmetry thesis applies only to events, and not the explanation of laws by theories, since one can't "predict" what is timeless. So much the worse for another objection from Scriven.[13] Finally, 'prediction' does not mean simply 'statement abut the future'. It is, of course, true that if one means that by 'prediction' then one can have predictions which are not explanations (e.g. those based on guesses), as Scriven points out,[14] but, contrary to Scriven's unfortunate conclusion, this in no way counts against the thesis of the symmetry of explanation and prediction entailed by the deductive model, which thesis is concerned with prediction in another, related, but more restricted sense, viz, "reasoned prediction".

So much for the boring objections. Let us turn to the claim that one can have (reasoned) predictions which are not explanations. The examples which Scriven and others use to attempt to establish this claim are of the following sort. Sunspots are such that when they occur so does radio interference on the earth. We may then use the latter to predict the former. But it is clear that we have thereby in no way explained the sunspots. Or so Scriven claims.[15] And that is all he does. It is perfectly open for the defender of the deductive model to counter-assert. Which this defender proceeds to do: it *is* an explanation. Therefore the symmetry thesis and the deductive model which entails it are not challenged by the example. Only, one must add, it is not a good explanation. The law which occurs in it and which makes prediction possible is *very* imperfect; it falls far short of the ideal of scientific explanation. This low explanatory power of the explanation accounts for the plausibility of Scriven's claim that example is an example of a non-explanation. But being a weak explanation and being a non-explanation are two things and not one, which means that though Scriven's claim is plausible it is nonetheless false. Perhaps the reason he accepts the false claim is that he fails to notice the defender of the deductive model has a criterion of better and worse for explanations which is a criterion in addition to the deductive criterion separating explanations from non-explanations. Certainly he gives no hint that such criteria of worth can be laid down by the defender of the deductive model.

Be that as it may, it is certainly true that *we know* the connection between sunspots and radio interference to be imperfect. *We* know sufficient (but still imperfect) laws, for example, to be able to assert the counter-factual, "if the earth were to disappear (and *ipso facto* radio interference) then the sunspots would not disappear". This counter-factual makes clear the way in which there can be sunspots without radio interference, and the way in which one can quite legitimately say that the former cause the latter but not conversely. It makes clear, too, the relevance of such conditioning factors as the existence of the earth. These two things we know on the basis of laws, and because *we* know them *we* recognize the weakness of the explanation provided by the sunspot/radio interference connection. Further, because *we can* do better we reject the explanation as adequate. Suppose we ask for an explanation. That is to ask for something we do not have; otherwise, why ask? In this context, it is to ask for an explanation better than any we have. We already have laws with greater explanatory power than that provided by the sunspot/radio interference law. So we reject the latter if it is offered to us. We reject it not because it fails to be an explanation, but because it fails to provide a better explanation than what we have. That is, in the context defined by what we already know, an explanation based on the sunspot/radio interference law would be useless as an instrument in the transaction of explaining. However — and this is the point — though useless for explaining, it would still be an explanation. On the other hand, if one were to not clearly distinguish being a non-explanation from being an explanation useless for explaining, then one might think the latter defines what it is to be a non-explanation. We have clearly distinguished those two things. It is equivalent to distinguishing explanations from the transaction of explaining. I shall suggest Scriven does not so neatly distinguish. Nor does Collins. That might explain why the latter so hastily dismiss the sunspot/radio interference example as non-explanatory.

Another sort of example that has been appealed to is the relation between weather and the records of a barometer.[16] It is suggested we can use the barometer to predict but not to explain the storm. With this one may again disagree. The inference from barometer reading to storm does have explanatory power. To be sure, it is not ideal, but one can offer less than ideal explanations and still explain. Nor is one enabled to control the process, if one knows the barometer-storm laws. But one must emphasize that inability to control does not entail complete absence of explanatory power. What one must realize is precisely what one is explaining in this case. There is a law connecting the state of the weather to the state of the barometer. It is here where one has an explanatory connection. We do not have an explanation of the weather process itself. To this last we *know* the barometer to be irrelevant: the weather process would be what it is even if the barometer were absent — this counterfactual is justified by appeal to known laws. And, of course, it is just these laws, just this counterfactual, which inform us that manipulating the barometer would fail to control the weather, that the explanatory connection does not yield a strategic connection enabling us to satisfy some pragmatic interest we might have in changing the weather.

If we look more closely at the example we find we have the following. There is (A) a rapid decrease of atmospheric-pressure, (B) a sharp fall of the barometer and (C) the occurrence of a storm. We may schematize this by considering three times t_1, t_2 and t_3. At t_i pressure is p_i, barometer height is b_i and weather is w_i. p_2 is very much less than p_1; p_3 approximately equals p_2. b_2 is very much less than b_1; b_3 approximately equals b_2. w_1 and w_2 are non-stormy weathers, w_3 is stormy weather. The air pressure decreases from t_1 to t_2 — we might suppose the air heats up and rises by convection. The surrounding atmosphere exerting pressure in all directions finds less resistance to its tendency to expand and moves into the area of decreased pressure and continues to do so until pressure is uniform throughout. This rushing in of air constitutes the storm. The suggestion is that (α) A causes B, A causes C, but B does not cause C, nor does it cause A; (β) A explains B, A explains C, but B does not either explain A or explain C; and (γ) on the basis of lawful relations A can be inferred from B, B can be inferred from A, and C can be inferred from B. From which it is concluded explanation and causation cannot be identical with lawful regularity. Now, on the basis of the law mentioned in (γ) we can say if b were b^* at t_1 then p would be p^* at that time, and if they were b^{**} and corresponding p^{**} at t_2, then if b^{**} were not much different from b^* then neither would p^{**} be much different from p^*; that is, had the barometer not sharply fallen then the pressure would not have sharply fallen. *Given the barometer readings we can use the law to explain why the pressure must have been as it was.* Briefly, I deny the last part of (β): we *can* use B to explain both A and C. (α) is more complicated. The decrease of pressure causes the storm. It is the causal relevance of B which is denied. Or rather, this must be inaccurate. For there *is* a causal connection between A and C on the one hand and B on the other. This, surely, must be granted once it is granted A causes B and A causes C. The idea is, then, that in spite of this B does not cause C. Which I take to mean that B is utterly irrelevant to the process which brings about C even though B and C result from A as a common cause. And this "irrelevance" I take to mean something like this, that if the barometer height were to be different then one still would not change C. Nor is the barometer height to be changed by changing A. Rather to the converse: A is to be changed by changing the height of the barometer. So the issue amounts to whether one could stop the decrease A by interfering with the barometer and thereby stop the storm. Could one increase atmospheric pressure by manipulating the height of the column of mercury in the barometer tube? This height can be manipulated in two ways (given the construction of the tube is constant), either by increasing atmospheric pressure – which we exclude since we are trying to manipulate that pressure through manipulating the height and not conversely – or by increasing the pressure in the closed portion of the tube above the mercury column. In fact the last is in principle a possible means, if we assume an absurdly small atmosphere and an equally absurd barometer. But in practice the atmosphere is so large and barometer so unable to withstand large internal pressure that any manipulation of column height by this second means would only with the utmost negligibility affect atmospheric pressure. So changes in the barometer cannot effect changes in pressure and are thereby in practice causally irrelevant to the occurrence of the storm. Still, they *do* explain it. The only possible strategy to control the storm is to go after the air pressure directly and not by manipulating the barometer. But the strategic irrelevance does not entail explanatory irrelevance.

A third sort of example that has appeared in the literature concerns the lawful relation between the length of a pendulum and its period, a law we all learn very early on in our study of classical mechanics. Given the length, one can deduce the period of the pendulum. Given the period, one can deduce the length. In both cases we can have prediction. But in only one do we have explanation. The length explains the period, but the period does not explain the length. We could reasonably answer, "Why does it have such and such a period?" with "Because it has so and so length," where we could not answer, "Why does it have so and so length?" with "Because it has such and such a period." Thus, the inference from period to length is predictive but not explanatory. The conclusion drawn is the rejection of the deductive model.[17]

Before discussing this example in detail some remarks about explanatory *transcation* will be in order. In such a transaction an explanation is offered to the explainee by the explainer. The latter has implicitly or explicitly asked a why-question, which is to say, he has some cognitive interest in receiving an explanation, an interest the explainer is endeavouring to satisfy. Now, the explainer might misunderstand that interest if it is poorly expressed by an explicitly asked why-question. Thus, if asked "*b* is *G*, why?" one might reply with

$$(x)(y)(Fx \& Rxy \supset Gy)$$

plus the initial conditions

$$Fa$$
$$Rab$$

providing an explanation for

$$Gb$$

The response then might be "What I had in mind was not, why is *b G*? but, why is *b of all things, G*?" To which the explainer might reply with; "Well, *b* and *b* alone was *R*-ing *a*", that is, with

$$(y)(Ray \supset y = b)$$

The last is not a further explanation, but, given the previously cited reason for b's being G, it could provide just the additional information the explainee also wanted or had a cognitive interest in (not all why-questions demand explanations as replies!). The question the explainee had in mind could also be, "How come b and only b came to stand in R to a?" This is a request for an explanation of the initial conditions, the givens, of the previous explanation. The appropriate response might be an explanation such as "Jones, in his experimental set-up, ensured that this guy b and no one else came into a's visual field." More complicated answers would be required by a question such as "How come the earth, of all planets, came to have a solitary natural satellite?" This point is of importance for the pendulum example.

I think one must indeed pay attention to questions asked if we are to understand explaining transactions. However, the anti-deductivist views often concluded from this point[18] do not follow, so far as I can see. Or, at least, the conversations reported are sufficiently incomplete that the conversations can be interpreted in a way conforming to the deductivist thesis. Consider an example of W. Ruddick designed to confute this thesis:[19] "The collapse was caused, not by the fact that the bolt gave way, but by the fact that the bolt gave way so suddenly and unexpectedly." The conversation is reported by Ruddick to be

Question: Why did it collapse?
Answer: (Because) the bolt gave way.
Objection: But there were safeguards.
Response: It gave way so suddenly and unexpectedly (that safeguards could not come into play).

The Answer will turn on an imperfect law of somewhat this sort:

$$(x) [Kx \supset (\text{Collapse } x \equiv : (Bx \mathbin{\&} (\exists f) (\mathscr{F}f \mathbin{\&} fx)) \vee S_2 x \vee \ldots .)]$$

The Answer picks out the first disjunct as relevant, implicitly denying the obtaining of the other sufficient conditions. It also implicitly affirms the other conjunct, since Bx, a bolt giving way, is not by itself sufficient for collapse. (The presence of these factors can be deduced from the *fact* of the collapse. Since we don't know what the other conditions are — the law is imperfect — we could not in fact predict the collapse, but given the collapse we can *ex post facto* explain it.[20]) The explainer may in fact have a less imperfect law available, one in which the $\mathscr{F}$s are more specifically detailed. But he, we might suppose — for the conversation so far as it has been described

THE REASONABILITY OF THE DEDUCTIVE MODEL

to us permits this — believes the explainer's cognitive interests will be satisfied by this knowledge. The objection is against this belief: what the explainee indicates is that he has further knowledge of some of the conditions gathered under $\mathscr{F}$, that he knows the disjunct is something more like

$$Bx \,\&\, \sim Cx \,\&\, (\exists f)(\mathscr{F}'f \,\&\, fx)$$

and that he believes C to be present from which we would infer there was present no sufficient condition for collapse. The response points out that C was in fact absent. The Response thus mentions a further initial condition and turns on the availability of a less imperfect law. One need not conclude the first answer did not provide an explanation; one need only conclude that what was desired was a better explanation than that originally provided. The conversation as sketched does not tell us whether the original answer is rejected because it fails to be an explanation or because it fails to be as good an explanation as cognitive interests demand. The deductivist has only to maintain the latter, and he has thereby replied sufficiently to the objection based on the conversation.

Consider another example of Ruddick's:[21] "Why is the train stopping here?" — "Because it always does". The questioner would not likely be satisfied with this answer. But not necessarily because it failed to be an explanation. Rather, one may say, mayn't one? that questioner is unsatisfied because on the basis of anyone's, including his, background knowledge it is a *thoroughly imperfect* explanation.

Or the case of Kepler's laws: Here we have laws yielding predictions about planetary orbits. The imperfection of this knowledge we know already. These laws enable us to predict position and velocity. We can also use them to say: if there were a planet here then it would move in an ellipse and its velocity at any point would be such and such depending on initial conditions and the eccentricity of the orbit. However, these laws do not support counterfactual assertions about what would happen were the planet's mass to be twice as great, or about what would happen if another planet were to be in another position or going twice as fast. That is, Kepler's laws leave undescribed certain connections we know to obtain among relevant variables. The less imperfect law Newton gave us for the solar system tells us about all these interconnections. Although Kepler's laws satisfy the deductivist model they do not yield all we might want by way of explanatory power. Newton's (approximation to a) process law does yield this. That is precisely what makes it better. But this has nothing to do with this law being explanatory *rather than* a mere "predictive regularity". It depends upon it mentioning more relevant variables — Newton, unlike Kepler, knew about mass. And it depends upon knowing more (indeed, all) of the details about the functional interconnections among the variables: if $x = f(y)$ is a (lawful) functional relation among variables then, even though x_1 and y_1 are the actual values, we can deduce that if y were y_2 then x would be x_2 and so on for all other non-actual but possible values. On the other hand, a process law, such as Newton's, no matter how powerful it is, will not answer *all* questions about the system in question. Thus, it won't answer, "Why are there but nine planets in this system?" nor "How come the earth, of all planets, came to have a solitary natural satellite?" In short, one must be clear on what is being asked for when an explanation is requested.

Now consider the case of the pendulum (see Figure 1). What is interesting about this example is that a process law is available. It therefore goes to the heart of our analysis. A simple pendulum is a particle of mass m suspended from a weightless string of (constant) length l, which is supposed to move in a vertical plane.

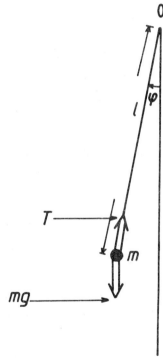

Fig. 1

The position of the mass can be measured by the angle ϕ (and the distance l from 0, but this is constant since the string is extensionless). Two forces act on the particle, the gravitational force mg and the unknown string tension T.[22] The path of the particle is a circle which 0 as its centre. The description of the situation is most easily done by resolving it into radial and tangential rather than the usual rectangular components. (Since l is constant, the radial component will always be along the string, and the tangential component will be perpendicular to the string. The force mg has component $mg \cos \phi$ radically to the path, and component $mg \sin \phi$ tangentially. The acceleration vector in space can be resolved into the normal = $\dot\phi^2 l$, and tangential = $l\ddot\phi$ components. Thus, radially force = $m\dot\phi^2 l$ and

$$T - mg \cos \phi = m\dot\phi^2 l$$

while tangentially force = $ml\ddot\phi$ and

$$-mg \sin \phi = ml\ddot\phi$$

(In this equation the force component on the left is written with a minus because the acceleration $\ddot\phi$ on the right is positive for increasing ϕ which we take to be towards the left, whereas the force component is to the right.) The radial equation enables us to calculate the unknown T once we know the motion, while the tangential equation determines this motion ϕ without involving T. In the latter, mass cancels (the motion of the pendulum is independent of its mass) and solving we have

$$\ddot\phi = -(g/l) \sin \phi$$

This cannot be solved by elementary means, since the solution involves elliptic functions which are known only in the form of infinite series. However, $\sin \phi$ may be developed in a power series

$$\sin \phi = \phi - \phi^2/6 + \cdots \approx \phi$$

and if all terms save the first are ignored the error involved is small for small angles. (So restricting ourselves amounts to assuming the radial and tangential are vertical and horizontal components, which is approximately true for small ϕ.) Thus, for small ϕ we have

$$\ddot\phi = -(g/l)\phi$$

which is the equation of motion for a simple vibrating system. The general solution of this is

$$\phi = \phi_0 \cos(\sqrt{g/l}\ t) + \dot{\phi}_0 \sqrt{l/g} \sin(\sqrt{g/l}\ t)$$

where ϕ_0 and $\dot{\phi}_0$ are initial ($t = 0$) positions and velocities. If it begins from rest the position is given by l (a constant) and by

(P) $\qquad \phi = \phi_0 \cos(\sqrt{g/l}\ t)$

and the velocity is given by radial acceleration (= zero, as the string is extensionless) and by the angular velocity

$$\dot{\phi} = -\phi_0 \sqrt{g/l} \sin(\sqrt{g/l}\ t)$$

sin and cos repeat themselves through $360° = 2\pi$ radians. Thus the time of one complete vibration is given when the angle $\sqrt{g/l}\ t$ increases through 2π:

$$\sqrt{g/l}\ t = 2\pi$$

or the time t for a complete vibration, called the period, is given by

(A) $\qquad t = 2\pi\sqrt{l/g}$

We have been able to deduce position and velocity for all times (at least to with a small error — which could be lessened by successive approximations). We assumed the system is closed and that the force law is that of Galileo's law. That is, we assumed the force-function for the earth-pendulum system is gravitational that given by (4.8) of Chapter 1. String is extensionless ($l = 0$). From this we deduced the law (A) relating length and period. It is clear (A) is imperfect relative to the fuller knowledge of the system just previously deduced, where exact details and not just the period of the motion were given expression. Though, note it is not imperfect for omitting a relevant variable, for holding only under certain conditions.

Assuming g to be constant, (A) supports both

(α) $\qquad$ If t were different, l would be different
(β) $\qquad$ If l were different, t would be different

On this basis surely one can quite correctly say that *given just this period we can explain why the pendulum must have just this length*. And, of course, conversely. Period and pendular length are mutually explanatory so far as the system itself is concerned. Not everything about the system in respect of its motion is accounted for by (A) – (A) is imperfect – but something is accounted for, or, rather, two things, namely, the length given the period and the period given the length.

In this system the relevant variables are mass, position and velocity, with mass disappearing owing to the fortunate features of Galileo's law. Period is not a variable, but rather represents the time in which patterns of behaviour exemplified by velocity and position repeat themselves overtime. Period is a function of one and only one variable, namely position. Position is given relative to 0 by the coordinates ϕ and l. Period is a function of the second coordinate alone. Therefore, if the period of the system is to be altered the proper strategy is to modify the length of the pendulum. (A) tells one by how much in order to achieve the desired period. Moreover, t can be changed no other way. Whence one cannot so to speak change t independently of l and by doing that bring about a change in l. Strategically, therefore, t depends on l but not conversely. But there is no explanatory asymmetry.

As I suggested, we do, it seems to me, explain length by appeal to period. This, though, is imperfect. In the process law, we assume $\dot{l} = 0$ throughout so that the change of position is given solely by changes in ϕ. That $\dot{l} = 0$ is part of the ideal process explanation of the behaviour of this system. Still, "Why is it l?" is correctly answerable by "Because the period is t" *at least when the question is understood in one way*. But we have noted questions are not entirely unambiguous. That why-question could also be understood as, "Why is the plendulum l, of all lengths?" which is rather different.

One answer to the latter might be "I made it so." This describes a causal process by which the system, *qua* having a pendulum of this length, came into being. This causal process requires an explanation, but one quite different than that which accounts for how the system works given that it exists. But what is crucial is that this second process is just that which occurs when one implements a strategy: "I made the length so in order to thereby make the period such and such", though, as we just saw, we could not say "I made the period so and so in order to thereby make the length such and such." We thus see that strategy issues are tied closely to certain styles of questions and also that those questions suggest strategies which may in turn introduce asymmetries.

But we must be clear just what these asymmetries are. In the first place one must notice that strategies inevitably introduce an ordering through the means-end relation. This relation is asymmetrical. This relation is a compound of an intentional action the aim of which is the achievement of the end, and a lawful connection relating some manipulable factor as sufficient for that end. The strategy of "changing x in order to bring about y" relates x and y as means and end but the lawful relation between x and y that makes this ordering possible need not be asymmetric, in the sense that while x can be manipulated to yield y the converse does not hold. In the length/period case there is, we saw, an asymmetry in just this sense. By way of contrast, however, for a case where this kind of asymmetry is absent consider the general solution for ϕ and its differentiation for velocity $\dot{\phi}$. Both ϕ and $\dot{\phi}$ depend on position ϕ_0 and velocity $\dot{\phi}_0$, so that either one may be manipulated in order to bring about a change in the other. Note, finally, that even where the relation between x and y is in the appropriate way asymmetric the asymmetry is deducible from the functional laws (regularities) describing the behaviour of the system. And functional connections are (we all know) inevitably symmetric, in that they are stated using an identity sign, "$y = f(x)$", and identity is a symmetrical relation. One does not really want to make much of this last remark, however, since the laws of which the functional relations are the tips of the iceberg might very well introduce some asymmetries or other, and it is the whole law which is crucial. And in any case, one must be careful about holding functional relations are purely symmetrical, for even these can record important asymmetries, as one realizes upon the thought of the laws of learning, or more generally, historical laws as represented in integro-differential equations, or, as in the equation (P) above, which relates the position ϕ of a pendulum to time and which associates with each time a unique position but associates with each position many times, the identity sign can be introduced in connection with a many-valued function which is inevitably asymmetrical. In many cases, however, it will remain true that the asymmetries enter with pragmatic and strategic considerations. All that it is important to remember is that the laws of the system determine all possible strategies, all one could want to know given any pragmatic cognitive interest.

Given the process law for the earth-pendulum system we are able to affirm (α) if the period were different then the length would be different. The law (A) stating the functional relation between length and period justifies, as we saw, our asserting this subjunctive conditional. When we assert (α), if the period were different, then the length would be different, on the basis of the law, we are making an assertion about the system as it was originally defined. After all, the process knowledge is knowledge about systems of a specified kind. In particular, it is knowledge with respect to systems in which the pendulum swings freely.

This becomes important as soon as one raises the problem about how one might interfere in the system so as to make the antecedent of the subjunctive conditional (α) come true. Because, if we could change the period, then we would thereby have changed the length. An easy way comes to mind: We will shorten the period by placing two walls either side of the pendulum in the path of its swing; these walls will cause the period to become shorter; therefore the length will become shorter. Call this the (α^*) method of shortening length.

Similarly, the functional relation of length to period supports the assertion of the subjunctive conditional (β) if the length were different then the period would be different. Here, too, we might think of interfering. We would attempt to make the antecedent of (β) come true, thereby changing the period. Thus: we shorten the length of the pendulum; the period then becomes shorter. Call this the (β^*) method of shortening the period.

Now, we know that the (β^*) method will work, but that the (α^*) method will not. In this sense, we can use the length to determine the period, but not conversely. From the viewpoint of pragmatic interest, strategic considerations, this is important. It is also of interest to idle curiosity, of course, but it is most often the former which determines our quotidien uses of 'cause', and I think it is that which leads us to the thought that the period is caused by the pendulum having the length it has. And that in turn leads us to the idea that the length explains the period but not conversely. It is this idea which is, I think, behind Scriven's conclusions from this example.[23] But how justified is the idea? The idea arises from pragmatic interest being directed at the (α^*) and (β^*) methods. That the one method works and the other does not is a matter of the laws of nature. The trouble is that these laws, which are relevant to (α^*) and (β^*), *are not laws about the original system about which the explanation vs. prediction question was first raised*. With respect to (α^*), the relevant kind of system consists of a pendulum swinging between two walls. In the original kind of system, with respect to with (α) was asserted, the pendulum swung freely. With respect to (β^*), the relevant kind of system is one in which there is a change Δl in the length of the pendulum, whereas in the original system with respect to which (β) was asserted, $\dot{l}$ is assumed to be = 0. The example which purported to show the asymmetry which destroys the deductive model involves systems of freely swinging pendulums in which $\dot{l}$ = 0. The idea which suggests length explains period but not conversely derives from other systems. It therefore remains, I suggest, an open question whether in the *original* example there is an asymmetry between explanation an prediction.

We tend to think of explanations as providing answers to "why?" questions.

Nor is this wholly wrong. On the other hand, of course, we must always bear in mind that such a thought could not bear the weight of an accurate description of the circumstances in which it would be appropriate to use an explanation. And in this case it is decidedly misleading. It is because we tend to think this way that we slide from talk about the original system, with respect to which (α) and (β) were asserted, to talk about the systems appropriate to the methods (α^*) and (β^*). The relevant "why?" question is "why does the pendulum have this length?" The suggestion it carries is "why this length rather than some other?" With respect to the original system, the reply, rather than the answer, to the suggested question is that it has no other length: for this kind of system $\dot{l} = 0$. Since it has no other length, the "why?" question which is suggested by the original "why this length?" question, that is, the question "why this length and not another?", is a question which has no answer, and, so far as the original carries this suggestion, neither does the original have an answer. And therefore, if explanations are answers to "why?" questions, there is no explanation of the length, so far as the original system is concerned. But the "why?" question with its suggestion *seems* reasonable, so we search for an interpretation of it which *would* require an explanation. This search leads us directly away from the original system and to the systems in which (α^*) and (β^*) were appropriate. In short, the "why?" question with its suggestion leads us away from the original example to another example without our noticing it; we then apply to the former thoughts which are relevant only to the latter, and come to think the original example violates the symmetry of explanation and prediction entailed by the deductive model, when in fact it remains an open question with respect to the original system whether the symmetry is violated.

Is that symmetry violated in the original system? Compare Newton's process knowledge for the solar system. It is part of this knowledge that the masses (m) of the ten objects do not change: $\dot{m} = 0$ for each object. The question "why do the objects have these masses (rather than some others)?" does not really arise, since they have no other masses. It does not follow from this that Newton provided no explanation of what went on in the system. To the contrary, he *did* provide such an explanation – the first (reasonable approximation) to the ideal of scientific explanation. Similarly, the process knowledge for the free-swinging pendulum system provides an explanation of the system – indeed, an instance of the ideal – in spite of the fact that it provides no answer to the question "why this length (rather than some other)?"

Just as there is a shift in the pendulum case from the original system to systems with respect to which the question "why this length (rather than some other)?" receives an answer, so there is a corresponding shift in the case of the solar system. Newton's process knowledge cannot answer the question "why just these masses (rather than some others)?" because it does not arise for a system like the solar system. It can be made to arise if we shift to talking about more complex systems, e.g., of sort which are discussed in the Kant-Laplace nebular hypothesis for the origin of the solar system.

It is not difficult to explain why we do not so easily make the shift from the original to the second systems in the solar system case as in the pendulum case. Solar system: mass is a relevant variable; how the other variables interact, their mode of interaction, depends upon the values which the masses assume. Pendulum: length is a relevant variable; how other variables (especially position and velocity, which define the period) interact depends upon the value of l. In both cases, if we are to change the *mode of interaction* of the remaining variables, it is necessary to change the values of those variables which remain constant, viz, mass and length respectively. Now, in the pendulum case, we can reasonably have a pragmatic interest in changing the mode of interaction (and thereby the period). This is because it is technologically possible to alter pendulum lengths. We cannot have a similar pragmatic interest in the solar system since it is not technologically feasible to alter the masses of the objects. But we saw it was pragmatic interest that led us to move from the original to the secondary systems in the case of the pendulum. Since pragmatic interests are absent in the solar system case, such a move is much less likely.

It is therefore at least an open question whether the inference from period to length is explanatory. Scriven has asserted one way, under the cover of certain misleading associations, and we shall assert the contrary: It *is* explanatory; it does *not* violate the symmetry thesis entailed by the deductive model; the latter is *not* refuted by the example. We add, of course, that better explanations are available. The length-period law is imperfect. It is easily derivable from the relevant process knowledge, but it is the latter and not the former which provides the best explanation. The process knowledge, in particular, gives the positions and velocities of the pendulum for each moment. The length-period law does not provide such information about the relevant variables. That is what makes it imperfect. And that imperfection is what makes it a worse explanation than that provided by the process knowledge. On the other hand, from the fact that it is not an explanation as good as others we could provide, it does not follow it is not an explanation. I conclude Scriven's example does not establish his case.

Let us now turn to the final example we will discuss. In this example one considers the shadow thrown by a building (see Figure 2). The building having a certain height both explains and enables one to deduce the length of the shadow. And from the length of the shadow one can deduce the height of the building. But the length of the shadow does not explain the height. So once again we are presented with a purported asymmetry where we have a predictive connection but not an explanatory one.[24]

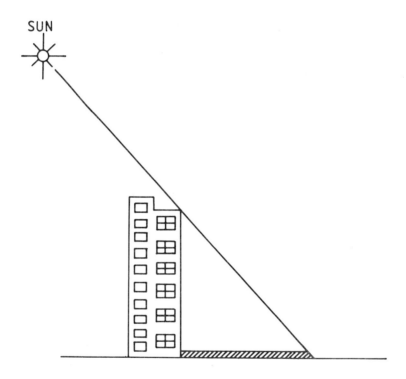

Fig. 2

In order to deal with this example it is necessary to turn to optics, and specifically geometrical optics.[25] Indeed, geometrical optics uncontaminated by wave optics is quite sufficient for what we are about.[26] I suppose this will be disappointing to those philosophers of science who expect quantum mechanics or what not to be trotted out at the earliest possible moment — even if it is to make the simplist points. But I fortunately suffer from a tendency to use the appropriate blade of Ockham's razor: complications in illustrations are not to be multiplied without necessity. So on to geometrical optics.

Some objects are by themselves visible. They are the self-luminous objects. Others are luminous only in the presence of the former. These are the dark objects. A dark object which is illuminated by a self-luminous object can illuminate a third dark object. The sum of the physical relations between one object and another determined by the feature of the visibility of the first object is the condition of illumination. In our example the self-luminous object is the sun. We have two dark objects which are illuminated by it, viz, the building and the earth. Only part of earth is illuminated. Part, viz, what we call the shadow of the building, remains unilluminated. The area of the earth illuminated (and therefore also the area not illuminated) is a function of the spatial relations of the source, the earth, and the intervening dark objects. Given the spatial relations one can determine the area of illumination by appeal to what is called the law of the Rectilinear Propagation of Light, though this law has nothing to do in any reasonable sense with "propagation". It is this same metaphor of "light" being "propagated" which leads then sun and other self-luminous objects being designated as "sources" of light. It was this pattern of metaphors pervading geometrical optics that led Mach to the comment that "the mechanism *imagined to be involved*, conditioned by the first object [and determining the rest of the conditions of illumination] is designated briefly as *light*." But it is still possible to use the law without taking this metaphor as more than that. The spatial relations and the Law of Rectilinear Propagation uniquely determine the area of illumination. For every set of spatial relations, there is exactly one area of illumination, one shadow. But the same shadow may be determined by several goemetrically distinct dark objects intervening between the earth and the source. However, the area of illumination and the law do determine certain *limits* on the geometry of the intervening dark object and on the location of the source. Thus, the sun, considered as a point source, must, according to the Law of Rectilinear Propagation, **lie** *somewhere* on the *straight line* extending from the end of the shadow, **the** edge of the illuminated area, through the topmost point of the interrvening dark object. If a building is then known to be perpendicular to the surface of the earth, then if the length of its shadow is known, then we may use the usual methods of geometry to calculate its height. But there may be portions higher than the edge if they lie behind the straight line connecting the edge to the source of illumination (and, **more obviously, there can be portions on top that are lower than the edge).**

It is still quite useful knowledge. With it we can support counterfactuals of the sort

> If the geometry (e.g., height) were thus and so then the area of illumination (equivalently, the shadow or non-illuminated area) would be such and such.
> If the area of illumination were such and such then the geometry would be thus and so.

That is, the shadow yields the height and the height yields the shadow. Here, as in the pendulum case, we have a symmetry, and I see no reason for saying we do not have an explanatory connection both ways: *given the area of iillumination (and the location of the source) we can explain why the geometry of the intervening dark objects must have just these features*; and, given the dark objects have these geometrical features (and the location of the source) we can explain why there must be just this area of illumination.

It is clear that (given the location of the source) if one wants to change the area of illumination one can change the location and geometry of the dark objects which intervene between the source and the illuminated object being considered. Can one conversely change the geometry of the intervening dark objects by changing the area of illumination? The answer is negative, for the reason that the laws establish that (given source and intervening objects) the only way to alter the area of illumination is by manipulating the opaque objects. One cannot change the area of illumination independently of the geometry of the intervening dark objects and by doing that thereby change the geometry. Strategically, therefore, shadow depends on height but not conversely. Strategic asymmetry exists, but no explanatory asymmetry, or, at least, from the strategic asymmetry one need not — and this one does not — conclude there is an explanatory asymmetry.

It is difficult to talk about these matters in pure generality, as Blalock for one so complacently does.[27] Still, we may hazard the following remarks on his distinction between "correlations" and "causal relations" where I understand both to be laws, save that correlations "merely predict" while causal relations are "explanatory". Let us suppose X and Y are correlated by virtue of having a common cause Z. If the examples discussed previously are not unrepresentative then there is no reason to suppose there is no explanatory connection between X and Y. Nor, of course — though here there is no disagreement — that there are no explanatory connections between Z and the other two. The difference between the $X-Y$ connection on the one hand and the $Z-X$ and $Z-Y$ connections on the other is not a matter of the non-explanation/explanation distinction but rather a matter of the strategically irrelevant/relevant distinction. Blalock's "causal relations", I am suggesting, are lawful relations which introduce strategic asymmetries. As our discussion of the examples has (I hope) made clear, strategic asymmetries *do* arise by virtue of the details of the functional interconnections among the variables. So the classification Blalock provides *does* reflect features of functional interconnections. The point is that one cannot conclude the "causal relations" are not functional connections, i.e., regularities; one cannot conclude these "causal relations" are different in kind from the regularities classed as "correlations", at least so far as being explanations is concerned. I conclude that so far as explanation is concerned Blalock's distinction is spurious.

Blalock suggests that "causal relations" should appear as axioms in a theoretical structure, "mere correlations" as theorems. There are good reaons for this, which are quite independent of any spurious distinction of cause vs. correlations. There are two important things one can do with a theory in the social sciences. One is to verify it experimentally. And the other is to apply it. In both cases the cognitive interest is pragmatic. So in both cases strategic considerations are of basic importance. Strategically relevant relations — Blalock's "causal relations" — ought therefore to be given emphasis as axioms.

I have suggested the so-called asymmetry of some causal relations can be understood as deducible from symmetric functional interconnections, and that the interest in such relations results from considerations of strategy. One of the difficulties I find in many discussions, e.g., of the pendulum example, is that the discussions do not look at the example *in detail*. It ssems to me people search for some "law" or other that will support their thesis, without ever looking *carefully* at whether it does or not. For ordinary situations the application of the "law" will be non-problematic, and we will be able to use it to predict, make counterfactual assertions, etc., even though we do not go consciously into all the details of the interactions involved. We rely, rather, on pre-conscious knowledge and fairly intuitive "know how". Because this knowledge is not conscious and articulate we tend to overlook it. But overlooking it leads to false impressions of explanatory asymmetries. What we should not do is leave this knowledge inarticulate, but rather raise it to consciousness by looking carefully at how the scientific explanations actually go through in all their detail.

2.3. IS CORRELATION LESS EXPLANATORY THAN CAUSATION?

This concludes the discussion of the examples designed to refute the deductivist thesis by purporting to be examples of predictions which are not explanations. What we have been discussing is the distinction between *causation* and *correlation*. The claim we have been examining is that correlations yield predictions but not explanations. Since the deductivist thesis entails that where one predicts there one explains, the claim thus aims to refute that thesis. Our response has been to argue that correlations do in fact provide explanations — though (we hastened to add) the explanations available when we have, as one says, "mere" correlations are imperfect, and they are therefore less adequate than one might reasonably desire. Here, "less adequate" means less adequate than one might desire out of idle curiosity or less adequate than one might desire out of some pragmatic interest. Where the interest is pragmatic one wants knowledge of how intelligently to interfere in order to bring out the end the desire for which defines the pragmatic interest. By way of diagnosis we suggested such pragmatic interests directed one's interest away from "mere" correlations towards less imperfect laws. The fact that the explanatory power of the latter was greater than that of the former would tend to lead us to overlook what explanatory power did in fact lie in the "mere" correlations.

The notion of *cause*, the usual contrast to *correlation*, is one of those concepts which is really a family of concepts blending one into another, and which for that reason remains thoroughly troublesome to philosophers. We have already remarked that Russell once argued, with considerable merit, that the notion of cause ought to be replaced by that of laws and, more specifically, by functional laws in which certain variables are mathematical functions of other variables. Certainly one must say that it is hard indeed to conceive a process law which did not involve functional relationships among variables. What the functional language enables one to do is state, efficiently and with logical perspicuity, a whole set of necessary and sufficient conditions.[28] In the less-than-ideal case one may well have statements of necessary and sufficient conditions that cannot be formulated in functional terms. Russell's point can be made equally well in these less-than-ideal cases by saying causal language should be replaced by talk of necessary and sufficient conditions. To achieve the explanatory ideal would therefore mean that our explanations would all be in terms of functional laws. Ideally, then, we would stop citing causes as explanations, and mention merely how one variable is functionally related to other variables. And in the less-than-ideal cases we should stop speaking of "causes" and speak only of "necessary conditions", and "necessary and sufficient conditions". The force of Russell's argument would therefore be that the deductivist need deal only with explanations in terms of functional relations or in terms of necessary and sufficient conditons; and that he may ignore the causal explanations these other ways replace. Still, if the deductivist thesis is correct, then, since citing causes does quite obviously explain, there must be laws involved in causal explanations. For the deductivist, the notion of *law* is fundamental, and all explanations involve laws, whether they are ideal or less than ideal, whether they are process laws, or causal laws, or "mere" correlations. Since the notion of law is what is fundamental, and since the ideal sort of law is functional rather than causal, the subtleties of our quotidien issue of 'cause' are not worth detailed exploration. And in fact, our everyday knowledge of lawful fact is thoroughly intermixed with judgements never justified by a methodical gathering of scientific evidence. We can therefore expect some irremediable confusions in our ordinary causal language. Specifically, it is worth mentioning that much of our causal language is irremediably anthropomorphic. The detailed exploration of that language would therefore not repay the effort. And effort it would be, since it would involve separating out the justified parts of causal discourse, those conforming to the deductive model, from the unjustified parts. This separation could not be done simply by looking at, e.g., the grammar of 'causes',[29] since one and the same concept or connective may be used in one explanatory context to convey a reasonable explanation (one involving a law, according to the deductive model) and in another context to convey nothing more than anthropomorphic pseudo-explanations. Thus, compare: "It was the striking of the match that caused the fire," and: "It was God's will that caused the accident." On the other hand, the deductivist should be able to account for at least the grosser features of our causal discourse. After all, it is evident we do succeed in explaining – reasonably explaining – quite a bit using causal language. And also, he should provide some account of those features of causal discourse which do not fit well with his thesis that all *justified* explanation involves laws. Thus, for example, he should be able to give some account of the anthropomorphism of much of our causal language. This latter task we cannot here undertake.[29a] But we must undertake to do the former, to show that much (just as much as is cogent!) of causal discourse can readily be understood within the framework of the deductivist thesis. This will show how Russell's programme, for replacing causal talk by a logically more adequate way of talking, could reasonably be implemented. Causal language would be replaced by another way of explaining. But not all the nuances of the former would appear in the latter. The unscientific portions would drop out; what would remain would be just those aspects which conformed to the deductivist thesis.

Here it must be emphasized once again that it is the notion of *law* which is fundamental. The deductivist is committed to the thesis that all explanation is in terms of laws. He is therefore committed to the thesis that all explanations in terms of causes involve laws. But not all need be "causal laws", laws which appear (perhaps implicitly) in causal explanations. Now, the point is often made that in the relation "x causes y" there is as part of what is meant an element of *priority*:[30] this relation holds in one direction, from x to y, but not in the other direction. In contrast, the relation "x is necessary and sufficient for y", does not preclude "y is sufficient for x" in the way in which the priority of x over y in "x causes y" precludes "y causes x". Thus, the germs are sufficient for the symptoms. And the symptoms may result only when those germs are present; i.e., the symptoms may be sufficient for the germs. But the germs cause the symptoms, and this precludes that the symptoms cause the germs to be there. Further, to say x is sufficient for y is to say that not-y is sufficient for not-x. But to say that x causes y is not to say not-y causes not-x. Heavy rain causes flooding, but the absence of flooding does not cause the absence of rain. Good rain causes wheat to grow, but wheat not growing does not cause a drought. The inference it is suggested be made from all this is that the language of conditions is not adequate to the analysis of causal relations. But this conclusion is not justified. To be sure, it does follow that the language of conditions cannot be identified with the language of causes. However, this is not to say the language of conditions cannot be used to analyse the (legitimate) parts of the language of causes. For, it could still be possible so to analyze the language of causes that causal explanations are a subclass of explanations in terms of conditions. It is, of course, this last to which the deductivist is committed. The question he must answer is this: what are the differentia which make causal explanations specifically different from other explanations in terms of conditions? The crucial thing these differentia must account for is the idea of causal priority. It is the element of causal priority which is often appealed to in order to distinguish causes from correlations. It is also used to distinguish causes from mere necessary and sufficient conditions, and from explanations in terms of functional relations. Correlations, conditions and functional connections can all be accommodated by the deductivist. Causal priority in this way tends to become a test of the adequacy of the deductivist position.

From our preceding discussions it should be clear where at least one

THE REASONABILITY OF THE DEDUCTIVE MODEL 109

asymmetry can enter, namely, via strategic considerations deriving from our pragmatic interests. The connection of these with human action is very close, in fact sufficiently close that some have defined the notion of causal priority in terms of human action. But, we shall see, this cannot be the whole story. And in any case, none of this tells against the deductivist thesis: actions, means-ends relations and causal relations all presuppose the notion of law. Or at least, they have that pre-supposition insofar as they are scientifically cogent. For, of course, much in this area is thoroughly anthropomorphic. Especially the concept of action is shot through with anthropomorphic features, and those who attempt to define causal priority in terms of action all too often fail to separate out these pre-scientific component.[31] This has the consequence that they miss that it is not the notion of action or of cause which is basic but rather the notion of law.

Action and strategy connect with causality through the idea that, by acting, people can interfere in on-going processes, and thereby cause to happen what otherwise would not have happened. Basic to such action is the fact that human beings can make bodily movements. This they do directly, not by doing anything else. Then, by making bodily movements, by moving arms, fingers, legs, or what, they can manipulate things. They can lift things, or, at least, certain things, hold these in certain positions, squeeze them, push them, rotate them, rub them against each other, or bang them together, and so on. Men discovered that whenever they manipulated certain things in certain ways in certain circumstances then certain things, perhaps desirable, perhaps undesirable, resulted. Sticks laid under a large rock and over a smaller when pushed upon raise rocks that persons cannot themselves move. Eggs break when squeezed. Striking certain kinds of stone together creates sparks which, when falling upon dry moss, can be blown up into a fire. Bashing a person on the head with a stone can crush his skull and halt his aggression. Men thus found out how to produce certain effects by manipulating things in certain ways. The two aspects of this – manipulation or action, and production or strategy – must both be looked at.

Collingwood,[32] quite correctly I believe, identified the most primitive notion of cause with human action. In this sense, that which is caused is the free and deliberate act of a conscious and responsible agent, and causing the agent to do what he does means affording him a motive for doing it. The Oxford English Dictionary quotes Sir John Fortescue, from 1460, about a monarch's acting in a certain way and "movyd therto by non other cause, save only drede of his rebellion". And Collingwood quotes a newspaper headline of 1936: "Mr. Baldwin's speech causes adjournment of House."

Neither the monarch nor the Speaker were compelled to act by that which is said to have caused them to do so. Rather, in each case, the cause afforded a motive upon which they freely acted. It is clear that for 'cause' in this sense one might substitute 'make', 'induce', 'urge', 'force' or even 'compel' according to differences in the kind of motive involved. With respect to causes of this sort one can distinguish two elements, which Collingwood designated as *causa quod* and *causa ut*.[33] The *causa quod* is a given situation, an already existing state of affairs, a standing condition, though, of course, such standing conditions need not in any way be stationary. The *causa ut* is a purpose or state of affairs to be brought about. Neither of these could function as a cause if the other were absent. The coming of darkness may cause a man to turn on the lamps, but the darkness would not cause him to so act unless he wanted the room to be light. A rumour would not cause him so to act unless he wanted to avoid a loss in the stockmarket. And, on the other hand, a desire to avoid falling over a precipice would not cause a man to walk more carefully if he knew there was no precipice about.

What the *causa quod* and the *causa ut* explain is an *action*, a bodily movement of some sort aimed at bringing about the desired state of affairs. We must not make the mistake of thinking that the *causa quod* is simply by itself explanatory. Rather, the standing conditions must be known or be believed by the agent to exist if they are to be explanatory. The agent must be aware of these conditions as part of his environment, and only if the agent is so related to his environment can those conditions function as causes. Even more importantly, the *causa ut*, the future state of affairs, is not as such explanatory of the action.[34] (The contrary is an important feature of the pre-scientific Aristotelian patterns of explanation.[35]) Rather, it is the state of affairs *qua desired* which explains the action. The future does not pull the present into it.[36] To the contrary, it is the present action which accounts for the coming to be (if it does come to be) of the future state of affairs.[37] And the action is to be accounted for in terms of the present desire. Of course, not every desire leads to action: we wish for many things, but only a few do we try to achieve. Sometimes a desire makes a difference to how we act into our environment, while other desires never reach expression in action. A full or fuller explanation would have to take into account the circumstances that determine why one desire issues into action where another does not. Such things as what the agent knows, how he deliberates, and so on, would be relevant. Nor could unconscious factors be omitted. And, what must be emphasized, not only must these various factors be mentioned — **the environment and the standing conditions in it, the perception of that** environment by the agent, his knowledge, his desires, the unconscious determinants of his behaviour — all these individual facts — but also it is necessary to mention just how these are relevant to each other and to the action. The mode of interaction must be indicated. Which is to say: *a law must be mentioned describing how the the various conditions interact to bring about the action.* This is so even where the action is done freely and the decision is freely made. Thus, the notion of cause we are now discussing, in which it is actions which are said to be caused, by some sufficiently motivating factor in some relevant environment, is a notion that presupposes the notion of law.[38] The deductivist thesis can therefore be defended as holding for the case of explanations in terms of this sense of 'cause'.

We must now look at the other aspect of strategic action, namely, production, the production of certain effects by manipulation.

What men have discovered is certain general manipulative techniques for producing certain kinds of effects. Thus, we have a general manipulative technique for making anything hot: we put it on a fire. We would reasonably say that putting an object an a fire *causes* it to become hotter. In this usage of 'cause', the cause of an event is that by means of which we can bring it about. If we want to produce or prevent something, and cannot do that directly (as we can directly produce or prevent certain types of movement of our own bodies), we look for its cause. To search for a cause of something is to search for a means by which we can produce or prevent it at will. Notice, now, that this involves discovering *two* things. In the case of our example, the *first* of these is that whenever an object is on a fire it becomes hotter. This is a generality, an imperfect law. The *second* is that it is in our power to put objects into fires. This, too, is a generality, asserting certain kinds of results follow upon certain kinds of bodily motion or action. Thus to speak of *causes* in this sense is to mention two laws, one stating what it is in our power to do by manipulation, the second stating what is consequent upon our doing what it is in our power of doing.

Collingwood defines this notion of 'causes' as follows: "A cause is an event or state of things which it is in our power to produce or prevent, and by producing or preventing which we can produce or prevent that whose cause it is said to be."[39]

Gasking offers a similar definition: "We could come rather [close] to the meaning of '*A* causes *B*' if we said: 'Events of the *B* sort can be produced by means of producing events of the A sort'."[40]

On the face of it these definitions are circular, since they define 'cause' in terms of itself, or at least in terms of such cognates from the language of causal explanation as 'power', 'produce' and 'prevent'. This point is made by Rosenberg who asks, what conditions must be fulfilled for a particular series of bodily motions to be considered a manipulative technique for bringing about some other event or state; and answers that one of the conditions must be that " ... the connection between the former events — actions or bodily movements — and the state of affairs which follows them is determined to be non-accidental, i.e. causal",[41] from which he concludes that the definition is circular. But, if those who propose such a definition are careful, it need not be circular. To be sure, Rosenberg is undoubtedly correct: the connection between our manipulations and the event these definitions call the cause, and the connection between this cause and its effect must both of them be non-accidental. But, contrary to Rosenberg's assertion, it does not follow from this, that the non-accidental nature of these connections must be characterized in *causal* terms. The proposed definitions are circular only if 'non-accidental' and 'causal' are to be identified. The correct contrast to *non-accidental* is *lawful*, not *causal*. So long as the concepts of 'power', 'produce', etc., are explained in terms of laws, then one has non-accidental connections and thereby a non-circular definition of 'cause'. But, as we saw, the definitions do presuppose that certain sorts of lawful connections obtain. We may therefore conclude the definitions of 'cause' given by Collingwood and Gasking are non-circular. And we may further conclude that explanations in terms of this sense of 'cause' are perfectly compatible with the deductivist thesis.

Gasking attempts to use his definition of 'cause' to make sense of *causal priority*. Consider the explanation that "This iron bar's now being at heat of 1200 °C is the cause of its currently glowing." According to Gasking the heating is the cause of the glowing even though the two are simultaneous, since " ... we have a general manipulative technique for making anything hot ... and we have no general manipulative technique for making things glow."[42] We can produce a higher temperature in almost anything we can move; the manipulative technique is that of putting it in a fire. But only a few things such as iron, glow when put in a fire. Indeed, in order to make such objects glow we must use the general manipulative technique which makes them hot, namely, putting them in a fire. Thus, "We speak of making iron glow by making it hot, i.e. by applying to it the usual manipulative technique for making things hot ..., which, in this special case, also make it glow. We do not speak of making iron hot by making it glow, for we have no general manipulative technique for making things glow. And we say that the high temperature causes the glowing, not vice-versa."[43] Gasking's criterion of causal priority thus involves there being a pair of laws. One says something like "Whenever A then B" — e.g., whenever an object is in a fire then it becomes hotter. The other says that for some cases of A not only does B occur but also something else. It is thus to the effect that "whenever A and C then B and D" — e.g., whenever an object is in a fire and is iron then it becomes hotter and glows. When we have a pair of laws like this we say that being A causes a C to B and being B causes the C also to D.

A somewhat similar account of causal priority, in terms of special features of the laws involved, has been given recently by Simon and Rescher.[44] The basic idea is simple enough. Suppose we had three variables, x_1, x_2 and x_3, for which we have the three laws

$$f_1(x_1, x_2, x_3) = 0$$
$$f_2(x_1, x_2, x_3) = 0$$
$$f_3(x_1, x_2, x_3) = 0$$

Of these three, it might turn out that x_3 plays no role in f_1 and f_2, that these equations are perspicuously represented as

$$f'_1(x_1, x_2) = 0$$
$$f'_2(x_1, x_2) = 0$$

and it may further turn out that f'_1 is the more perspicuous

$$f''_1(x_1) = 0$$

In this case, x_1 can be determined independently of x_2 and x_3; and both x_1 and x_2 can be determined independently of x_3. Then, according to Simon and Rescher, we would have x_1 as causally prior to x_2 and x_2 causally prior to x_3. This ordering could be represented by

$$x_1 \longrightarrow x_2 \longrightarrow x_3$$

On the other hand, if neither f'_1 nor f'_2 could be reduced to a formula like f''_1, of a single variable, then neither x_1 nor x_2 would be causally prior to the other, but both would be prior to x_3. This ordering could be represented by

$$\begin{matrix} x_1 \searrow \\ x_2 \nearrow \end{matrix} x_3$$

For example, consider a mass of gas in a cylinder with a movable piston. The gas laws tell us that

$$PV = KT$$

114   CHAPTER 2

We may suppose the weight on the piston can be varied by the experimenter. This means

$$P = \bar{W}$$

And let us suppose the temperature is equal instantaneously to the temperature of the environment and that the latter, too, can be varied at will by the experimenter. This means we also have

$$T = \bar{T}$$

We then have the causal ordering

$$\begin{matrix} P \searrow \\ T \nearrow \end{matrix} V$$

which we would read as saying something like, "In order to decrease the volume of the gas in the cylinder, increase the weight on the piston, or decrease the temperature of the environment." Clearly, the scheme can be extended to the case where we have n variables and n functions. Causal ordering depends upon which variables *do not* appear in certain functions. In this respect, it is analogous to Gasking's less sophisticated principle of ordering. There is another close analogy. In the gas example, it is assumed P and T can be determined (predicted) independently of temperature. We guaranteed this assumption to hold by identifying these as exogenous variables. More generally, we need equations like

$$f_1''(x_1) = 0$$

In a system of several variables these can be obtained by assuming some one or more of the variables are exogenous, by assuming their values can be determined independently of the values of the remaining variables in the system. One way of achieving this is by assuming, as we did in the gas case, that these variables can be determined by an experimenter, that is, by manipulation. Thus, the causally prior variables will be those the values of which can be determined by some manipulation, just as with Gasking. But it is more general, because considerations other than manipulation might well be introduced. For example, rain (R) causes wheat to grow (W):

$$f_1(R, W) = 0$$

and from this we can establish the causal order

$$R \rightarrow W$$

if we have a theory

$$f_2(R) = 0$$

about the weather alone which enables us to predict the rainfall independently of the growth of wheat. The assumption that such a theory exists is clearly one of lawful fact. That there is such a theory would follow from the fairly safe generalization that the behaviour of any system involving very large quantities of energy (e.g., the atmosphere) is practically autonomous of the behaviour of variables involving very much less energy (e.g., wheat growing). It was, of course, just this sort of assumption concerning lawful connections that we appealed to in our discussions of the weather and barometer cases. Once again we have an account of causal priority in terms of laws. Once again we have an account of causal language which is compatible with the deductivist thesis.

The virtue of the Simon and Rescher definition of causal priority over that of Gasking is that it divorces causal priority from manipulating, from what it is in the power of men to do. We do speak of causal priority where the idea of manipulation technique is quite inapplicable. Thus, this account would make it improper to speak of knowing the cause of cancer if we could not use our knowledge to prevent it. Collingwood accepts this consequence of his view,[45] but Gasking suggests a way in which it may be avoided.[46] By way of example he considers the case of the melting of the polar ice cap causing a rise in the mean sea level during a certain geological epoch, and not *vice versa*. Gasking suggests we speak of causal order here because we have a manipulative technique for producing events of the same *sort* as the cause: we can produce a similar effect, a rising water level in a bucket, by manipulating ice in a similar way, by melting it into the container. In order to extend the manipulation criterion to situations of a cosmological scale we must be able to sort out all events into classes such that in each class some events are equipped with manipulative techniques and are sufficiently similar to other members of the class not so equipped that we can determine analogically which among two causally connected cosmological events is the cause. But in virtue of what constitutive properties of events could we determine their respective classes? Clearly, they must be properties which enter into a theory, a body of laws, which apply indifferently to the manipulable and cosmological cases. The appeal to the notion of law here does not render circular Gasking's account of causal priority, though Rosenberg has suggested otherwise. Again, the crucial point is that the notion of law does not presuppose the notion of cause or of causal priority. Rosenberg's objection rests upon the mistaken assumption that the presupposition does hold, that the notion of law is essentially that of causal law. But even if Gasking's extension of his definition of causal priority to cover cosmological events is not circular, it is still inadequate to those events. Consider planetary systems. Uranus does not follow its predicted path. This leads to the suggestion that the unexplained perturbations are caused by the gravitational force of an as-yet-undiscovered planet further out in the solar system beyond Uranus. Such a cause was of course discovered, and was named Neptune. The point is that there is nothing we ever manipulate that corresponds sufficiently to a solar system to yield a verification of the inverse-square law. The closest systems are objects falling freely and projectiles; but, though these may be manipulated, at the time of Newton, and into the 19th century, technology was not sufficient to decide whether these verified a strictly constant force attracting such objects towards the earth, or a force that varied as the inverse square of the distance from the earth. (That it was inverse square was considered confirmed on the basis of the success of the gravitational theory at the non-experimental planetary level.) Yet, though the science was non-experimental, Newton provided scientists with a body of laws quite sufficient to predicting the existence, position, velocity and mass of Neptune. The axioms of classical mechanics can be put to experimental tests; that is, there are systems which are analogous to the systems with which celestial mechanics deals. But the law of gravity is quite another thing. Newton's work on the solar system was totally non-experimental (which is not to say there were not certain test situations). By tying causal priority to human powers to manipulate, Gasking is unable to account for our use of the language of causal priority in the context of cosmological events. And, in contrast, Simon and Rescher provide an analysis of causal priority which is quite capable of application to cosmological events.

Unfortunately, however, though it is applicable to cosmological events, it still fails to provide an adequate account of causal priority in that area. Consider again the case of Uranus and Neptune. Here we have a nine-body system, the sun, six planets between the sun and Uranus, Uranus itself and Neptune. (We may pretend Pluto does not exist.) The mass, position and velocity of each of these objects constitute a complete set of relevant variables. For these there is a process law. This law enables us to compute the values of the variables at any time, given that we know their values at one time. So far as concerns position and velocity, there is *total interaction*: the value of any *one* of these variables depends upon the values of *all* the

THE REASONABILITY OF THE DEDUCTIVE MODEL 117

other variables. (Mass, of course, is constant.) By virtue of this feature of total interaction there is no way in which the value of one can be determined independently of the rest. In particular, the position and velocity of Neptune cannot be determined independently of the masses, positions and velocities of the other objects in the system. These variables therefore do not fit the condition of being exogamous, according to the model of Simon and Rescher. And hence, if Simon and Rescher are correct, the motion of Neptune cannot cause the perturbations of the orbit of Uranus. But we do in fact say that Neptune causes those perturbations, and imply that it is the motion of Neptune which is causally prior. So Simon and Rescher have not, after all, accounted for this aspect of our causal discourse.

We do say correctly that if Neptune were absent then the perturbations would not occur. But this will not account for the causal priority of the former over the latter. The ability to support subjunctive conditionals is the mark of lawful connections generally, not just of those special lawful connections which are causal. After all, even in the weather-barometer case we can reasonably assert the conditional that if the barometer were different then the weather would be different, even though the direction of causal priority goes from the weather to the barometer. In the present case, of Neptune and Uranus, a similar subjunctive conditional can be asserted, that if Uranus were not perturbed Neptune could not be there, even though causal priority is from Neptune to the perturbations.

There is a further puzzle about this case, that must also be mentioned, and be accounted for by any analysis of causal discourse. It is the fact that the perturbations, the effects Neptune is said to cause are not so much Uranus positively having certain properties, but rather its lacking certain other properties. What is explained is not the presence of certain properties but the absence of others. Uranus has a certain orbit. This orbit is determined by all the objects in the solar system. To speak of the perturbations caused by Neptune is to contrast the orbit Uranus actually has with the orbit it would have if Neptune were not to exist. The effect Neptune accounts for is why Uranus is *not* elsewhere, specifically, not where it would be were Neptune non-existent. The specification is important. It is a very definite property the absence of which Neptune explains. Now, there are very many properties which Uranus lacks, and among these many orbits it does not follow. Knowing the place where Uranus is, and its velocity – knowing these positive facts – we can deduce where it is not and what velocities it does not have. From the positive facts we can deduce what negative facts

118   CHAPTER 2

obtain. For, it is part of the laws of Kinematics that an object is at exactly one place and has exactly one velocity at any given time. In this sense, all these negative facts are explained once we have explained, via the process law, the positive facts of Uranus' position and velocity. But from all these negative facts thus accounted for, we single out one group as of special interest, and worthy of a different sort of explanation, namely, just those that are caused by the presence of Neptune. Why should just these negative facts interest us and not the rest? Why should certain absences demand causal explanaton and not others? Any account of causal explanation must provide an answer to these questions.

2.4. IS CAUSATION INSEPARABLE FROM ACTION?

Another approach to causality and causal priority is that of von Wright.[47] He considers event-types, what he calls generic events. He understands "P is a sufficient condition for Q" as either "Whenever P then Q" or as "As a matter of lawful necessity, if P then Q" but where the latter entails the former. Such statements of conditions are statements of law. He recognizes that causal relations have an asymmetry which an analysis in terms of conditions alone seems incapable, by itself, of capturing. He proposes an account which does capture that asymmetry.

Von Wright considers a world in which there are n types of event, or basic states. A possible state of the world is represented by a conjunction of n terms such that each basic state or its negation appears as a term. A given point in time is an occasion. On any occasion 2^n different states are logically possible. Over m occasions 2^{mn} different successions or total histories of length m are logically possible. Thus, suppose there are two basic states P and Q. The occasion O_1 has four possible states

Diagram 1

$P \& Q$

$P \& \sim Q$

$\sim P \& Q$

$\sim P \& \sim Q$

THE REASONABILITY OF THE DEDUCTIVE MODEL 119

O_1 will have O_2 as its successor

Diagram 2

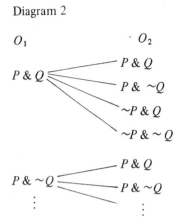

Corresponding to each of the $2^2 = 4$ possible states at O_1 will be four possible states at O_2, yielding $2^{2 \times 2} = 2^4 = 16$ histories of length 2. However, all that is logically possible is not lawfully possible. The number of lawfully possible histories of length m may be less than 2^m. Instead of

Diagram 3

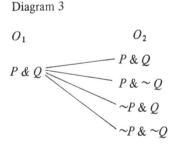

we will have, say

Diagram 4

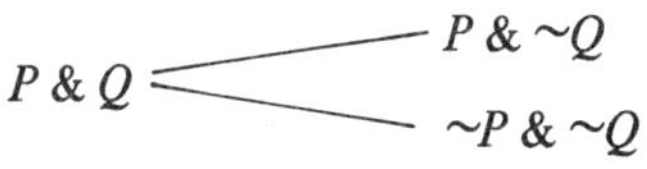

$$P \& Q \begin{array}{c} \diagup P \& {\sim}Q \\ \diagdown {\sim}P \& {\sim}Q \end{array}$$

if it is a law that, whenever P then (on the next occasion) $\sim Q$. In general we will be able to give the following sort of tree-diagram for the lawfully possible histories:

Diagram 5

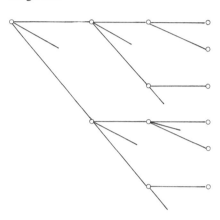

The diagram begins with the first occasion on the left and history moves to the right. Each node located at a given occasion represents the possible alternative total states consequent upon the total state of the preceeding occasion to which it is connected by a line. The one node on the first occasion is that possible state which is actual on that occasion. Progression along a line is a lawfully possible history. The branchings at any particular node indicate the various alternative developments which are immediately open to the world the total state of which is represented by that node. Von Wright now distinguishes the *natural* from other developments of the world after the actual state of it on occasion one. (He adopts the convention that the natural development after any given node is represented by the topmost layer of nodes in the branch of the tree which has the given node as its apex. In the above diagram, the horizontal lines represent the various possible natural developments.) The natural development is the course that nature *would* take unless interference with the course of nature takes place. The natural course of events is " . . . the course of future developments if nature is 'left alone', 'untouched', 'to itself', to continue its course from any given point. It may have come to this point either 'of itself', or from some previous point in the same 'natural' history, or thanks to some interference. This last means that the development from some point in another history was 'deflected' so as to become the starting point of a new history."[48] Thus, turning to Diagram 4, if we have P & Q then naturally we have P & $\sim Q$, unless interference occurred so that $\sim P$ & $\sim Q$ happened instead. Von Wright now defines causation by reference to the interference of agents: P is a cause relative to Q, and Q an effect relative to P, if and only if by doing P we could bring about Q or by suppressing P we would remove Q or prevent it from happening. "What makes P a cause-factor relative to the effect-factor Q is ... the fact that by *manipulating P*, i.e. by producing changes in it 'at will' as we say, we could bring about changes in Q."[49] Von Wright's definition of causation is thus very similar to Gasking's and to Collingwood's. It is open to the same change of circularity as were theirs, and can be defended in the same way, by indicating it presupposes in the notions of *bringing about Q by doing P* the notion of law and not the notion of cause. Von Wright's definition is open to the same objection as Gasking's that it does not apply to causation at the level of cosmological events. Von Wright defends his account as follows: "Causation, needless to say, operates throughout nature independently of agency, also in regions of the world forever inaccessible to human interference. But the test procedures characteristic of causal laws, including those whose operation is far removed from us in space or time, belong to the scientists' laboratories – and they belong there *essentially*, because of their conceptual connection with the mode of action we call experiment."[50] But this is not adequate to the same example for which Gasking's account was not adequate, namely, the law of gravitation which explains the behaviour of the solar system: at the time when Newton advanced that explanation, the idea of *experimental* verification of that law in a laboratory would have been rather more than ludicrous. We may therefore judge von Wright's account inadequate, however elaborately he may spell it out using special "logics" so-called.

However, unlike Gasking, who simply argues from usage, von Wright gives a serious argument for his definition. Von Wright considers a world which is naturally $\sim P$. We now interfere in the situation and make a P-world out of this which, but for the interference, would remain a $\sim P$-world. This newly brought about P-world now transforms itself, without further interference, into a Q-world. This will confirm for us that the regular sequence of P and Q we may have observed in the past was no mere accident but signified a causal tie between the two factors. However, it possible Q would have come into being even if we had refrained from interference, letting the original world continue as a $\sim P$-world. We cannot check this *post hoc*. But we might obtain, either by waiting for, or deliberately creating, a situation in which P is not present, and will remain there unless there is interference. We *then* deliberately abstain from interfering. If it turns out that Q is not there directly after, or at least is not there on some such occasions, then we have further strengthened our belief in the causal connection.

Of these two operations, viz. that of producing p and always finding q on the next occasion and that of letting p remain absent and not always finding q on the next occasion I shall say that they come 'as near as logically conceivable' to the verification of the counterfactual statement that on the past occasions when p was not there, q would have immediately followed had p been there. It is no proof, of course, that the counterfactual is true. But it is what makes us believe this. It confirms our belief that an observed regularity ammounts to a causal or normic connection.[51]

Von Wright recognizes the subjunctive and counterfactual aspect of causation.

Continued observation ... is tied to the 'surface of reality' so to speak. What is required is a peep under this surface, somehow; a dive into the depths of unactualized possibilities for the sake of making sure that even if the actual course of events had been different from what it was, whenever p might have obtained, q would have followed.[52]

There is another way of expressing what I take to be substantially the same craving or requirement: we want to 'substitute' for a given world in which p is *not* the case another world in which p is the case, so as to be able to show that had p been true of the given world, q would be true of the following one.[53]

Von Wright's two operations, then, attempt to get "below the surface" of reality, to the level of unactual possibilities, without, however, actually going there. For, after all, there really is no such "depth" to the world. What von Wright substitutes for such impossible plumbing of the depths is a "shift in reality": an active intervention that substitutes one reality for another than would have happened had the interference had not occurred. "Such deflections [from what would otherwise naturally have happened] can happen, in principle, at any point in any of the natural histories. They could also be called 'shifts in reality'. They are by no means impossible nor even uncommon happenings. They occur whenever agents act".[54] Unactual possibilities, the subjunctive aspect of lawfulness, are thus "tied conceptually" to the interferring activities of agents, according to von Wright. The operations are designed to establish two points. The first is that *in a world naturally ~P, and P could cause Q*. The second is that *in a world naturally ~P, only interference could cause it to change to P*.

This is not a matter of simply confirming certain regularities. To be sure, the first operation involves a P being succeeded by a Q, and this confirms

$P \supset Q$. The second operation does not confirm this regularity, however, though it does confirm something closely related. Think of excluding that, Q would have happened even if P has not happened, by excluding the pair Q & $\sim P$. This is logically equivalent to excluding $\sim(Q \supset P)$. In turn this is equivalent to confirming $Q \supset P$. In the operation, we have a $\sim P$-world. The instantiation of $Q \supset P$ for this world would then have a false consequent. If Q turns out to be present in this world, then the antecedent of the instantiation of $Q \supset P$ would be true, and the generality $Q \supset P$ therefore false. If, on the other hand, Q turns out to be absent from this world, $Q \supset P$ is not falsified. However, this is to say the second operation tends to confirm that P is necessary for Q, not that P is sufficient for Q. But von Wright clearly has in mind that the second operation is relevant to establishing a causal relation, whereby P causing Q he understands a relation of sufficiency.

The two operations are designed to exclude certain counterfactuals. They do this, not so much by confirming certain laws (though that is part of it), but by rejecting other generalities that could be used to support the counterfactuals to be excluded. Von Wright considers two (singular) counterfactuals that could be asserted of a situation in which we had interfered to make it a P- rather than a $\sim P$-world, and in which the P had been followed by a Q. Such a situation occurs when the first operation is performed. Such a situation confirms the regularity that $P \supset Q$ but this by itself does not establish that $P \supset Q$ is a causal relation unless these counterfactuals remain unsupported. One of these counterfactuals is that

(C_1) Q would have happened, in this situation, even if P had been absent

or, what amounts is the same, that

If P had been absent, Q would nonetheless have happened.

This counterfactual could be asserted if there was a law that $R \supset Q$ and if R is present in the situation. In this situation Q has a more remote cause, so that Q would have been there in any case, whether or not P happened to be inserted into the situation immediately before Q occurred. If there was such a remote cause we should not say P was the cause of Q. The other counterfactual about the situation in which the introduction of a P has been again followed by a Q is that

(C_2) Q would have happened, even if we hadn't interfered.

This counterfactual could be asserted if there was a law that $R \supset P$ and if R was present. In that case, P would have been brought about by R, and then that P would have brought about Q. In this situation P and Q have a common cause, and, if there is such a common cause, we should not say that there is a causal tie *between* P and Q.

We have two situations, that of the first operation and that of the second. In these situations there is the conditioned property Q and the two conditioning properties P and R. In the first operation, one interferes in a P-world to make it P and Q follows. The only change is from $\sim P$ to P. *One now assumes that the world remains otherwise the same.* This means that any R, whatever it may be, that happens to be present will continue to be present. In this situation, then, Q, P and R will all be present. In the second situation one has a world otherwise like the previous one. In particular, this world will be naturally R and naturally $\sim P$. One now abstains from interfering, and Q does not occur. In this situation, R will be present, and P and Q will be absent. These data are summarized in the table:

	R	P	Q
first situation	T	T	T
second situation	T	F	F

The first situation confirms that $P \supset Q$. The second establishes that $R \supset Q$ is false. *It also establishes that $R \supset P$ is false*. Thus, the second of von Wright's operations serves to exclude any laws which might support the counterfactuals (C_1) and (C_2). We can conclude from the first of the two points von Wright wished his operations to establish, that in a world naturally $\sim P$, only P could cause Q.

The second of von Wright's points was that in a world naturally $\sim P$, only interference could bring it to be P. We can see how he argues for this by noticing that von Wright does not describe the case of (C_2) or common cause in quite the way we have. We have just seen that it is the second of von Wright's two operations which excludes the possibility of a common cause. Von Wright himself argues that it is the first operation which does this:

... [the possibility] ... of a common cause, is eliminated by the experiment in which we shifted to a new succession of natural developments. For, at the point where the shift took place, the world would, we feel confident, have continued to be in the state not-*p* had we not 'steered' it to the state *p*, and the possibility of 'steering' the world rules out the existence of an antecedent cause which would have taken the world to *p*, i.e. produced *p*, and thereby also rules out the possibility of a cause responsible for the succession of *q* upon *p*.[55]

THE REASONABILITY OF THE DEDUCTIVE MODEL 125

What this passage makes clear is that von Wright is not so much interested in excluding some common cause R as interested in establishing that *in the context of that world the "steering" is the **only** sufficient condition for P*. Our data therefore is intended to concern not just the properties P, Q, and R, but also the property S, of "steering". The two situations in which the interference occurs and does not occur is meant to provide not merely the data recorded in the table above but also the data

	P	S	R
first situation	T	T	T
second situation	F	F	F

From these data *if one assumes that the world remains otherwise the same*, i.e., if one assumes all such factors R remain present (or remain absent), *then one can conclude S is the only sufficient condition for P*. The operations which establish that the regularity $P \supset Q$ is causal are thus operations which establish that "steering" or "interference in the natural course of events" is the only cause of P coming to be in a world that naturally would be $\sim P$. The notions of 'steering' and 'interference' are, of course, concepts closely associated with the concept of action. So the idea of "cause" cannot be separated from the idea of "action".

Von Wright concludes:

> In an important sense, therefore, the causal relation can be said to be dependent upon the concept of (human) *action*. This dependence is *epistemological*, rather than ontological, because it has to do with the way causal relations are established and distinguished from accidental regulates. But this dependence is also ... *logical*, because it is connected with features which are peculiar to the *concept* of action.[56]

It is this logical connection to action that distinguishes causal relations from other cases of sufficient or necessary and sufficient condition, and that gives to causal relations that asymmetry which is typical of those relations. Note that the logical apparatus is that of elimination. Action itself is treated as a case of sufficient conditions, and its relevance in certain contexts established by the eliminative methods characteristic of experimental research. Thus, the notion of causation is explained by von Wright in terms of that of law, of necessary and/or sufficient conditions. Causal explanation thus turns out to be a special case of explanation in terms of laws. On von Wright's account, then, there is nothing about causal explanation that is incompatible with the deductivist thesis concerning the nature of scientific explanation.

The question that confronts us, therefore, is not how to defend the deductivist account of explanation, but rather whether von Wright's analysis of cause and causal asymmetry establishes a conceptual connection between these and the concept of action. We argued above that there were in fact cases where causal language was appropriate but where the concept of action was obviously out of place. Specifically, we had the case of Neptune causing the perturbations in the orbit of Uranus. The (less imperfect) law which explains these causal assertions is the law of gravity, and this law could reasonably be asserted, and the causal connections it accounts for could reasonably be asserted, long before any *experimental* test of the law was conceivable. So von Wright's account is inadequate. Where, then, does his argument go wrong?

The eliminative mechanisms establish that common and non-proximate causes of Q are absent, and that the "steering" or act of interference was the only sufficient condition present to bring about the effect Q. They do this only on the assumption that the world *continues* otherwise to be the same, that the world *continues naturally* to be the same save in the respect which interference changed. This assumption is one of *lawful knowledge*. That such knowledge of laws is required is hardly suprising: the eliminative mechanisms cannot enable one to conclude from a set of observation data to an assertion of sufficient and/or necessary conditions unless some such background knowledge is assumed as an additional premiss. It is the feature of natural behaviour, and its contrast, that of changes wrought by interference, that von Wright sees to be the crucial link-up with concept of action.

We feel confident − presumably on the basis of past experience, or for some other reason − that p will continue to be absent from the world on the next occasion unless we produce it, i.e., change the world into one that contains the state p . . . by and large, the type of confidence to which I am here referring is trustworthy. Were it not so, *action* would not be possible. It is a conceptual feature of the utmost importance of that which we call action that certain changes in nature would not have occurred had we not produced them − or have happened had we not prevented them.[57]

Now, it may well be that the notion of "cause" is tied to the contrast natural development vs. interference. And it may well be that the concept of "action" makes no sense without that contrast. It does not not follow, however, that "cause" must be tied to "action". It would follow only if one assumed that the contrast connected with both notions must be defined in terms of "action". Von Wright's case for connecting "cause" with "action" makes just this assumption. If, therefore, we can successfully challenge it, we shall have shown where his argument goes wrong.

The concept of interference can best be approached in our terms by means

THE REASONABILITY OF THE DEDUCTIVE MODEL 127

of the idea of closure, or rather the breaking of closure. One has a closed system. Call it a. a develops according to the (hopefully process) law that describes its behaviour over time. According to this law, a will be $\sim P$ at t_n. Another object b comes into relation R with a at t_{n-1}. One now has a *compound system* consisting of a and b in relation R. The law for the development of this compound system will be different from the law for the isolated a. In the law for the compound system, there being something R-ing a is a relevant variable. In particular, the law predicts that when it is present then a is P at the next succeeding occasion. So, for the compound system that has come into existence, we have the prediction that a will be P at t_n, rather than $\sim P$ as it would have been had not the compound system come into existence at t_{n-1}. R-ing will, in general, not be the only relevant variable. Thus, for example, manipulating brings about changes only with respect to certain sorts of objects and with respect to certain sorts of changes. Where a is of one sort, then the laws assert a will continue to be $\sim P$ as it would had it not come to be R-ed by b. If it is of another sort, then the laws assert that a will come to be P upon being R-ed by b rather than remaining $\sim P$ had it not been R-ed.

In the case of action, b is the actor, and R-ing a consists of manipulating a in some particular way or other. Where action can bring about changes and where it cannot is something that has been learned. "We feel confident ... that p will continue to be absent from the world on the next occasion unless we produce it ... because we have *learnt* (been taught) how to do it."[58] What we have learnt is certain laws of nature, about how our bodily activities are relevant to the presence and absence of properties in other objects. But there is nothing in the notion of interference that requires the interfering object to be one that learns, i.e., comes to know, and acts upon what it has learned. The solar system is closed. A hitherto distant star suddenly approaches. Its gravitational effect becomes appreciable. That is to say, closure is broken. The solar system no longer behaves as it would have behaved had the interference not occurred. Here b is the approaching star, a is the solar system, and R-ing a consists of coming within a certain distant of a.

Or consider another example, this one due to Taylor, and which raises some other issues also.

Consider ... a locomotive that is pulling a caboose .... Now here the motion of the locomotive is sufficient for the motion of the caboose, the two being connected in such a way that the former cannot move without the latter moving with it. But so also, the motion of the caboose is sufficient for the motion of the locomotive, for given that the two are connected as they are, it would be impossible for the caboose to be moving without the locomotive moving with it. From this it logically follows that, conditions being such as they are – both objects are in motion, there are no other moves present, no obstructions to motion, and so on – the motion of each object is also necessary for the other . . . .

What, then, distinguishes cause and effect in the foregoing [example]? . . . It is not any difference in the relations of necessity and sufficiency, for these are identical in both ways. But there is one thing which . . . appears to distinguish the cause from the effect; namely, the cause acts upon something else to produce some change. The locomotive *pulls* the caboose, but the caboose does not *push* the locomotive; it just follows passively along.[59]

This example is, in certain respects, like the barometer example we discussed previously. It provides an example of a law, a relation of necessity and sufficiency which yields predictions but not explanations: at least, that is the suggestion. The motion of the caboose is necessary and sufficient for the motion of the engine, yet the former does not explain the latter. In contrast, the motion of the engine does, it is said, explain the motion of the caboose. To this the reply is, we know, that the motion of the caboose does have explanatory power *vis-à-vis* the motion of the engine: just as, given the barometer readings we can use the barometer-atmospheric pressure law to explain why the pressure must have been as it was, so also, given the motion of the caboose we can use the law that this suffices for the motion of the engine to explain why the motion of the engine must have been as it was. Still, one can give *better* explanations than this. The correct contrast is not, as Taylor would have it, between a non-explanation and an explanation but between an explanation and a better explanation. We know that the motion of a train and of a caboose can begin and can be continued only if there is a source of power. The *laws* are such as to support the subjunctive conditionals that if there were no source of power than the motion of the caboose could not begin, and that if there were no source of power then the motion of the caboose would not continue. And we further know that the source of power is in the engine. So we know that if there were no engine then the motion of the caboose would neither begin nor continue. From the point of view either of idle curiosity or of pragmatic interests these laws are far better than those that permit the inference of engine motion from caboose motion.

However Taylor can also be construed as contrasting not explanation to non-explanation but causal explanation to other sorts of explanation. His criterion for causal explanation would then be that where one has *action* rather than passion: causes are active, effects are what is passively consequent upon such activity. Clearly, upon this account the typical asymmetry of the causal relation is easily accounted for.

Clearly, there is some plausibility to Taylor's suggestion, but there are clear cases which establish its inadequacy as an analysis of causal language. Mill gives a couple of examples that do the job neatly.

> ... would it be more agreeable to custom to say that man fell because his foot slipped, or that he fell because of his weight? for his weight and not the motion of his foot, was the active force that determined his fall. If a person walking out on a frosty day stumbled and fell, it might be said that he stumbled because the ground was slippery, or because he was not sufficiently careful; but few people, I suppose, would say that he stumbled because he walked. Yet the only active force concerned was that which he exerted in walking; the others were mere negative conditions; but they happened to be the only ones which there could be any necessity to state; for he walked, most likely, in exactly his usual manner, and the negative conditions made all the difference.[60]

Causes are thus not necessarily active. But, like some actions and activities, they do interfere with the normal or natural course of events. Failures and omissions — he did *not* walk with sufficient care — are among the events cited as causes. These are particularly important where men have discovered routines capable of *preventing* harm. Men have discovered not only that nature can be harmful *if* we interfere but also that it can be harmful *unless* we intervene. In order to avoid such harmful consequences men and societies have invented routines, regular patterns of activities aimed at systematically counteracting the regular course of unattended nature. Thus, the effect of drought is regularly neutralized by governmental precautions in conserving food or water; disease is neutralized by innoculation; rain by carrying umbrellas; icy patches by walking with care. Now, in these cases *failing to act* can be cited as a cause. The famine was *caused* by the government's failure to store adequate amounts of food. He caught cholera *because* he neglected to be innoculated prior to his trip. I got wet *because* I left my umbrella at home. He slipped *because* he did not take care. That failure to act can be a clause makes clear that the idea of an "active force" is not an essential ingredient of the notion of cause. Taylor has therefore failed to characterize adequately the nature of the causal relation.

On the other hand, both the example of Taylor and the examples of failing to act fit well into an account of causation in terms of a contrast between "natural development" and "interference". In the case of the caboose we have a heavy object resting on the ground. Such an object will, we know, continue in that state unless some force is exercised. The force, the source of power, lies in the engine. The "natural state" of the caboose is rest. What interferes and changes the state of the caboose from what it would naturally continue to be, is the pulling of the engine. We therefore call this the cause. In the case of the famine, the cholera, the becoming wet, the "natural development" is not that of uncontrolled nature, but rather nature as modified by these customary and regular activities of men which aim at preventing the harm implicit in uncontrolled nature. Such man-made natural developments are established and regulated by certain norms and standards of performance. Deviations from these norms will come to regarded as exceptional and so to be counted as the causes of the harm that results.

There is one objection to treating omissions as causes that is worth noticing.[61] A gardener whose duty is to water the flowers fails to do so and they die as a consequence. It could be argued that one can treat the gardener's omission as the cause of the flowers dying only if we are prepared to say that the "failure" on the part of everyone else to water the flowers was equally the cause. But the latter hardly fits into our ordinary causal discourse, and therefore neither could the former. This *modus tolens* can be countered by challenging the premiss, arguing that one can, after all, accept the antecedent and reject the consequent. One does this by distinguishing the nature of the gardener's failure from the nature of the "failures" of other persons. The "failure" of other persons is what is naturally to be expected; it is not their business, not a duty with which they have been charged, to water the flowers. But the gardener's failure to water the flowers is quite different. It is part of what is naturally to be expected of gardeners that they will water the flowers. The gardener's failure is therefore not just a breach of duty, but a deviation from an established routine, from a set of man-made natural developments. So his failure, but not that of other persons, is, as it goes in ordinary discourse, the cause of the flowers dying.

We may draw two conclusions now. One is that non-human entities may *interfere* in the course of natural developments and be reckoned as causes. The other is that omissions may also be counted as *interference* in the course of natural developments and be reckoned as causes. Together these make quite clear that, contrary to von Wright, while action is a case of interference, interference does not itself conceptually presuppose the notion of action. Failure to recognize this may simply be a matter of undue preoccupation with those cases where human manipulation or alteration of the environment is the cause. Or, more likely, it may be due to a, perhaps unconscious, commitment to an anthropomorphic model of causation. Certainly, one wants to insist that in our ordinary causal language there is a great amount of anthropomorphism. One must discipline one's reflections upon ordinary causal discourse with the recognition that all explanation, including causal explanation, *ought* to be in terms of laws. Unless one so disciplines oneself, one is liable to be misled by the non-scientific, anthropomorphic features of ordinary causal language into assimilating all causes to the model of human action.

Von Wright attempts to link causes to action through his contrast of "natural behaviour" to "interference". We have just seen that the latter is not tied to the concept of action. What, now, of the other half of von Wright's contrast: Does the notion of "natural development" conceptually presuppose the notion of action?

We may begin by noting that the distinction between natural and interfering factors is relative to context. Consider, first, a famine consequent upon a drought. The cause of the famine may be identified by the peasant as the drought, but the World Food Authority may identify it as the government's failure to build up adequate reserves of food. It was no doubt such examples as this that led Collingwood to insist that the identification of a cause among the other — "mere" — conditions is always dictated by some pragmatic interest, attributing the relativity of the distinction to the varying means at the disposal of different people to produce results. But the peasant can do nothing about the drought, though he nonetheless cites it as the cause. Similarly, lightning causes fire, though we have no way of controlling lightning. Indeed, when we do learn to establish techniques for controlling these things we may cease to look upon them as the cause and shift to identifying the cause as the failure to use established technique[62] — in the case of famine and drought, food reserves and reservoirs for water; in the case of lightning, the placing of lightning rods. Not liking the harm that results from uncontrolled nature, as we learn how to control events, we expect such knowledge to be implemented, and routines instituted, that such harm might be avoided. We expect man-made natural situations to come into being as our capacities to anticipate and control nature increase. The government may be assumed to have the capacity to act to prevent the famine. Its failure to act is therefore the cause of the famine. The peasant would no doubt act to turn the government out of office if he knew, as does the World Food Authority, that the government has the capacity to prevent the famine; for he would see, then, that his government's failure to act was the cause of his miserable condition. The difference between the peasant and the World Food Authority is not, as Colingwood would have it, a matter of pragmatic interests alone, but a matter of knowledge. The World Food Authority knows what the peasant does not know, that the government could have but failed to act. This makes clear one way in which the identification of causes depends upon context: it depends upon what those who explain in terms of causes happen to know. But this has to do with the pragmatics of explaining, not with the logic of explanation, or to logic of cause.

There are other situations, however, in which one and the same law is cited in explanations, but in which one identifies two different conditions as *the* cause.[63] In most cases where a building is destroyed by fire, the natural situation will include the presence of oxygen, the dryness of the building, the presence of combustible material, and so on. It will be the dropping of the lit cigarette that will be counted as the cause.[64] On the other hand, suppose one has a laboratory or a factory in which special care is exercised to exclude oxygen during part of an experiment or part of a manufacturing process, success depending upon safety from fire, while at the same time there is quite likely to be present static electrical sparks. In that context, if a fire breaks out then the presence of oxygen might well be cited as the cause of a fire. In both these cases a relatively full explanation will have to mention the presence of oxygen, the presence of an open flame or spark, the presence of combustible materials, and so on. These are all lawfully relevant factors. This lawful knowledge is used in both cases when one distinguishes what is naturally present and what is the interfering factor. The distinction between condition and cause lies elsewhere.

In order to find out what this is, we must look at features of the two situations which are other than those which enter into the explanations of the events in the two situations. In the first case oxygen is present in the world which developes naturally, without interference, and in the world which is "steered" towards the fire. In that context, then, it could not be the cause of the fire; the eliminative mechanisms to which von Wright directed our attention would not pick it out as the only sufficient condition present. In the second case, the presence of oxygen (as contrasted to, say, the dryness of the building, or the presence of static electricity) is *not* a feature common to the disaster and the natural situation. We thus have in each case, systems of some kind, say K, and the regularities which hold in these systems are the same. Thus, we might have laws of the sort

$$(x) [Kx \supset ((P_1 x \:\&\: \sim P_2 x) \supset Qx)]$$

K then divides into two subclasses. In o e, P_1 is naturally present. With respect to this class of Ks, the eliminative mechanisms von Wright introduces will establish $\sim P_2$ as the only cause of Q. In the other class of Ks, P_1 is naturally absent. With respect to this class, the eliminative mechanisms will establish that P_2 is the only cause of Q. For all Ks, any *explanation* of the presence of Q will mention the presence of P_1 and the absence of P_2. But in those Ks in which P_1 is naturally present, the absence of P_2 will be *the cause* of Q, while in those Ks in which P_2 is naturally absent, the presence of P_1 will be *the cause* of Q. The question is, what grounds could we have for expecting a K to be naturally P_1 or naturally $\sim P_2$? It is clear that we must go beyond the factors, P_1, P_2 and Q, if we are to have grounds for expecting one K to be P_1, another K to be $\sim P_2$. There will have to be characters of Ks, H_1 and H_2 say, which exclusively and exhaustively divide K, and such that we say

normally, H_1s are P_1

and

normally, H_2s are $\sim P_2$

Such judgements may be based on simple statistics: e.g., "the vast majority of H_1s are P_1." Or other considerations might be relevant. H_2 might be the feature that someone has a duty towards a K in his charge to ensure it remains P_2. Given the assurance that men normally do their duty, we will be able reasonably to expect H_2s to be $\sim P_2$. This describes the circumstance of man-made natural developments. In either case, however, what is crucial is one's expectations with respect to a K. Some Ks will be expected to be one way, other Ks will be expected to be the other way. *Judgements of causation will made in terms of these expectations.*

If this tentative conclusion is acceptable then von Wright's category of "natural development" in no way presupposes the notion of action. We may therefore accept von Wright's case that "cause" is to be understood in terms of the contrast between "natural development" and "interference", while rejecting his claim that the notion of "cause" presupposes that of "action". But our conclusion is tentative. We must look at the case we have thus far made the touchstone for any analysis of cause, the perturbations in the orbit of Uranus caused by the presence of Neptune.

Under what conditions would one cite Neptune as the cause of the perturbations of Uranus' orbit? Surely in the context where someone was expecting Uranus to have a certain orbit, discovered it did not, and then wondered why his expectations were not fulfilled. One thinks here of the astronomer who calculated Uranian orbits only to find that planet was elsewhere than he had predicted it to be. He then would look for those omitted relevant factors the presence of which accounts for Uranus being where it is, or, what here is the same, why his expectations were not fulfilled. This interpretation is certainly plausible. Plausible enough for John Stuart Mill to adopt it, when he connected the distinction between "cause" and "conditions" primarily with the contrast between what a person inquiring into the cause already knows at the outset, and what he may require to be informed of as previously unknown to him.

The reviewer [of an earlier edition of the *System of Logic*] observes, that when a person dies of poison, his possession of bodily organs is a necessary condition, but that no one would ever speak of it as the cause. I admit the fact; but I believe the reason to be, that the occasion could never arise for so speaking of it; for when in the accuracy of common discourse we are led to speak of some one condition of a phenomenon as its cause, the condition so spoken of is always one which it is at least possible that the hearer may be required to be informed of. The possession of bodily organs is a known condition, and to give that as the answer, when asked the cause of a person's death, would not supply the information sought. Once conceive that a doubt could exist as to his having bodily organs, or that he were to be compared to some being who had them not, and cases may be imagined in which it might be said that his possession of them was the cause of his death. If Faust and Mephistopheles together took poison it might be said that Faust died because he was a human being and had a body, while Mephistopheles survived because he was a spirit.[65]

And we may take this interpretation to be confirmed by the fact that it is endorsed by those against whom we are arguing most vigorously. Thus, Toulmin discusses unexpected deviations in the orbits of Jupiter, and relates them quite neatly to our ordinary causal discourse:

Suppose, for example, that we look out the window, into the street. One car travels steadily down the road comes into sight, passes our window, and goes on out of sight again: it may well escape our attention. Another car comes down the road haltingly, perhaps jerking and backfiring, perhaps only stopping dead and starting up again several times: our attention is immediatley arrested, and we begin to ask questions – 'Why is it behaving like that?' From this example it is only a step to the case of a practical astronomer, for whom the continued motion round its orbit of the planet Jupiter is no mystery: but for whom questions would immediately arise if the planet were suddenly to fly off along a tangent to its orbit and out into space: 'What made it do that?'[66]

He elsewhere makes the same point in more general terms:

'Deviations' – as soon as one begins to characterize phenomena, the very ink in one's pen becomes saturated with revealing words like 'deviation', 'anomaly', and 'irregularity'. All these imply quite clearly that we know of a straight, smooth, regular course of events which would be intelligible and rational and natural in a way that the 'phenomenon' is not. And this is just the conclusion we are now prepared for: the scientist's prior expectations are governed by certain rational ideas or conceptions of the regular order

THE REASONABILITY OF THE DEDUCTIVE MODEL 135

of Nature. Things which happen according to these ideas he finds unmysterious; the cause or explanation of an event comes in question ... through seemingly deviating from this regular way; ... and, before the scientist can be satisfied, he must find some way of applying or extending or modifying his prior ideas about Nature so as to bring the deviant event into the fold.[67]

Except, of course, that we do not, where Toulmin does, identify explanation with causal explanation. Doing so misleads Toulmin decisively into a systematic misinterpretation of a good deal of science. What it appears to be is that Toulmin has been misled into the identification by a commitment to some of the non-scientific, anthropomorphic patterns of explanation implicit in much of ordinary language.[67a]

In much the same vein, another of our opponents, Scriven, has remarked upon our tendency to characterize as laws certain generalities that are only approximately true.

One of the reasons for our willingness to accept an approximation is that we hope to explain its deviations from accuracy and that it gives us a basis from which to work. The nature of explanation is such that it is greatly simplified if a standard pattern of behaviour can be found, in a particular field, to which the actual behaviour *approximates*. For it is then possible to provide (or seek) a single explanation of the standard pattern of behaviour, no matter how often exemplified, and to provide accessory explanations only for the residual departures from this standard. The big step, for the theoretician, is the first step – getting the approximation .... This, I believe, is why we happily refer to propositions as laws which are known to be inaccurate. It is not mere historical habit, but a recognition of these propositions as framework claims valid for certain limiting cases and known-to-be-false assumptions which are nevertheless *related to* actual conditions.[68]

What Scriven here makes clear is how naturally the language of cause fits into the *research* context. One has a piece of imperfect knowledge describing what is naturally to be expected. One then looks for causes which determine deviations from this norm. The search after causes in the research context is thus equivalent to the search after less imperfect knowledge. But what must equally be emphasized is that this less imperfect knowledge which the research yields need not consist of causal laws. It is might turn out to be better than knowledge of causal laws. It might, for example, turn out to be process knowledge. *The search after causes may well yield knowledge which is better than causal, knowledge which absorbs and explains causal knowledge as the less imperfect absorbs and explains the more imperfect.* More strongly, since process knowledge is the ideal of scientific explanation, *it is strongly to be hoped that any search after causes will yield knowledge which is better than causal*. However, to emphasize this does not mean that one cannot also recognize that much research is appropriately described as a search after causes, and that much of causal language is correctly applied, in particular the distinction between standard or natural behaviour, and the deviations from it that are to be accounted for by, as yet unknown, factors which interfere with the natural developments.

I am not at all sure it is possible to *prove* the natural-interference distinction turns upon expectations. But we have made cosmological events the touchstone of any analysis of causation. And when we find our opponents agreeing with us on our interpretation of this case we may, I think, take that to be as much proof as could reasonably be desired.

What it is important to recognize, what Toulmin and the others against whom we are arguing do not recognize, or do not recognize clearly, is that at no point is one to say causal explanations are *the best* one can do. The case of Uranus makes clear that the causal explanations are but part of a better explanation, an explanation in terms of process knowledge. But once causation comes to be understood in terms of the natural-interference distinction, it is clear that the notion of law is basic, that causal explanations are but a species of explanation in terms of laws, and that there are explanations in terms of laws that are better than causal explanations.

We began this discussion of causation by a consideration of examples, produced by those opposed to the deductivist model, which purport to be cases of predictions which are not explanations. Our discussion of predictions which are not explanations must now come to an end. Not all examples that have been used have been discussed. However, I think that, save for one group, they introduce nothing essentially different from what is involved in the above examples. To discuss these examples would therefore become tedious, and so I pass them by. The one remaining group are those which involve statistics, and these we must leave until later.[68a]

2.5. ARE THERE EXPLANATIONS WITHOUT PREDICTIONS?

Let us now turn to the examples the opponents of the deductivist thesis advance in an attempt to show we can have explanations without predictions.

The first example we shall look at is this. Consider the destruction caused by an earthquake.[69] The occurrence of the latter explains the occurence of the former. However, the latter could not be used to predict the former, since we could not predict the occurrence of the destruction until the simultaneous occurrence of the earthquake. Thus, Scriven. The example fails to make its

THE REASONABILITY OF THE DEDUCTIVE MODEL 137

point, however. That there is explanation is agreed. Further, it is by deduction that we know that earthquakes cause damage. Scriven holds there could be no prediction here. His only reason seems to be that the destruction occurs *simultaneously* with the earthquake, and since predictions are about the future, the destruction can't be predicted from the earthquake. However, for the defender of the deductive model, the essential idea of 'prediction' is not that of an inference to the future, but the more general idea (of which the inference to the future is but a special case) of an inference from the known to the hirtherto unknown. It is consistent with the more general idea that the inference-prediction be to a simultaneous event. The example therefore fails to establish its point. Of course, it is true that in all likelihood, if we come to know about the earthquake, e.g., by feeling the earth trembling, then we shall simultaneously come to know about the destruction. In that case, since both facts are known, it is not possible for there to be an inference from the known one to the unknown one and therefore no prediction in the broad sense. There would still be the explanation. However, the fact that the explanation would not *in that context* be used to predict does not entail that the explanation *could not have been* used to predict in another context, one in which the earthquake was known and the destruction not known.[70] The fact that it would not be used does not touch the symmetry thesis which says only that where explanation is possible, there prediction is also possible. Not actual, only possible — that is, possible in the appropriate context. I conclude Scriven's example completely fails of its purpose.

He also appeals to what we shall call *the ink-bottle example*: " ... your knee catches the edge of the table and thus turns over the ink-bottle, the contents of which proceed to run over the table's edge and ruin the carpet. If you are subsequently asked to explain how the carpet was damaged, you have a complete explanation. You did it, by knocking over the ink."[71] But there could be no prediction because no law is available which would permit such prediction: "You could not produce any true universal hypothesis in which the antecedent was identifiably present (i.e., which avoids such terms as 'knocks hard enough'), and the consequent is the effect to be explained."[72] Since the antecedent cannot be clearly identified, there can be no prediction. On the other hand, there is an explanation. Scriven concludes against the deductive model. Let us see.

We grant him that an explanation has been produced. However, it is an explanation because there is a law available. And this law permits prediction. So his example does not count against the deductive model after all. It is true that the law used in the explanation and prediction is only imperfect. But that does not count against the deductive model. Unfortunately Scriven overlooks this point. He wrongly thinks the laws used in explanations have to be "exact". Indeed, he tells us "the deductive model ... requires the exact truth of the premises and exact deduction."[73] What exact as opposed to inexact deduction is, I cannot say. Deduction is either valid or invalid, with no degrees. Certainly, the model requires valid deduction. But there can be valid deductions from imperfect laws, so no problem arises. As for "exact truth": again, statements are either true or false, with no degrees. And again, the deductive model requires the premises to be true. But, also again, we can have true statements of imperfect laws, so neither is there a problem here.[74] The force of 'exact' has to be something like 'if the phenomenon is quantifiable then a determinate numerical value must be predicted.' Only a determinate value is "exact"; if one predicts a determinable range of such values then the prediction is "inexact". Such is the only sense which I can attach to Scriven's claim. But then the deductive model does *not* require "exact" predictions. A law may state that if a factor is within such and such a range of values then so and so will occur. Initial conditions state that the factor is within the determinable range. One deduces and explains and predicts that so and so will occur. Or a law may state that if a factor is within such and such a range then another factor will be in so and so a range of values. Initial conditions state the first factor is within the relevant range. One deduces and explains and predicts that the other factor is within the so and so range of values. Nothing "exact", but still, contrary to Scriven, deduction — and therefore explanation and prediction. Of course, the laws used in the deduction (explanation, prediction) are imperfect. The fact that the prediction is *not* exact is a mark of that imperfection. Only with perfect knowledge are all such imperfections overcome. To suggest that the deductive model requires "exact" truth, is to suggest that the only premises the deductive model permits are those which involve perfect knowledge. But this is not so. Of course, the event we want to explain may be one which has the factor taking on a determinate value. Then our imperfect law explains not this event but another (the description of) which is implied by (the description of) the event in which we are interested. The latter is the factor taking on a determinate specific value; from that it follows the factor takes on a generic value within the so and so range (that it takes on the value 5 implies it takes on a value in the 4–6 range). The fact that all we succeed in explaining is the so to speak generic event whereas we want an explanation of the specific event means simply we are disatisfied with what we have been offered, and desire another, more perfect, explanation. [So, to advert to our example in Section 1.2, we desire to replace (1.2) by (1.3).] Such disatisfaction is simply a mark of the fact that process knowledge does, where imperfect knowledge does not, tell us everything we could possibly want to know about a system. On the other hand, while we do want explanations in terms of process knowledge, while we do want "exact truth", often we do not have it, and settle for less. The result is the use of "inexactly true" laws, i.e., imperfect laws, as the premisses of our deductive arguments and therefore explanations and predictions. And this is what we have and have settled for in the ink-bottle example. A blow with an air-bourne feather will not knock the bottle over; a rapidly moving elbow, squarely striking will suffice. Below a certain point, blows are not sufficient. Above that point they are sufficient. The exact dividing point is unclear. This is because there are a number of relevant contributing factors — the strength of the blow, the direction, the nature and situation of the ink-bottle, and so on. That is, we have an "inexact truth", an imperfect law. But we can use it to predict. "He's going to knock the ink over," we might say, recognizing the direction in which his elbow is moving, and recognizing its speed is above the critical point. An "inexact prediction" to be sure, but, for all that, still a prediction, and, indeed, a reasoned prediction involving a deductive from initial conditions and an "inexact", imperfect, law. I conclude that, contrary to Scriven's claims, the ink-bottle example does involve an explanation, the explanation *does* include a law, and the law *could* be used to predict.

The crucial thing in the ink-bottle example that must be noticed if the deductive model is to be defended is the imperfect law which occurs in the explanation. That the generality is imperfect is marked by the fact that we express it with such phrases as "knocks hard enough" and "knocks sufficiently hard." There are a number of such expressions, of which Scriven makes much,[75] which mark imperfect knowledge. According to Scriven, such expressions do not mark statements of law, imperfect laws, but rather statements which are "truisms".[76] With respect to such a truism, Scriven asks, "who could deny it, but who would bother to say it?" Laws presumably, could be denied. Since truisms cannot be denied, they are not laws. What figures in the ink-bottle case is not a law but a truism. Hence, the deductive model is false — not so much for insisting upon deduction, but for insisting that the premisses include at least one law. This is a different objection, but one with which we must deal. The question to be asked is this: Does the fact that a "truism" cannot be denied render it not a law? Now, one may be unable to deny a statement for either one of two reasons. First, it is analytic. Second, it is synthetic but evidence which is available to all is overwhelmingly and conclusively in its favour.[77] "An ink-bottle will fall if knocked sufficiently hard" *may* be understood, given the appropriate interpretation for 'sufficiently', as analytic. In the present case, it is not being understood this way. As we saw, such a use of 'sufficient' or 'hard enough' serves to mark a certain range with an indefinite boundary. Definite occurrences are excluded as not "hard enough". An analytic statement would excluded nothing. So the statement is not analytic. However, the ordinary quotidien experience of everyone (who lives in a cultural ambience which makes things like ink-bottles familiar objects) provides overwhelming evidence that the statement is true. It is this overwhelming evidence which renders the statement a 'truism": It can't be denied (by a rational person) because that would be to fly in the face of conclusive evidence to the contrary. It is, of course, true that the statement has, as Scriven puts it, a "selective immunity" to counter-examples. This is by virtue of its hazy border-line. But it excludes certain things, and therefore does not have complete immunity to possible counter-examples; from which it follow it is not analytic. But while it is not immune to *possible* counter-examples, the overwhelming evidence for its truth constitutes overwhelming evidence that there will be no *actual* counter-examples. Because the evidence available to all is conclusive in its favour, "who could deny it?" That is, it is therefore a truism. But being a truism in this sense is compatible with being synthetic. A law must be synthetic. Nothing prevents its also being a truism. Indeed as Scriven himself says of his "truisms", they are "trivial though not empty", and "the truism tells us nothing *new* at all; but it says *something* and it says something *true*, even if vague and dull". There is no need for laws used in explanations to all have the intellectual excitement of the general theory of relativity; there is no reason why they might not be vague and dull. All the deductive model excludes from the premises are non-laws. It does not exclude laws which are truisms. So the fact that an explanation may use a law, that is, an imperfect law, which is also a truism does not in any way show a difficulty in the deductive model.

The difficulty with Scriven is that he simply takes a response to a sentence, viz, "could not deny it", and elevates it into a logical category.[78] Responses, however, are not logical categories. The latter must be features of sentences defined in terms of the sentences themselves, intrinsic to the sentences. Elevating responses to the level of logic replaces the latter by descriptive psychology. The latter way be interesting but it is not philosophy. For it ignores the critical, *normative* question: *ought* he to have responded as he did? If the response defines the category then this question cannot be asked. Scriven's procedures embody the abandonment of critical intelligence.

THE REASONABILITY OF THE DEDUCTIVE MODEL   141

Let us turn now to a slightly different case. In fact, it is an extension of the previous one. The "knocks hard enough" covers a border-line area which is hazy. Some cases are neither clearly within nor clearly without. This is what gives it its "selective immunity" to counter-examples. This is simply a special case of a law (imperfect, of course) with "exceptions".

> The normic statement says that *everything* falls into a certain category *except* those to which *certain special conditions* apply. And, although the normic statement itself does not explicitly list what count as exceptional conditions, it employs a vocabulary which reminds us of our knowledge of this, our trained judgment of exceptions ....[79]

> ... if exceptions were few in number and readily described one could convert a normic statement into an exact generalization by listing them. Normic statements are useful where the system of exceptions, although perfectly comprehensible in the sense that one can learn how to judge their relevance, is exceedingly complex.[80]

There is a great deal in this to which we shall have to return. The immediate point with which we want to deal is this: since there are exceptions the nature of which we do not know (if we did know them, we would list them, and use a generality rather than a normic truism) — though we may "judge" when we have an exceptional case — since there are these unknown exceptions, it follows we cannot *with knowledge* employ the truism to predict; i.e., we cannot make reasoned predictions using it. On the other hand, under certain circumstances, we *can* use it to explain. Hence explanation without reasoned prediction, contrary to the symmetry thesis, and so much the worse for the deductive model.

Once again Scriven's familiar pattern. It begins from a sound position. There *are* imperfect laws with "exceptions" the nature of which we do not know. (We shall give examples below.) "Judgment" *does* play a role — though we have to figure out what this "judgment" is. The significant question is whether any of this touches the deductive model? Or is Scriven wrong once again? Wrong once again, I am afraid we shall have to conclude. What is crucial is understanding what it is about the logical structure of laws, imperfect laws, that permits us to explain where we cannot predict.

We can begin by looking at our example (2.1) from Section 1.2:

$$(x)(Kx \supset (Fx \equiv Gx))$$

Suppose we know this to be false in all generality. That is, we suppose it has been used in successful prediction in some cases, but that it has failed in others. If (2.3)

$$(x)[Kx \supset (Cx \equiv (Fx \equiv Gx))]$$

were known, we would know the conditions under which it (2.1) is successful viz, which Ks are Cs, and those under which it fails, viz, which Ks are not Cs. But let us also suppose (2.3) is *not* known. We are therefore supposing we know (2.1) is sometimes useful, but that we do not know the specific conditions under which it is useful. In all liklihood, our background knowledge and theory would make reasonable certain limitations upon precisely what the conditions are under which it is useful. These conditions will be of some type C. In such a situation the following hypothesis would suggest itself as reasonable [81]

(5.1) There is a factor of sort $\mathscr{C}$ such that for Ks its presence is a necessary and sufficient condition for the join presence or absence of F and G.

which, as a first stab at formalizing, we might represent by after the fashion of (2.2):

(5.2) $(\exists f) \{ \mathscr{C}f \& (x) [Kx \supset (fx \equiv (Fx \equiv Gx))] \}$

As discussion proceeds we shall find (5.2) does not adequately represent (5.1) *if* (5.1) is to be understood in such a way that we can use it to explain where we cannot predict. The discussion shall proceed in such a way as to attempt to bring out what, logically, is involved in the pieces of imperfect knowledge that permit explanation without prediction.

We should begin by noting the order of the quantifiers in (5.2). Compare

(5.3) $(\exists f)(x)(\mathscr{C}f \& fx)$

with

(5.4) $(x)(\exists f)(\mathscr{C}f \& fx)$

(5.3) says "there is at least one colour which every thing has"; it asserts at least one determinate colour is common to every thing. (5.4) asserts only the weaker claim that every thing is coloured, has some colour or other. Since '$\mathscr{C}f$' contains no 'x' the universal quantifier in (5.2) can be shifted to yield the equivalent expression

(5.5) $(\exists f)(x) \{ \mathscr{C}f \& [Kx \supset (fx \equiv (Fx \equiv Gx))] \}$

which makes the analogy to (5.3) clear. What (5.2) asserts is that there is a determinate $\mathscr{C}$ exemplified commonly by all x satisfying certain conditions.

THE REASONABILITY OF THE DEDUCTIVE MODEL 143

It asserts that there is a determinate $\mathscr{C}$, common to all Ks, the presence of which is necessary and sufficient for the joint presence or absence of F and G.

As an example of (5.1) we might have 'F' to be hunger as defined in terms of length of deprivation of food, and 'G' hunger as defined in terms of the strength of the tendency to prefer food to some alternative, say, a female. Then background theory suggests there is some condition f which is physiological in nature – '$\mathscr{C}$' = the property of being a physiological factor – which is necessary and sufficient for the simultaneous presence and absence of F and G. This specific physiological condition (the $\mathscr{C}$ of (5.1)) is, we know, the amount of sugar in the blood. We are now supposing, however, a situation in which this factor is not specifically known. All we are supposing to be known is that it exists, rather than specifically what it is. We are supposing simply that (5.1) is reasonably believed.

If (5.1) is reasonably believed then scientists would begin to seach for the factor (5.1) asserts to exist.[82] Prior to its discovery, however, scientists might reason as follows. "We know, by (5.1), that such a factor exists. Let us call it 'C^*'." On that basis they would begin to use a statement of the form

(5.6) For Ks, being C^* is necessary and sufficient for the joint presence or absence of F and G.

This can be formalized as

(5.7) $(x) [Kx \supset (C^* x \equiv (Fx \equiv Gx))]$

which is, of course, parallel to (2.3) of Section 1.2. But it must be emphasized that (5.7) is *not* (2.3). In (2.3), we assumed C was known. In (5.7), 'C^*' is the label of a factor we do not know but which we *hope to discover*. When we use 'C' we are using the concept of a known relevant factor; when we use 'C^*' we are using the *hope* of a known relevant factor. 'C' marks knowledge; 'C^*' marks ignorance, though ignorance which, we hope, will be replaced by knowledge. Since we do not know specifically what C^* is, we can have no direct confirmation of (5.7) = (5.6). We may, however, have indirect evidence. As the physiology example makes clear, background theory may make it quite reasonable to affirm (5.1) and therefore (5.6) = (5.7).

Now, once we have (5.6) = (5.7), then it is possible to offer explanations by citing the presence of C^*. These would go as follows. Suppose system a is a K-system. Then

(5.8) Ka

is true. Suppose, further, that this is a K-system which is F and also G. In that case

(5.9) Fa

and

(5.10) Ga

are both true. Since we do not know what C^* is, we obviously could not have predicted this would be a system which was G. On the other hand, since we have reason to suppose (5.1) and therefore (5.2) are true, we can use (5.3), (5.9) and (5.10) to deduce from the *synthetic truth* (5.7) that

(5.11) C^*a

holds. We can say: since (5.8), (5.11) and (5.9) obtain, and since (we have reason to believe) (5.7) obtains, we deduce (5.10). In other words, we can explain why a is G. *We can explain where we could not predict.* But, as always, we must be careful.

In this explanation we have explained Ga only so to speak *ex post facto*. The event which we explain, Ga, is itself the evidence which we use to deduce the existence of the explaining event C^*a. Indeed, since 'C^*' is not a concept but only the hope of concept we could *not* have independent evidence for its occurrence. On the other hand, the fact that Ga is the evidence for C^*a in no way prevents us from using C^*a to explain Ga. The deductive model does not place any restrictions on how we acquire knowledge of the explaining event, and, in particular, it does not exclude our using as evidence the event to be explained.

I believe this pattern is exemplified throughout science. We see this in our example in terms of the germ theory of disease. We shall look at it also in terms of tipping over ink-bottles and of immunity to poision. Another example we could have used is psychoanalytic theory and its application in the therapeutic context. On this theory, psychoneuroses are due to an inner conflict between an impulse and the interconnected rest of the personality. This interconnection is a mode of interaction, and in particular, a way of interacting with the world. The interconnections are such as to adjust the person to reality. They are referred to as the "ego". So psychoneuroses are due to conflicts between impulses and the ego. Now, a common form of complaint against psychoanalysis is that it is circular, in this way, that the existence of the conflict is inferred from the neurosis, and the conflict is

THE REASONABILITY OF THE DEDUCTIVE MODEL 145

then used to explain the neurosis. We now see that there is no circularity in this procedure, that the complaint is groundless.

We must, however, recognize explanations of the sort we are now examining as essentially imperfect. An explanation using (2.3)

$$(x) [Kx \supset (Cx \equiv (Fx \equiv Gx))]$$

and

(5.12) Ca

is less imperfect. In the previous explanation we used (5.7), or, what is the same thing, (5.1). (5.1) is more imperfect than (2.3), because (5.1) says only *there is* a relevant factor without specifying *what it is*, whereas (2.3) does specify what it is. We use (5.1) because of failure of completeness. We do not have a complete set of relevant variables; we use (5.1) because we know there are other relevant variables but not what they are. What (5.11) says is only that, whatever the relevant factor may be, it is present. In contrast, (5.12) says a definite factor is present. Since our knowledge of (5.11) depends on our knowledge of (5.7), i.e., (5.1), (5.11) so to speak partakes of the imperfection of that law. Thus, although we can explain where we could not predict, the explanation is essentially imperfect.

(5.7), insofar as it involves *knowledge*, asserts nothing more nor less than (5.1) which justified the introduction of 'C^*'. It follows that where we explained using C^* and (5.7), we should be able to explain in a similar fashion using (5.1) alone. The difference would be that in place of (5.11) we would assert an existentially quantified sentence of the sort

(5.13) $(\exists f)(\mathscr{C}f \& fa)$

'C^*' enters the language of science in (5.7). But there it enters not as a concept, but simply as a piece of shorthand which acts as proxy for the "there is" part of (5.1). We should therefore expect contexts involving 'C^*' to be systematically replaceable by contexts involving a "there is" statement of the sort found in (5.1), and for which 'C^*' is acting as proxy. In particular, we should expect an explanation involving 'C^*' to be replaceable by an explanation involving such a "there is" statement. So we should expect to be able to understand (5.11) in terms of something like (5.13).

This line of argument is certainly just. As we pursue it, we shall discover something about the logical form of (5.1). Specifically we shall find out that (5.2) is not an adequate formalization of (5.1). If we use (5.2) then inferences do not go through as the line of argument just traced indicates they should go through. Those inferences will go through only if (5.1) is understood to have a somewhat different logical form. Once we recognize that this form is we shall also gain further insight into the nature of '$C*$'.

Take, for the moment, (5.2) to be the formalization of (5.1), and let us try to generate the explanation we generated using (5.7). We proceed by existential instantiation. Using this pattern we infer from (5.2):

(5.14) $\mathscr{C}C* \;\&\; (x) \{Kx \supset (C* x \equiv (Fx \equiv Gx))\}$

where '$C*$' is functioning as the arbitrary constant of the inference pattern. We can now detach the two conjuncts from each other

(5.15) $\mathscr{C}C*$
(5.16) $(x) \{Kx \supset (C* x \equiv (Fx \equiv Gx))\}$

We can now instantiate (5.16) with the individual constant 'a'

(5.17) $Ka \supset (C*a \equiv (Fa \equiv Ga))$

This 'a' is the 'a' of (5.8)–(5.11). We there assumed

(5.8) Ka
(5.9) Fa
(5.10) Ga

were all true. With these, we can infer from (5.17)

(5.18) $C*a$

which can be conjoined to (5.15):

(5.19) $\mathscr{C}C* \;\&\; C*a$

We can now use the rule of existnetial generalization to infer (5.13). (5.13) replaces (5.11).

When we had (5.11) we turned around and appealed to it in order to explain why a was G. From (5.7), (5.8), (5.9) and (5.11) we deduced (5.10). This was our explanation where we could not predict. The analogous inference with (5.13) would be as follows: (5.2) would replace (5.7); (5.8) and (5.9) remain the same; (5.13) replaces (5.11); and as before (5.10) is to be deduced. Let us see. From (5.2) we proceed by existential instantiation to

THE REASONABILITY OF THE DEDUCTIVE MODEL 147

(5.14). We then separate (5.15) and (5.16). By instantiating 'a' we obtain (5.17). Using (5.8) and (5.9) we can obtain

(5.20) $C^*a \equiv Ga$

If we can now infer 'C^*a' from (5.11), the other permitted premiss, then we have deduced (5.10), as we wanted. Now, from (5.11), it is possible to move by existential instantiation to

(5.21) $\mathscr{C}C^{**} \& C^{**}a$

which by detachment yields

(5.22) $C^{**}a$

But (5.22) is *not* the 'C^*a' we have in (5.20). The reason is the restrictions on the use of the inference pattern called existential instantiation. If an arbitrary constant has been introduced when the pattern is used once, then the same constant cannot be used when the pattern is applied again. If it is true premisses can be made to yield false conclusions. Thus, from

There is something which is red = $(\exists x)(fx)$

we infer by existential instantiation to

fa

and from

There is something which is green = $(\exists x)(gx)$

we infer to

ga

If we now conjoin these we can conclude by existential generalization that

$(\exists x)(fx \& gx)$

which says something exists which is both red and green. It is to prevent such illicit inferences that one introduces the usual "flagging" rules for existential instantiation. These rules require us to use one constant, 'C^*', when we instantiate (5.2) and another constant, 'C^{**}', when we instantiate (5.13).

The explanation involving 'C^*' was unobjectionable. We now seem, however, unable to reproduce it. Or rather, we cannot reproduce it without some additional assumption. It is clear what it must be. *The uniqueness of the explaining property must somehow be guaranteed.* There must be a restriction which guarantees that 'C^{**}' coincides with 'C^*'.

What this means is that (5.2) must somehow or other be supplemented. Additional informational premisses, beyond those mentioned, must be present if (5.7) is to be used in the way our line of argument suggested it could be used, to explain where we cannot predict. (5.2) secures that there is at least one C^*. What the additional information must secure is that there is at most one such C^*. This already makes more evident the role of 'C^*' in (5.7): our *hope-for-a-concept is in effect a definite description; it is a definite description of the species or type of factor we hope to discover.*[83]

There are at least two ways in which this uniqueness can be secured. Both types occur in science. And both types are used by Scriven and Collins in their attempts to confute the deductive model. The more complicated case is where the law itself states uniqueness. We will deal with this first. A rather complicated example of Collins is particularly relevant. But there are, I shall suggest, other examples of law of the relevant sort. The other case is where uniqueness is asserted as a subsidiary premiss. We find this pattern exemplified in the ink-bottle example of Scriven.

The first way in which uniqueness can be secured involves modifying (5.2) to include, not just the condition that there is at least one f such that . . . , as it now does, but also the condition that there is at most one f such that . . . This involves our attributing to the "there is" of (5.1) an implicit condition of uniqueness. This would never be implausible. As Strawson has been at pains to show, the "there is" of English often carries logical freight not explicit in the existential quantifier of the logicians. (Compare his example: "Someone fell over the cliff" — which clearly implies uniqueness.) What that shows is that in English grammatical form and logical form do not coincide. One must look carefully at what is said and what inferences are drawn from it in order to make explicit the logical form of an English sentence. The way we are now considering, by which uniqueness may be secured, requires us to understand (5.1) as asserting uniqueness, which may plausibly be done in many circumstances. So understood, (5.1) would be formalized not as (5.2) but rather as

(5.23) $\quad (\exists f) \, \{ \mathscr{C}f \, \& \, (x) \, [Kx \supset (fx \equiv (Fx \equiv Gx))]$
$\quad\quad\quad \& \, (g) \, [\mathscr{C}g \, \& \, (x) \, [Kx \supset (gx \equiv (Fx \equiv Gx))] \supset g = f] \, \}$

As an example of (5.23), consider the class K as persons, F and G as disease symptoms. C is the genus *germs*. The law (5.1) = (5.23) states that there is a unique species — exactly one species — of germs which accounts for the presence of the sysptoms. This example shows how plausible it is in certain contexts to understand a statement like (5.1) to involve an assertion of uniqueness.[84]

To make matters typographically simpler and therefore more perspicuous let us reduce '$Fx \equiv Gx$' to 'Fx'. And let us assume that the variable 'x' in both its occurrences ranges over all and only Ks. In that case (5.23) becomes

(5.24) $(\exists f) \{ \mathscr{C}f \,\&\, (x)(fx \equiv Fx) \,\&\, (g)[\mathscr{C}g \,\&\, (x)(fx \equiv Fx) \colon \supset g = f]\}$

If we write

(5.25) φf

as short for

(5.26) $\mathscr{C}f \,\&\, (x)(fx \equiv Fx)$

then (5.24) becomes

(5.27) $(\exists f)(\varphi f \,\&\, (g)(\varphi g \supset g = f))$

which will be recognized as the uniqueness ("E-Shriek") condition

(5.28) $E! \,(\imath f)(\varphi f)$

for the definite description

(5.29) $(\imath f)(\varphi f)$

We may predicate ψ of (5.29)

(5.30) $\psi(\imath f)(\varphi f)$

(5.30) contains the defined iota-operator. It may be expanded into

(5.31) $(\exists f)[\varphi f \,\&\, (g)(\varphi g \supset g = f) \,\&\, \psi f]$

If in particular we have

(5.32) $\varphi(\imath f)(\varphi f)$

the expansion analogous to (5.31) is

(5.33) $(\exists f)[\varphi f \,\&\, (g)(\varphi g \supset g = f) \,\&\, \varphi f]$

Given that '$p \equiv (p \,\&\, p)$' is a tautology, (5.33) is logically equivalent to the uniqueness condition (5.27). It is important that (5.32) is *not* analytic. We just saw (5.32) is logically equivalent to (5.27). Since the latter is synthetic, so is the former.

(5.29) is a definite description. It denotes a species or type of factor. To assert (5.32) is to assert, of the species denoted by (5.29), that (5.26) holds. Which is to say that it asserts of that species that that species is of genus $\mathscr{C}$ and is necessary and sufficient for the presence of F. That is to assert a conjunctive statement; this is clear from 'φ' merely abbreviating (5.26). We may detach the second conjunct, and simply assert of the species denoted by (5.29) that it is necessary and sufficient for the presence of F. If we abbreviate (5.29) by the expression 'C^*', then what we have asserted is

(5.34) $(x)(C^* x \equiv Fx)$

This, too, is synthetic. (5.29) and therefore 'C^*' can function in all inferences as a constant just in case it is a successful definite description, i.e., just in case the uniqueness condition (5.27) is fulfilled. But (5.27) is the same as (5.24) which, given our typographical simplifications, is in turn the same as (5.23). (5.23) is the formalization of (5.1). Since we are supposing the latter to hold, we are thereby supposing (5.29) to be successful. Which is to suppose 'C^*' can function in inferences as if it were a predicate constant. Given our typographical simplifications, (5.34) turns out to be the same as (5.7). All the inferences we made with the latter, treating 'C^*' as a predicate constant, turn out to be justified.

In particular those inferences where we explained but could not predict turn out to be justified. Given (5.1) = (5.23) we will be able to explain by appealing to the presence of C^*. The individual fact (5.11)

$$C^* a$$

is what is cited in these explanations. 'C^*' here is functioning as the definite description (5.29). That means (5.11) can be expanded in the usual way, eliminating the definite description. If we write

$$fa$$

as

$$\theta f$$

then (5.11) becomes

$$\theta(\imath f)(\varphi f)$$

which expands into

$$(\exists f)(\varphi f \& (g)(\varphi g \supset g = f) \& \theta f)$$

which is

(5.35) $\quad (\exists f)(\varphi f \& (g)(\varphi g \supset g = f) \& fa)$

This is the correct reading of '$C*a$' if we are to be able to explain using the hope-of-a-concept or definite description '$C*$'. We see how far off was our supposition that the correct reading was (5.13). The more complicated reading results from our treating '$C*$' in (5.7) as a definite description.

Let us apply the above considerations to the following examples of Collins. This discussion will also apply to some of Scriven's examples.

> Suppose a man takes poison and dies. We will think that taking the poison was the cause of the man's death and we will cite having taken the poison in explanation of his death. Of course we may be wrong. But supposing alternative accounts to be investigated and ruled out, confidence in this explanation can reach virtual certainty without our being able to formulate any law or other conditions from which we can deduce a statement asserting the man's death, or make a selection of descriptions of the event, the poison-taking, and the circumstances in which it took place, with the help of which we have any hope of getting a universal generalization. We may be unable to manage a description of the man, the dose, and the circumstances for which the correlation of poison-taking and death is higher than, say, twenty-five per cent ... we [may] nonetheless believe that there is a law:
>
> [*] Whenever a man with such and such [presumably determinate] characteristics in such and such circumstances takes such and such poison, that man dies.
>
> Which covers our case. But we do not wait to find the law [i.e., I take it, such a law] in order to be able to explain the event, or to be sure that the explanation is correct, or to be sure that the causal assessment on which it rests is correct.[85]

I take it [*] is meant to mention only determinate characteristics. That makes it like our (2.3) of Section 1.2. So Collins takes all laws to be like (2.3). It follows he ignores imperfect laws like (5.23). This comes out in another remark of his that "from the viewpoint of the law-deduction theory I cannot be sure that A is one of the several causal factors and conditions unless I know what the others are."[86] This is simply false, and the law-deduction theorist is not committed to it, as our discussion of imperfect knowledge has made clear. Again he tells us that with respect to the law-deduction theory, "What is wrong is the line of thought that argues: if we cannot be sure of all the causal factors and necessary conditions for a particular event, then we cannot be sure of any of them." It is indeed a wrong line of thought. But no law-deduction theorist is committed to it by his defence of the deductive model. What the remarks show is that Collins identifies, wrongly, the law-deduction theory with the use in the premisses of the laws involved in perfect knowledge. It shows he, wrongly, will not let the law-deduction theorist include imperfect laws among his premisses. These latter, however, are relevant to his poison example, since it is essential to the example that we know that there are relevant factors without knowing what those factors are. Collins uses his example to establish we can have explanation without prediction, and also that we can have explanation without laws. Since he ignores imperfect laws, all he establishes is that we can have explanation without perfect laws. As for whether there is explanation without prediction, I think this case is simply a complicated instance of the previous example involving (5.1) and (5.7). If that can be shown then the remarks about explaining via (5.1) apply equally to this case, with the consequence that this case equally fails to seriously challenge the deductive model.

The relevant systems, K, are human bodies. We apply poison, P, to them. P by itself is not sufficient for death D, since D does not invariably follow P. We believe that there is a definite further condition, the specific nature of which we do not know, which is such that P and it is jointly sufficient for death. The poison P works only in the presence of a certain definite physiological state, present in some persons, absent in others. How do we know *there is* such a factor and that it is unique? Our background knowledge of physiology and medicine *very strongly* supports the hypothesis there are such factors. These factors will be of a physiological nature. (This limitation corresponds to the $\mathscr{C}$ of (5.1).) Background theory makes the hypothesis (subjectively) worthy of acceptance. Call these factors C^*. The background theory justifies our using this definite description. These are other sufficient conditons of death, say, F_1, and F_2, and F_3. The sum of all sufficient conditions will yield the complete necessary condition. We obtain the imperfect law

(5.36) $(x) \{Kx \supset [(Px \,\&\, C^*x) \vee F_1 x \vee F_2 x \vee F_3 x) \equiv Dx]\}$

It is imperfect because C^*, as in (5.7), signifies that there is a relevant factor, of some generic sort determined by the background theory, but that the specific nature of this factor is not known (though we hope to discover it!). (There is a further imperfection: the list F_1, F_2, etc., will be one which is vague in the sense that the borderline between the set of things sufficient

THE REASONABILITY OF THE DEDUCTIVE MODEL 153

for death and the set not sufficient will be hazy — cf.: how hard a blow is sufficient to kill him? — but we have discussed such a point above, and so need say nothing more about it.) Now consider person b. We know for him that

(5.37) Kb

From (5.37) and (5.36) we obtain

(5.38) $[(Pb \,\&\, C^*b) \vee F_1 b \vee F_2 b \vee F_3 b] \equiv Db$

Suppose he takes poison

(5.39) Pb

and dies

(5.40) Db

Did the poison cause the death? Well, we are, with Collins, "supposing alternative accounts to be investigated and ruled out," i.e., that

(5.41) $\sim F_1 b \,\&\, \sim F_2 b \,\&\, \sim F_3 b$

(5.38), (5.40) and (5.41) yield

(5.42) C^*b

We can now use (5.42) and (5.39) as initial conditions (together with (5.37)) to yield an explanation of the death on the basis of the imperfect law (5.36). There is no actual prediction since C^* cannot be known independently to obtain. But, as we know, this does not affect the claims of the defender of the deductive model. Conversely, if death does not occur, i.e., if

(5.24) $\sim Db$

then from this and (5.38) we obtain

$\sim [(Pb \,\&\, C^*b) \vee F_1 b \vee F_2 b \vee F_3 b]$

which is equivalent to

$\sim (Pb \,\&\, C^*b) \,\&\, \sim F_1 b \,\&\, \sim F_2 b \,\&\, \sim F_3 b$

Simplification yields

$\sim (Pb \,\&\, C^*b)$

which is equivalent to

(5.43) $\sim Pb \vee \sim C^*b$

But he did take poison; i.e., (5.39) obtains. From (5.39) and (5.43) we can deduce

(5.44) $\sim C^*b$

We can now use (5.44), (5.39) and (5.41) as initial conditions (together with (5.37)), and these yield an explanation of the event of his not dying, on the basis of the imperfect law (5.36). Again, explanation without prediction, but only without it in that innocuous way which does not touched the deductive model. Collins tells us

> When there are a limited number of ways in which *B can* come about and one of them, *A*, obtained prior to the occurrence of *B*, then there is a strong *prima facie* case that *A* not only obtained but also operated in effecting *B*. [Agreed.] The most compelling reason for supposing this to be the case is that *B* happened, after all. [Agreed – cf. our use of (5.40) to obtain knowledge of the initial conditions.] And we can be completely convinced that *A* caused *B* by discovering that the other ways in which *B* ordinarily *can be* brought about did not apply. [Agreed – cf. our use of (5.41).] This realistic bit of reasoning is inaccessible to the law-deduction theory because it employs our knowledge that *B* did happen in order to explain why it did. The law-deduction theory, seeking premises from which to deduce that *B* happens, must, so to speak, feign ignorance of the fact that it did happen.[87]

Agreed until the point where it is claimed the reasoning is inaccessible to the law-deduction theorist. Far from being inaccessible that reasoning actually *involves* a law, viz, (5.36). It is an imperfect law for all that. However, we saw Collins in [*] unfortunately identify laws with perfect laws and so he misses this elementary point. The law-deduction theorist could not, suggests Collins, in any case have an explanation since the explanation is in terms of initial conditions, the knowledge of which depends upon the explained event, upon the fact that *B* did happen. This suggests that "since *A* explains *B* and *B* evidences *A*, therefore *B* explains *B*", in which case the law-deduction theorist is trying to explain *B* in terms of itself. If this be Collins' thought, it is simply misguided, as the "therefore" just doesn't follow. And, I feel it must be Collins' thought, as I can't see any other reason why he should suppose, which is completely unsupported, and which seems to be patently wrong, that "the law-deduction theory, seeking premises from which to deduce that *B* happens, must, so to speak, feign ignorance of the fact that it did happen."

Bromberger introduces a distinction between a "general rule" and an "abnormal law".[88] The general rule is a lawlike statement — it must be capable of supporting subjunctive conditionals — but need not be true. It would be of the sort "Poison causes death". An abnormal law which is said to "complete" such a general rule in effect gives a whole set of "unless conditions": "Poison causes death unless . . . ". If there were no imperfection in (5.36) — either with respect to C^* or with respect to the list of F_is — then that would constitute the abnormal law which completes the rule that poison causes death. Bromberger then gives a rather complicated account of what constitutes an explanation. It is proposed to explain why a person did not die when he took poison. The explanation involves, first, a general rule. The initial condition of his taking poison and this general rule entail a conclusion contrary to the event to be explained, viz. that he did not die. The same initial condition, *plus others* — he was not hit by a car and he has within him the appropriate physiological mechanisms — together with the abnormal law completing the general rule entail the event to be explained. It is these other initial conditions that constitute the explanation of why the person died.

Bromberger has unfortunately adopted an approach based on linguistic theory, the context of communication, rather than ours. It is this which leads him to characterize just some of the initial conditions rather than the totality of the initial conditions plus law as the explanation. This difference is mainly one of the approaches. At least, it makes no difference to the point I am about to make. Bromberger is quite correct in his suggestion that an argument based on what he calls the general rule cannot be an explanation. But he fails to note that often neither is his abnormal law. The latter has none of our C^*s in it, only Cs, the predicates available after the elimination of the imperfections the former involve. We may just not know what the physiological conditions are the absence of which prevent the poison causing death. And if they are not known then on Bromberger's account we are unable to explain. But, of course, Scriven and Collins are quite right in this matter: we often do explain in these situations. Only, contrary to Scriven, Collins, and Bromberger, it is by means of a law, an imperfect law — and the imperfection is what one must notice. Bromberger fails to recognize imperfect laws and therefore is quite unable to reply to Scriven and Collins on these matters. Bromberger trots out a good deal of the apparatus of the linguistics that is currently fashionable. Yet he misses this simple point we have been emphasizing. The moral, I suppose, is that one ought to pander to fashion rather less and to think rather more.

We may use the poison-non-dying case to remark about a thesis of Scriven.

We explain the fact or event (5.43) (the event (5.43) is about), using the law (5.36). Presupposing (5.37), we use the facts (5.39), (5.45) and (5.41) as initial conditions for the deduction (explanation). However, if we were talking as we ordinarily do we should select one of these items as *the* cause of the person's not dying when the poison was administered. Each of the conditions in (5.41) is relevant, yet they are only "conditions", not "the cause" of b's not dying upon the administration of the poison, Pb. "The cause" of b's not dying upon the administration of the poison, is the absence of the crucial factor C^*; i.e., "*the* cause" which provides "*the* explanation" is given by (5.45) alone, and not by (5.45) conjoined with (5.41). When we speak in this everyday fashion and offer (5.45) as "the explanation" we select, out of the set of causally relevant antecedents, the one antecedent which is *relevant* in the context in which the explanation is given. Scriven uses this, as we might expect, to go after the deductive model. To have predicted (5.43) upon the knowledge of (5.37) and (5.39) we would have had to know both (5.45) and (5.41). Yet the explanation uses only (5.45). Hence, we have an explanation without prediction, in violation of the requirements of the deductive model. Let us see. To omit (5.41) from our explanation is, in effect, to omit F_1, F_2 and F_3 from the list of relevant variables. In other words, it is to use

(5.46) $(x) \{Kx \supset [(Px \ \& \ C^*x) \equiv Dx]\}$

rather than (5.36). Since (5.46) omits to mention relevant variables which are mentioned in (5.36), (5.46) is imperfect relative to (5.36). What we offer in the everyday contexts as "the explanation" is imperfect, relying on (5.46), and is an explanation which is not as good as we could provide, were we to want it, namely, the explanation in terms of (5.36). To use (5.46) rather than (5.36) is like using (2.1) instead of (2.3) – to go back to our example of Chapter 1. Nor is that wrong. We can use (2.1) instead of (2.3) so long as the relevant conditions obtain, so long as the Ks to which we apply (2.1) are also as a matter of fact Cs. Similarly, we can use (5.46) instead of (5.36) so long as the Ks to which we apply it are also Ks which (like b, according to (5.41)) are neither F_1 nor F_2 nor F_3. Thus, since b satisfies this condition, it is permissible to use (5.46) in the explanation, and therefore only (5.45) (presupposing (5.37) and (5.39) as given). We know from the discussion in Chapter 1 of (2.1), (2.2), and (2.3) that the use of imperfect knowledge does not violate the deductive model, so long as the imperfect knowledge is used in appropriately restricted circumstances. The same comments apply to the use of (5.46) rather than (5.36). Contrary to Scriven, then, the deductive model is not violated. Still another point can be made. Suppose both (5.36) and (5.46) are known. It is still easy to see that in ever so many quotidien contexts people would be prepared to accept (5.46) rather than (5.36). It is this fact which makes Scriven's example seem rather more plausible than it is. Can we explain the fact? The appropriate answer to this can be obtained if we ask which of the two would be desired by an explainee who was merely idly curious, who merely wanted to know as far as possible what antecedent facts or events necessitated b's not dying. Since the facts (5.41) are in fact relevant, as (5.36) makes clear, the idly curious explainee will want the explanation to be in terms of (5.36) rather than (5.46), and in terms of (5.41) *and* (5.45) rather than (5.45) alone. An explainee would settle for (5.46) only if his interest was other than idle curiosity, i.e., only if his interest were some pragmatic interest. As we know, pragmatic interest is often satisfied with less knowledge than is idle curiosity. It is just such an interest which is present in quotidien contexts: one wants to know what it is that *prevents* the operation of the poison, and other relevant factors are not of interest; one wants to know what prevents the death, not the other factors that might have caused it; and the reason one wants to know this is the very general interest in preventing death, which is, of course, a pragmatic interest. It is, I suggest, the presence of this pragmatic interest, and these strategic considerations, which in the everyday contexts makes explainees in general satisfied with an explanation in terms of (5.46) rather than (5.36). By virtue of this pragmatic interest, the absences of F_1, F_2 and F_3 are not relevant facts; only the absence of C^*, i.e., only the truth of (5.45), is relevant. Hence, as Scriven correctly points out, the explainer picks this out as being, given the context defined by the explainee's pragmatic interest, *the* relevant causal antecedent, i.e., *the* causal antecedent knowledge of which, alone among the several, is relevant to the explainee's pragmatic interest. But, of course, as we have seen, none of this touches the deductive model.

This concludes our discussion of the first way in which uniqueness can be secured, by building it into the law (5.1). Let us turn now to the second way. We can deal with it more briefly.

This way construes (5.1) as having the logical form (5.2). As before we use the facts to be explained in order to deduce the presence of the unknown condition. From

(5.8) Ka
(5.9) Fa
(5.10) Ga

158   CHAPTER 2

we deduce from (5.2) that

(5.13) $(\exists f)(\mathscr{C}f \,\&\, fa)$

The idea is to now explain why a is G by appeal to (5.13). So we must use (5.8), (5.9), and (5.13) as initial conditions, and (5.2) as a law, to deduce (5.10). As it went before, from (5.8), (5.9) and (5.10) we could deduce

(5.20) $C^*a \equiv Ga$

where 'C^*' is the arbitrary constant of the existential instantiation. From (5.13) we were able to obtain

(5.22) $C^{**}a$

where 'C^{**}' is a different constant of existential instantiation. We can deduce (5.10) from (5.20) if we can establish

(5.45) $C^* = C^{**}$

This will follow if we have an additional premiss of the sort

(5.46)   For all x, if x has at least one property of sort $\mathscr{C}$ then it has at most one property of that sort.

or, in symbols,

(5.47) $(x)(f)(g)\,[\mathscr{C}f \,\&\, fx \,\&\, \mathscr{C}g \,\&\, gx: \supset g = f]$

What (5.46) = (5.47) asserts is that properties of sort $\mathscr{C}$ form a class which mutually exclude each other.

As an example of this sort of case, consider our reading of (5.1) = (5.2) in terms of a physiological basis for the presence of the behaviourally defined hunger properties. The law (5.2) leaves it open whether there is more than one physiological state necessary and sufficient for the presence of the behavioural properties. (And therefore each of these physiological states would be necessary and sufficient for each other.) Thus, we could have stomach contractions as well as blood sugar. And this may be satisfactory in a qualitative sort of way, except that the scientists might be striving for something more adequate, namely, a reliable quantitative association. Behaviourally, we have length of time that an animal has not eaten, and the strength of its tendency to pursue food rather than a mate. Both these are quantified, and the search would be for a similarly quantifiable physiological dimension, one in which

THE REASONABILITY OF THE DEDUCTIVE MODEL 159

the facts permit good sense to be made of the notion of "amount". It might then be the case that physiologists have reason to suppose – given their general knowledge of physiology – that there will be only one such indicator for which the notion of "amount" can be defined. The scientists may not know quite what this factor is, but they might reason as follows. "Nourishment is carried throughout the body by the blood. We should therefore be able to measure the amount of nourishment available to the body at any moment by finding some factor in the blood. The amount of this factor will correspond to the time since food has been taken in. It is likely there is only one such mechanism by which quantities of nourishment get systematically conveyed to all parts of the body." The last comment would mean that these physiologist-psychologists would have reason to believe a law of the sort (5.46) = (5.47).

In both the cases we have discussed we have worked in terms of (5.1), understood either as (5.2) or as (5.23). With the latter, uniqueness is built into the law being used. This permits its formulation as (5.7), and explanation proceeds in the usual way. Where the law (5.1) is understood as (5.2), one is unable to formulate directly a statement of the sort (5.7). This is because the 'C^*' of (5.7) involves a uniqueness claim which the law (5.2) does not justify. However, there is, we supposed, an auxiliary hypothesis of the sort (5.47). This auxiliary hypothesis permits one to explain where one could not predict.

2.6 EXPLANATION AND JUDGMENT

The preceding cases are not the only ones where one can have explanation without prediction – but without violating the deductive model. Both turned upon certain features of (5.1). We chose that example deliberately because several examples of Scriven and Collins turn on there being laws of that sort. Two things made it logically possible to explain where we could not predict. Throughout, prediction is not possible because one does not know *specifically* what type of event it is that is cited as the explaining event. Rather, one refers to the type or species indirectly, by means of a definite description, or a term which is the hope-of-a-concept. The *existence* and *uniqueness* of the *type* is secured by the laws involved. In addition, one must, first, be able, on the basis of observed individual facts, to deduce the presence of an event of that type, and, second, be able, on the basis of the presence of an event of that type, to deduce and thereby explain the observed facts. In the examples there is a certain event to be explained. Its occurrence is observed. On that basis one deduces the presence of the event which is to do the explaining. For this inference to go through the event to be explained must be sufficient for the explaining event, the latter necessary for the former. Once the presence of the explaining event comes thus to be known one then uses it to explain the other event. For this second inference to go through the explaining event must be sufficient for the event to be explained. In short, the explaining event must be necessary and sufficient for the explained event. Or, at least, this is how the examples go. And, in general, something like this sort of thing has to be involved wherever we can explain without being able to predict. Laws must secure the uniqueness of the explaining type, and laws must permit one to infer the explaining event from the observed event to be explained, and conversely. But these conditions need not be fulfilled in precisely the way they are in the examples based on (5.1). Another way in which they can be achieved can be seen if we turn once again to Scriven's ink-bottle example.

Let us begin to look more closely at the law which enables us to make determinable predictions. Let 'Kx' represent that a system is of the same specific mechanical sort as the ink-bottle Scriven so carelessly tipped over. Let 'Tx' represent that x tips over. Let 'f' be a variable ranging over forces. Finally, let '$\mathscr{C}f$' represent that f is within a determinable range of forces. (We shall return to both '$\mathscr{C}$' and 'K' directly.) Then the law is this

(6.1) $(x) \{Kx \supset [(\exists f)(\mathscr{C}f \& fx) \equiv Tx]\}$

which states that so far as ink bottles of that sort are concerned, a necessary and sufficient condition of their tipping is that a force of the determinable sort be exerted upon them. If we know that

(6.2) Ka

and observe that

(6.3) Ta

we deduce the determinable fact

(6.4) $(\exists f)(\mathscr{C}f \& fa)$

We can now use (6.4) to explain why, given a is K, that a is T. For, (6.1), (6.2) and (6.4) entail (6.3). Notice that this explanation is in terms of a determinable event, an event of a determinable rather than a determinate type. No determinate force is as such necessary for the occurrence of tipping. That means on the basis of the fact of tipping one cannot deduce the presence of any determinate force. One can only deduce the presence of a necesary condition for tipping, namely, that there is present a force falling within a

THE REASONABILITY OF THE DEDUCTIVE MODEL   161

certain determinable range. It seems reasonable to believe a determinate event exists the presence of which is sufficient to explain the tipping. It seems we should be able to cite that event. Yet on the basis of the law (6.1) we are unable to do so. The difficulty is that (6.4) does not secure the uniqueness of the type. Nor, since many forces are sufficient for tipping can we expect uniqueness at this point. An additional premiss is needed.

If we can represent the determinate event by

(6.5) C^*a

where

(6.6) $\mathscr{C}C^*$

then we can deduce (6.4) and from that via (6.1) we can deduce (6.3), thereby explaining why a is T. The difficulty is to arrive at (6.5) from (6.3). (6.1) and (6.3) yield (6.4) but we cannot go from there to (6.5). Or rather, if we are to arrive at (6.5), it is here that we need the further premiss. (6.4) asserts there is *at least* one force of type $\mathscr{C}$ which is present in a. (6.5) and (6.6), assert there is *exactly* one such force present in a. What we need a premiss to the effect that *at most* one such force is present in a. And such a premiss is in fact available. We know that forces form a mutually excluding class. If the (total) force exerted on an object is such and such a value then there is no other (total) force being exerted on that object. Which is to say, we know a law of the form (5.46) = (5.47) applying to (total) forces. This law permits us to infer (6.5) from (6.4). Having thus obtained (6.5) from (6.3), we can now turn around and explain (6.3) on the basis of (6.5). Again, We have explained where we could not give a reasoned prediction based on determinate properties present in the situation.

(6.1) is imperfect insofar as there is in it an existential quantifier, and involves reference to a determinable rather than a determinate type of event. In the sorts of situations Scriven envisages the law used will actually be even more imperfect. Imperfection will enter in both the definition of '$\mathscr{C}$' and of 'K'. Consider '$\mathscr{C}$' first. '$\mathscr{C}$' represents the range of forces sufficient to cause tipping in Ks. For each force E in this range there will be a law of the form

(6.7) $(x) \{Kx \supset (Ex \supset Tx)\}$

Forces can be ranged in an order of greater and less. Thus, any pair E_i and E_j of forces satisfying (6.7) will be such that

(6.8) $E_i \leq E_j$

162   CHAPTER 2

or conversely; let us assume (6.8). We know there is a force which is *the* least force satisfying (6.7). That is, if we represent the fact that E satisifes (6.7) by 'φE', then what we know is that

(6.9) $\quad (\exists E)\, [\varphi E\ \&\ (E')\,(\varphi E' \supset E \leqslant E')]$

This statement is a synthetic generality. It asserts the existence and uniqueness condition for the definite description

(6.10)   the least force sufficient for tipping

which we may abbreviate as

$\quad E_L$

E_L is a sufficient condition for tipping. To asserts this is to assert E_L is φ. This assertion is equivalent to the assertion of (6.8), which is synthetic. We use 'E_L' to define the range of forces $\mathscr{C}$ mentioned in (6.1):

(6.11)   '$\mathscr{C}f$' is short for '$f \geqslant E_L$'

Thus, '$\mathscr{C}$' in (6.1) involves the definite description 'E_L' which functions in (6.18) much as 'C^*' functions in (5.7).

The other point at which imperfection enters is in 'K'. What is crucial to the tipping of the ink bottle upon the imposition of external force are the mechanical properties of the bottle. 'K' represents those mechanical properties. 'Ka' represents that a exemplifies these properties. Now, we may not know *specifically* what these properties are. But we can give them a definite description:

(6.12)   'Kx' is short for 'x exemplifies the mechanical properties exemplified by a'

The definite description will be successful just in case mechanical properties form a mutually excluding class. They will do this if there is a law of the sort (5.47). In the case of mechanical properties there is such a law, so the definite description is successful. Background knowledge from mechanics yields an assertion of the following sort:

(6.12)   For all kinds of mechanical properties, there is a least force such that for all objects exemplifying a given set of mechanical properties that force defines a determinable range the presence of which is necessary and sufficient for tipping.

THE REASONABILITY OF THE DEDUCTIVE MODEL 163

(6.18) is obtained from (6.12) by instantiating the first universal quantifier: one instantiates with the successful definite description (6.18).

Note the many and mixed quantifications in (6.12). *Even everyday knowledge of laws is knowledge of quantificationally very complex propositions.* I suppose that if Scriven has done nothing more he has at least driven this point home to us. When so many people, from the Popper to Hempel, take '$(x) (fx \supset gx)$' to be *the* form of scientific laws Scriven's point is worth emphasizing. On the other hand, the moral that Scriven draws does not follow.

He concludes that, since we can explain but not predict in these cases, therefore explanation and prediction are not symmetric, as required by the deductive model; and concludes so much the worse for that model. Let us see. Our explanations throughout involve a reference to an event we called 'C^*a'. This appears in (5.11) [understood as (5.35)], (5.20), and (6.5). In each case 'C^*' functions as a definite description for a *type* of event. Laws in each case guarantee the existence and uniqueness of the type. What we must show is that where we use 'C^*' in explaining where we cannot predict, prediction does not fail for a reason which violates the requirement of the deductive model. All the deductive model requires is that prediction be possible, that one could have predicted. All this possibility requires is that the statement of initial conditions be objectively worthy of assertion, i.e., true. The possibility does not require those conditions to be subjectively worthy of assertion, i.e., that we know the initial conditions to obtain, prior to one's coming to know the event to be explained. The actuality of prediction requires such knowledge of the initial conditions, but not the possibility. Now, (5.11) = (5.35) is essentially an existential statement. It says that there is a relevant variable such that (5.23) holds of them and that this variable is instantiated in a. Existential statements are objectively true or false. They can, therefore, be used in making rational predictions — provided their truth is known. The difficulty is we cannot know them as such to be true directly. We cannot observe that there are lions; we observe only particular instances of lions, and infer the truth of the existential statement from this knowledge. Similarly we do not observe there are properties in a; we observe definite determinate properties in a, and infer the truth of the existential statement from this knowledge. We can obtain the truth of C^*a from observation only if we acquire by observation knowledge of Ca and infer the truth of C^*a from it. (This is only a necessary, not a sufficient condition for the inference. We also need the universal premiss that C is unique in order to derive 'C^*a' = (5.35).) However, by hypothesis we do not have knowledge of Ca, since we have not yet succeeded in discovering the relevant factor C. We therefore cannot acquire the knowledge of (5.11) which would enable us to actually predict Ga, though in principle we could so predict, since C^*a is (we are assuming) objectively true. Thus, the nature of the case is such that prediction is always possible but never actual. Still, it *is* always *objectively possible*, and that is all that is required by the symmetry thesis entailed by the deduction model.[89] Once again that model escapes unscathed.

However, if the explanation is to be subjectively worthy of acceptance some reason must be available, some evidence to suppose true the laws guaranteeing the success of the definite description. This evidence will not be direct. It will therefore be supplied by background theory. If such theoretical evidence is not available the charge of *ad hocery* will not be avoidable.[90]

Let me make the following remarks in terms of but one of our examples, viz, (5.1) understood as (5.23). Analogous remarks will hold for the other cases.

Suppose we explain the presence of G by citing the presence of C^*, as we supposed we could explain (5.10) by citing (5.11). The law which provides the explanatory connection is (5.7). (5.7) is simply (5.32), or equivalently (5.33). But the latter is the same as (5.23). Thus, (5.23) both provides the explanatory connection *and* guarantees the existence of the property used to explain. The law which provides the explanatory connection is the very same law that justifies the use of the definite description 'C^*'. 'C^*' is thus used to explain the very same effects that *so to speek define it*. Its use in an explanation will therefore apparently be quite *ad hoc*. For all that appearance, however, the explanatory connection is in fact synthetic, and the explanation perfectly adequate, deductive, and based on a matter of fact generality. The appearance of *ad hocery* is generated by the idea that the effects *define* the property cited in their explanation. However, it is not a definition that is involved but a definite description, the legitimate use of which depends upon certain lawful generalities being true. It is these laws – here: (5.1) = (5.23) – that guarantee the explanation is not tautologous, and not *ad hoc*.[91]

There is, however, one situation in which appeal to C^* could be justly characterized as *ad hoc*.[92] That is when there is no evidence in favour of the law (5.1) = (5.23) that justifies its use and provides the explanatory connection. The evidence rendering this law acceptable cannot be by way of its instances. For, such instantial evidence would require one to cite facts of the sort (5.12). But if such facts were known then we could cite these facts rather than C^* as explanatory. Or, to say it differently, we could give a better explanation through the use the less imperfect (2.3) of Section 1.2 rather than (5.7). So the evidence must be indirect, by way of a more general theory. Thus, we give our definite description of the (species of) germs that cause this disease, these symptoms. This will be based on the law that such symptoms are caused by one and at most one species of germs. This law will have to be justified in terms of a more general law: this disease is of some generic sort, and all diseases of this generic sort are caused by some unique species or other of germs. This generic law could be justified by citing the actual isolation of the appropriate germs in the case of other diseases of the generic sort. This indirect evidence justifies accepting the law justifying the definite description in the case of the particular disease we are now interested in. Or, we can speak of *the* temperature function for a system. This will be justified by citing the axioms, the theory, of thermodynamics. The evidence for this theory will all indirectly justify our speaking of *the* temperature function, even when we have not determined just what that function is. So, where there is no such background theory the use of C^* in an explanation is reasonably characterized as *ad hoc*. But where there is such a theory, C^* is available for explanatory purposes, though, as we saw, in such cases prediction is not possible where C^* is used.

Given reason for supposing the law (5.1) to be true, we will thereby have reason to believe a certain line of research will be successful.[93] We will be able to formulate a series of hypotheses of the form

(6.13) $\quad C_1 = C^*$
$ C_2 = C^*$

and so on. Research will gradually eliminate all but one. That is, eliminate all but one *if* and only if the definite description 'C^*' or (5.29) is successful. Which is to say, eliminate all but one if and only if the hypothesis (5.1) = (5.23) asserting the uniqueness condition is a true hypothesis. In this whole discussion we have been supposing (5.1) = (5.23) to be true. But it is synthetic and a generalization, so it might be false, after all. In that case none of the hypotheses (6.13) would be true. Or several might be true. But until that research is undertaken, our reason for accepting (5.1) will justify our using 'C^*' in explaining where we cannot predict.

We noted earlier that Scriven makes much of the fact that the normic statements he talks about, the truisms, i.e., the pieces of imperfect knowledge, which we often use in explanations, employ a vocabulary which reminds that they have exceptions, which exceptions we perhaps do not know, but which trained judgment enables us to locate. Scriven says that the use of such truisms requires a "trained judgment", and that the theorists of the deductive model ignore this feature. "This capacity for indentifying causes is learnt, is better developed in some people than in others, can be tested, and is the basis for which we call judgments."[94] We recognize possible causes easily and reliably. The " ... historian's principles of judgment are most nearly, though inadequately, expressed in the form of truisms."[95] If one accepts our discussion, this capacity becomes less mysterious than Scriven would have it. Even though *antecedently* we cannot discover the truth of C^*a, *ex post facto* evidence is available, namely the event to be explained. Since historians explain *ex post facto*, we are not surprised at their ability to locate causes. Such location simply employs the imperfect law which, after the location succeeds, is used in the explanation. The locating and explaining are, so to speak, simply two sides of the same coin.

There is a role for "judgment", however, which Scriven overlooks. He holds such judgmental ability enters the picture when it comes to explaining where there is no actual prediction. We just saw this is not so. In those cases there is no element of judgment, over and above ordinary inferential abilities, which enters into the picture.

What is important about these cases is that the explanation is in terms of C^*, which since it is not a concept but the hope of a concept, cannot be known (e.g., by observation) to be present antecedently to the occurrence of the event it is used to explain. Prediction was not possible owing to the fact that its presence cannot be known antecedently. Or, at least, *reasoned* prediction is not possible. On the other hand – and this is the point which Scriven misses – *non-reasoned* prediction often *is* possible in such cases. Scriven is so concerned to establish there is explanation without prediction that he fails to notice in these cases of explanation where there is never actual reasoned prediction, there *is ex post facto* explanation and also *antecedently non-reasoned* prediction. Reasoned prediction requires one to know the initial conditions and to then use the law to infer the as yet unknown future state. Reasoned prediction is not possible since the initial conditions cannot be antecedently identified. Thus, we can explain why b fails to die, but cannot give a reasoned prediction of the fact because we cannot antecedently identify the absence of the crucial physiological condition C^* which prevents the poison from working. However, in point of fact trained doctors are often able to predict in such cases that the patient, b, would not die. That means they give non-reasoned predictions of such facts, where only *ex post facto* explanation is possible. *This is where trained judgment enters.*

Let us look again at (2.1) from Chapter 1.

$$(x)\,(Kx \supset (Fx \equiv Gx))$$

We are supposing it to be false in all generality and to hold only under the conditions (2.3)

$$(x)\,[Kx \supset (Cx \equiv (Fx \equiv Gx))]$$

indicates. That means the imperfect knowledge (2.1) can be used to predict successfully only where Ks are Cs. This is so whether one can identify Ks as Cs or not. Suppose one does not know C is among the relevant variables, i.e., that one cannot identify Ks as Cs. Then, in the first place, one cannot make reasoned predictions, since that would require knowing both C and (2.3). Nor, in the second place, can one distinguish the circumstances in which (2.1) does hold from those in which it does not hold, since those conditions are a matter of being C the relevance of which is not know; which means one cannot identify the circumstances in which one can and in which one cannot legitimately employ that imperfect knowledge (2.1) which one does have available. The relevant variables, the ignorance of which makes our knowledge imperfect, are not known, and that being so, when one attempts to apply the imperfect law in order to predict, one does not *know* whether the application is legitimate or whether this is one of the exceptional cases, and therefore, illegitimate, yielding false explanations and false predictions. We must now distinguish two things: in the *first* place, there is the ability of the scientist to identify C, in the sense of being able as a matter of conscious thought to recognize a thing as a C. In such a case, a reasoned prediction based on C is possible. If identification is not a matter of conscious thought, *reasoned* prediction, that is, consciously made inferences from the known to the unknown, is not possible. If we cannot identify as a matter of conscious thought some relevant variable, then we are ignorant of that variable. For, the goal is, after all, consciously articulable knowledge. There is often available, however, a substitute for such conscious knowledge which will enable us at least partially to overcome such ignorance. This brings us to what we must notice in the *second* place, namely, the ability of the trained scientist to know, on the basis of unconscious cues, to which Ks he can legitimately apply (2.1) and those to which he cannot do so. On the basis of unconscious cues the scientist A separates Ks into Cs and non-Cs, even though he cannot consciously separate Ks into Cs and non-Cs, even though he cannot *consciously* separate Ks into those classes. Because he has this trained ability,

168   CHAPTER 2

A can use (2.1) to predict successfully. With respect to scientist A, we can say that scientist A applies (2.1) to a K if and only if that K is also C. That this generality holds of A is a consequence of his training. And it holds even though he cannot consciously say what criteria he uses to divide the class of Ks. We may call this ability, following Scriven, the trained "judgment" of the scientist, a judgmental ability which he acquires as he acquires his expertise in an area. In medicine most knowledge is in fact imperfect. But we train doctors so that they acquire sufficient intuitive "know how" that they can, on the basis of their judgmental ability, successfully apply such imperfect knowledge correctly and make successful predictions. For example, on the basis of unconsciously noted cues, the trained doctor will notice the absence of C and will thereupon apply (5.36) and successfully predict b will not die even thought the poison has in fact been administered. This ability to make successful judgments is what makes the doctor's prognosis so useful in respect of his and our pragmatic interest, and so much more reliable, as a basis for decision-making, than pure guess-work or coin-tossing and chance.[96]

We may say then, that trained judgment is indeed part of one's capacities as a scientist. If the above is correct then one thing such capacities enable one to do is to make non-reasoned prediction possible, where explanation is not possible until after the occurrence of the event predicted, and where reasoned prediction can never be actual. To put it briefly, its point, contrary to what Scriven holds, is to make prediction possible where explanation is not (yet) possible. Only, one must add, since the prediction is non-reasoned, this in no way touches the deductive model and the symmetry thesis.

But the impression should not be given that such judgment abilities are restricted to scientists. Nor the impression that such abilities relate to only the presence or absence of relevant variables. For, we all have such abilities, and we all use them all the time. This can be made clear if we return to the ink-bottle example.

What we considered in that case were the forces exerted by certain objects upon ink-bottles. Let us assume normal circumstances, excluding such cases as the bottle being glued down, and so on. This is as we excluded F_1, etc., in the poison-non-dying case. So we have only to consider the forces, and the tipping of ink-bottles. This enables us to schematize the situation into laws of the general form "whenever A then B" or, which is the same, "$(x) (Fx \supset Gx)$". These are related in the obvious way to (6.7). Let us refer to the tipping by "T" and to the several forces as "E_a", "E_b", "E_c", and so on. For each force ther will be a general statement to the effect that whenever that force is exerted then the bottle tips:

(6.14) $(x)(E_a x \supset Tx)$
(6.15) $(x)(E_b x \supset Tx)$
(6.16) $(x)(E_c x \supset Tx)$
(6.17) $(x)(E_d x \supset Tx)$

Now, it is possible for scientists to quantify exactly these forces, and put them in a cardinal ordering in which they are assigned numbers according to exactly how much greater and less any given force is when compared to all others. But to determine exactly which amount of force is exerted on any given occasion requires measurements to be taken with instruments. These are not present in most situations. Nor, in ordinary situations, do we need them. For, we can get along for the most part with just a rank ordering into greater and less, without reference to how much greater or less. In fact, we can often recognize *by simply looking* roughly where an exemplified force falls on this scale of greater and less. This recognition may be gross, but for most purposes it suffices. Thus, some forces will not be sufficient to tip the bottle. Suppose E_a is one of them. In that case the generality (6.14) will be false. Other forces will be sufficient to tip, say E_d. Then (6.17) will be true. We will have the ordering

$$E_d > E_a$$

Among the forces there will be (we know, from mechanics) a least force capable of tipping the bottle. We may well not be able to identify such a force as such if we see it exemplified. But we may well recognize limits, an interval. Let us suppose E_d is the upper limit of the interval, E_a the lower limit. One would then be able to recognize a force being exemplified as greater than E_d, and thereby infer that it is sufficient to knock the bottle over, or, what is the same, predict the tipping. And one could recognize other forces as less than E_a and withhold any prediction of tipping. And finally there would be certain force with respect to which one could neither assert nor deny sufficiency for knocking the bottle over. These last would be the forces between E_d and E_a. We may suppose E_c is such a case. It is, as one says, a borderline case.

Suppose a exerts E_c (at t). Then we have

(6.18) $E_c a$

170   CHAPTER 2

We do not know whether (6.16) is true or not, since we are supposing E_c is between E_d and E_a. Now, it could happen the bottle does tip when struck by this force. We would then have

(6.19) Ta

(6.18) and (6.19) thus provide confirming evidence that (6.16) is true. Nor would the inference of the truth (6.16) from (6.18) and (6.19) be an induction from a single instance. Background knowledge lends its support. If the bottle is to move, certain forces, e.g., those of friction, must be overcome. If the exerted force is greater than those resisting forces the bottle will move, otherwise it will not. Thus, any force is either sufficient for tipping or it is not. With this from background theory – or rather, everyday knowledge, for what was just described hardly ventures the advanced sort of mechanics we needed Newton to articulate to us – with this from background knowledge the existence of the one instance (6.18) & (6.19) suffices for one to conclude conclusively to the truth of the law (6.16). We examined the general pattern of such inferences in our discussion of Ducasse's criticism of Hume in Section 1.2. So upon seeing an exerted force tip an ink-bottle one can explain via an immediately available law the tipping by the exertion of the force. However, prediction will still not be possible. This will be so because by our hypothesis one is unable to distinguish those forces in the interval E_d–E_a into those which are sufficient and those not sufficient to tip. If one does in fact see such a force tip the bottle one can infer directly to its general sufficiency. This is what we just noted. The point is that one cannot reidentify that force, if it is presented again, in any way save in the E_d–E_a interval, and therefore if it is presented again one cannot say for sure whether it is or is not sufficient to tip – until the tipping occurs.

We do not have a general set of names E_a, etc., for all the forces. Once quantification occurs, we can use numerals as names, but all that is not available in the situation we are now looking at. Rather, we tend to use demonstratives – "*that* force is sufficient" – or definite descriptions – "*the* force being now exemplified by (exerted by) a". So, instead of explaining the tipping (6.19) by means of a sentence such as (5.23), the explanation might well be expressed via demonstratives – "that knock caused the ink to tip" – or definite descriptions – "the bottle tipped because of the push Jones gave it". The use of these ways of referring to types of forces – the same types as are referred to by such general terms as 'E_a' – these special ways of referring may well obscure the role the law plays in the explanation. Perhaps

THE REASONABILITY OF THE DEDUCTIVE MODEL 171

some of this helps to account for Scriven's or Collins' inability to see these matters clearly.

Judgment thus enters not only to detect the presence and absence of variables, but also to classify them roughly along such dimensions as that of greater and less. We looked at a simple case. We can imagine more complicated cases: "Just another pinch of salt will make it a very good soup"; "His heart is not quite strong enough! Quickly! Do so and so!", "Are those all the pills he took? That's not quite enough to kill him! He'll sleep it off!" What we do is train scientists – and cooks – in such a way that these judgmental abilities are increased. As the discussion just above showed, its point may be to make non-reasoned prediction possible. Or, as the discussion of the ink-bottle showed, its point may be to classify properties into those we can make reasoned predictions with (any weight outside the $E_d - E_a$ interval) and those we cannot make reasoned predictions with. In any case, however, the deductive model and the symmetry thesis are quite compatible with scientists needing and using trained judgment.

In connection with intuitive expertise, trained judgment, one further point is worth making here, in spite of its not bearing directly on the symmetry thesis. The scientist who offers (non-reasoned) predictions based on his intuitive expertise will have a certain success rate. This rate can be measured statistically.[97] Of course, if we knew the situation at which he is looking, and the unconscious perceptual mechanisms – the unconscious cues and how he systematically responds to them – generating his judgment, then we could predict perfectly his success rate. In other words, if our knowledge here were less than imperfect we would not need the actuarial assessment of his reliability: we could replace that imperfect knowledge with less imperfect knowledge.[98] But then in that case we would not need his trained judgment either! Supposing we do need his intuitive expertise then we can at least assess its reliability in the usual actuarial way. Now, it may turn out that our scientist can divide his intuitive judgments into two kinds, confident and non-confident. It could then be that the success rate of the former is higher than that of the latter. The statistics will tell us just how risky it is to rely upon a judgment of one kind or the other. In this matter the ideal would seem to be something like the following. The scientist has a whole set of feelings of confidence, going in degrees from low to high. To these, numbers between 0 and 1 are assigned. He then reports his degree of confidence using some number n/m. Then the statistical probability of a judgment with this degree of confidence being correct is n/m. But this is clearly ideal. It is hardly likely anything beyond a confident/non-confident distinction will be useful.

However, all this is but a digression, and cannot be pursued further, interesting as it is.

This concludes our discussion of the issue of whether one can discover cases where there is an asymmetry between explanation and prediction. It has not been possible to deal with all the supposed counter-examples that have been presented in the literature to refute the symmetry thesis, and thereby the deductivist thesis that entails the symmetry. I do believe, however, that the above discussion has introduced all the material that one needs to successfully handle those other supposed counter-examples and to show that after all they are not counter-examples. We may therefore rest this aspect of our case in defence of the deductivist thesis, and pass on to other topics.

CHAPTER 3

EXPLANATIONS AND EXPLAININGS

The attack on the deductive model has many fronts. Some we have examined in detail already. None of these seriously challenged the idea that if laws were available, then explanations could proceed as deductions from those laws. The main point of the present chapter is to challenge a somewhat different line of attack, that argues that even if laws are available, they are *not part of* the explanation but rather of its justification. The argument for this claim involves the exploration of the idea of explaining as well as that of explanation. Another line presupposing the same sort of exploration of the idea of explanation is that it is possible to explain even where no law is available. It is to these matters that we now turn. What will turn out is that in fact these lines of argument against the deductive model are plausible only because there has been insufficient exploration of the ideas of explanation and of explaining. Interestingly enough, as we shall see, some recent formalist criticisms of the deductive model also are vitiated by an insufficiency of attention being paid to those same ideas.

3.1. EXPLANATIONS IN THE CONTEXT OF COMMUNICATION

Earlier we distinguished an *explanation*, in the sense of a set of sentences, from the transaction of *explaining*. Whether a set of sentences is an explanation is a matter of its satisfying certain objective criteria, objective in the sense that they depend on neither explainer nor explainee. In the transaction of explaining, the explainee wants but does not have an explanation. Exactly what sort of explanation he wants depends on his interests. Those interests might be idle curiosity or they might be some pragmatic interests. Let us suppose that an explainee has a quizzical look on his face whenever he wants but does not have an acceptable explanation. The explainer offers him a set of sentences. He accepts the sentences and the quizzical look disappears. There are two possibilities. (1) The explainer has presented the explainee with a set of sentences which constitute an explanation and which are adequate to supplying (so far as evidence, etc., makes it possible) what the explainee wants by way of explanation. (2) The explainer has presented the explainee with a set of sentences which are either not an explanation or are not adequate to what he wants. In both cases, whether or not the sentences constitute an explanation and whether or not they are adequate to his wants is something which is independent of whether he *in fact* accepts the sentences and of whether the quizzical look disappears. Now, we are supposing that in both cases (1) and (2) he accepts the sentences and the quizzical look disappears. In case (1), this acceptance and the concomitant disappearance of the puzzled feeling is justified. In case (2) it is not justified. In case (1) he ought to have accepted, and did. In case (2) he ought not to have accepted but did. In case (1) the explainee did have what he wanted. In case (2) he did not have what he wanted, but *thought* he had what he wanted; this false thought led him to the unjustified acceptance of the set of sentences as an adequate explanation. We can make this basic contrast between the two cases because our objective criteria for being an explanation and for being better or worse as an explanation enable us to judge, *normatively*, whether acceptance, and the disappearance of the quizzical look, is or is not justified.

The intention of the explainer in presenting the set of sentences to the explainee is *to explain*, that is, to explain to the explainee whatever the latter wants an explanation of. The attempt to carry out this intention is the transaction of explaining. We may define *two* senses in which such a transaction may be said to be successful. *One*: The transaction of explaining is successful just in case the explainee accepts the set of sentences he is offered and the quizzical look leaves his face. *Two*: The transaction of explaining is successful in the first sense, and furthermore the acceptance of the sentences by the explainee is justified.

Let us call a set of sentences an "explanation$_B$" (relative to a particular context) just in case the set is used in an explaining transaction which is successful in the first sense. To say a set of sentences constitute an explanation$_B$ is to say that they constitute a verbal stimulus which evokes a certain response in the explainee, viz, acceptance and the disappearance of the quizzical look. Thus, the concept of being an explanation$_B$ is essentially a concept of behaviouristic psychology (whence the "*B*"). As a matter of fact there are regularities connecting verbal stimuli of certain kinds to behavioural responses of certain kinds. These regularities are laws, imperfect, to be sure, but still laws. The concept of being an explanation$_B$ is useful because it enables us to state these laws, these matter-of-fact regularities. To "explore" this concept is to examine these regularities.[1] For, it is to ask for the "criteria" which make a set of sentences into an explanation, in the sense of asking for what characteristics of verbal stimuli are sufficient to evoke the response definitory of explanations$_B$. But that is to ask what the S–R connections are, what the laws are. In short, to "explore" the concept in this sense is to do armchair psychology. And, in fact, whether something is an explanation$_B$ is completely irrelevant to the philosophy of science.

An explanation$_B$ is of interest to the philosophy of science only if it is successful in the second sense, that is, only if it is an explanation in our sense, i.e., according to the deductive model, and only if it is good enough, on the scale defined by the ideal of process, to satisfy the explainee's interests. An explanation$_B$ is of interest to the philosophy of science only if it is worthy of the explainee's acceptance. Let us say that an explanation$_B$ which satisfies this condition is a "w-explanation$_B$" (w=worthy). But even a w-explanation$_B$ is of interest to the philosophy of science, *not* because it is an explanation$_B$ but rather because it is *an explanation worthy of acceptance*. What makes the set of sentences of interest to the philosophy of science is not the fact that the explainee accepts them but the fact that they are *an explanation worthy of acceptance by one who has the explainee's interests*.

It is true that in the explaining situation we want to present to the explainee sets of sentences which (a) are explanations and are worthy of his acceptance, given his interests, and (b) he will in fact accept. (b) may be rephrased as saying we want to present him with explanations$_B$. Given what we are trying to do, we hope to present sets of sentences satisfying condition (a) and which are also explanations$_B$. Whether we succeed in the latter is a practical consideration, one of immense importance. It is, however, a question of the psychology and sociology of communication. It is of no interest to the philosophy of science, which is interested in explanations and their relative worth, and not in whether they can be successfully communicated to another.

I emphasize all this, which seems rather elementary, because the critics of the deductive model of explanation unfortunately fail to notice much of it at all. I mentioned that some lines of attack on the deductive model depend upon an exploration of the idea of explanation. That was inaccurate. What the attacks depend upon is an exploration of the concept of being an explanation$_B$. But that concept is irrelevant to the philosophy of science and the concerns of the law-deduction theorist. In other words, the attacks are simply off their target, and are so right from the beginning. Or, so I propose to argue.

I do not want to say that the approach through psychology or linguistics, the approach through communication, can never be illuminating. We noted in Chapter 2 that some such as Collins and Scriven raise the silly objection to the deductive model that explanations of the rules of chess are not deductive.[2] As if the law-deduction theorists were doing anything other than consider scientific explanations! Bromberger[3] has suggested how this confusion might have arisen. To make us aware of the ambiguity of 'explanation' he asks us simply to

> ... notice that 'explanation' may refer to the answers of a huge variety of questions besides why-questions, the only requirement being that their *oratio obliqua* form fit as grammatical object of the verb 'to explain' and its nominalization 'explanation of', e.g., how-questions, what-is-the-cause-of-questions, what-corresponds-at-the-microscopic-level-questions, etc. Yet, the issues raised by these other types call for considerations peculiar to each type and different from those called for in the case of why-questions. Confusion is therefore likely to ensue and is apt to be further compounded if we allow ourselves to forget that 'explanation' may also refer to things not readily specified as answers to a specific class of questions.[4]

I suppose forgetting the ambiguity is what is behind Scriven and Collins. Certainly, confusion ensued. Yet I doubt Bromberger's elaborate explanation is really needed for one to notice the ambiguity. I should have thought some rather more simple reflections on a couple of examples would suffice. As Hempel and the rest who have stated the deductive model have also thought. On the other hand, those simple reflections did not work in the case of Scriven and Collins. Perhaps Bromberger's approach will. Certainly, one must admit there is a certain superficial sort of mind that will accept a point that could be put simply only if it is in fact put in somewhat fashionable and elaborate jargon. They are the Mrs. Generals of philosophy, ultimately interested only in the varnish on an object. And if Scriven and Collins are of this sort then, indeed, perhaps they will accept the simple point now that Bromberger has put a thick enough coat of lacquer on it.

Bromberger gives a rather elaborate definition of what, according to him, is to count as a why-question.

> We will mean by a *why-question* a question that can be put in English in the form of an interrogative sentence of which the following is true: (1) the sentence begins with the word *why*; (2) the remainder has the (surface) structure of an interrogative sentence designed to ask a whether-question, i.e. a question whose right answers in English, if any, must be either "yes" or "no"; (3) the sentence contains no parenthetical verbs, in Urmson's sense.[4a] A why-question put as an English sentence that satisfies (1), (2), and (3) will be said to be in *normal form*. By the *inner question* of a why-question we will mean the question alluded to in (2) above, i.e., the question reached by putting the why-question in normal form, then deleting the initial "why" and uttering the remaining string as a question. By the *presupposition* of a why-question we will mean that which one would be saying is the case if, upon being asked the inner question of the why-question through an affirmative interrogative sentence, one were or reply "yes", or what one would be saying is the case if, upon being asked the inner question through a

negative sentence, one were to reply "no". Thus, "Why does copper turn green when exposed to air?" is a why-question in normal form; its inner question is "Does copper turn green when exposed to air?"; and its presupposition is that copper turns green when exposed to air.[5]

After all this Bromberger then restricts his attention to only some why-questions. He proposes to ignore "why-questions whose normal forms are not in the indicative." He proposes to ignore "why-questions whose presupposition refers to human acts or intentions or mental states." And finally, he proposes to ignore "why-questions whose correct answer cannot be put in the form 'because-p', where p indicates a position reserved for declarative sentences."[6] So *now* it all comes out! What Bromberger wants to talk about are causal explanations. I should have thought Bromberger could have stated what he was going to talk about in rather less elaborate fashion. I guess some persons just take a long time to get to the point.

What should be noticed in the long quotation is the phrase 'right answer in English'. Now, if the reference to 'in English' be taken seriously, an answer will be right just in case it conforms to the syntax and semantics of English. That would leave certain *bad* explanations as possible answers to why-questions, since English, the language, can hardly judge such explanations to be bad. However, it would be unfortunate for the reader to take this thought too seriously. For, very shortly we are told that the third of his qualifications as to which why-questions he is going to attend to is a qualification which " ... affects ... in particular why-questions with false presupposition and why-questions whose inner question admits of no answer, e.g., 'Why doesn't iron form any compounds with oxygen?' and 'Why does phlogiston combine with calx?' More may be ruled out ... "[7] – but, we are given to understand, these at least are certainly ruled out. The former is ruled out because its presupposition, that iron forms no compounds with oxygen, is false; while the latter is ruled out because its inner question, Does phlogiston combine with calx?, admits of no answer, which is turn is presumably because there is no such thing as phlogiston. Now, that iron does combine with oxygen and that phlogiston does not exist are both *factual* truths. So there are no explanations associated with those why-questions because certain facts obtain. Thus, whether an answer to a why-question is "right in English" turns out to be not simply a matter of conforming to grammar but of the way the world is. I do not find this last objectionable. After all, we ourselves defined something to be an explanation only if the law-premiss was true. Only then would it be objectively worthy of acceptance as an explanation. So I am quite prepared to say we do not explain iron's failure to compound with oxygen nor phlogiston's combining with calx, and that we do not explain because, given certain factual false-hoods, there is nothing to be explained scientifically. More strongly, Bromberger is to be strongly commended for entering into these normative considerations. He should, perhaps, distinguish, as we have, between objective and subjective worth. Otherwise, Bromberger will end up convicting of irrationality and bad science those who, with good reason, accepted the phlogiston theory, and so accepted with reason the truth of the presupposition that phlogiston combines with calx.[8] But this is minor. What dismays one is Bromberger's clearly held idea that the failure of these why-questions to admit of answers somehow has to do with what is "right in English". As if it were a rule of English that decreed the non-existence of phlogiston, or the existence of iron oxide! Bromberger completely runs together two ever so different reasons for rejecting a why-question as having "no correct answer", viz, on the one hand, it is ungrammatical and, on the other hand, certain states of affairs as a matter of fact do not obtain. We suggested above that one who adopted with respect to explanations the descriptive-psychological approach via explanation$_B$ would be likely to fail to distinguish reasons for accepting (or rejecting) an explanation. One who adopts the same approach to why-questions will be equally likely to fail to distinguish reasons for rejecting why-questions. In other words, Bromberger's ostentatious adoption of techniques from contemporary linguistics, his adoption of the descriptive-psychological approach, leads him to run together precisely what must be distinguished if philosophy of science is to get on with its normative tasks.

I believe the descriptive-psychological approach also leads Bromberger astray at another point. He connects the use of why-questions to contexts in which persons are puzzled. In this he is no doubt correct. He defines[9] someone as being " ... in a *p-predicament* ... with regard to some question Q, if and only if on that person's views, the question Q admits of a right answer, yet the person can think of no answer, can make up no answer, can generate from his mental repetoire no answer to which, given that person's views, there are no decisive objections." And a person is " ... in a *b-predicament* with regard to a question Q if and only if the question admits of a right answer, no matter what the views of the person, but that answer is beyond what that person can think of, can state, can generate from his mental repertoire." (I note that any scientist who had puzzles about phlogiston was in neither a *p*-predicament nor a *b*-predicament. Bromberger's failure to have a category here is simply his failure to draw the distinctions we insisted upon in the last paragraph, and therefore a consequence of his descriptive-psychological approach. The blurs the latter introduces lead to the approach itself not being worked out fully! Actually, as we shall see, Scriven is rather better on these things. Which is to say he is even further than is Bromberger from the normative issues philosophy of science should be concerned with.) Bromberger proceeds:

Let us say furthermore that a question Q is *unanswerable relative to a certain set of propositions and concepts C* if and only if anyone who subscribes to these propositions and limits himself to these concepts must be in either a *p*-predicament or *b*-predicament with regard to the question Q. The search for and discovery of scientific *explanations* ... is essentially the search for and discovery of answers to questions that are unanswerable relative to prevailing beliefs and concepts. It is not, therefore, merely a question for evidence to settle which available answer is correct, it is a quest for the unthought-of.[10]

A scientific explanation is thus an answer to a question arising from a certain context. This is presumably a comment on the grammar or use of 'scientific explanation'. Let us accept it as correct, that is, correct as a descriptive-psychological account of 'scientific explanation'. It then turns out that a good deal of scientific research is *not* aimed at producing scientific explanations. Very often, research is necessary even where thought-of answers are available to why-questions; thus, Newton answered why the solar system behaves as it does. Yet a good deal of research went into determining, what Newton had not determined, indeed, had not needed to determine — namely, the universal constant G of gravitation. Or again, it was known certain entities called "penicillin" caused the death of germs and effected thereby the cure of disease, prior to those entities having actually being identified. Events were explained by appeal to penicillin. Research still had to proceed, however, to replace this C^* with its C.[11] What was being sought had, however, been thought-of, talked about, generated from mental repertoires (whatever they are) as answers to questions. So the research was not after the unthought-of, and was therefore not, according to Bromberger, directed at finding a scientific explanation. Now, the two examples just given involve a search for a *better explanation*, the elimination of imperfection. This, of course, is *our* use of 'explanation' — which is *not* descriptive-psychological. The point is that for Bromberger these are *not* scientific explanations. Bromberger is thus effectively prevented from seeing that explanations (our use) can be better or worse. His descriptive-psychological approach prevents him from seeing the distinction between perfect and imperfect knowledge. We noted earlier (Chapter 2)[12] that he did not make this distinction. We now understand why.

Let us now turn to Collins, who tells us that "To explain something is to make that thing intelligible, understandable, or clear to someone, perhaps to oneself."[13] Collins is here clearly talking about explanations$_B$. We are told about these that "What succeeds in making something intelligible depends on what kind of thing it is that we want to explain and on the context in which the need for an explanation arises."[14] We ask about the "success" here. Is it merely the sort of success which any explanation$_B$ has (but might not deserve)? Or is it the sort of success a w-explanation$_B$ has (and deserves)? Collins never tells us. However, he does make the point that the law-deduction theory "prescribes" a pattern for explanations and that by virtue of this prescription " ... investigators are enjoined to improve themselves in directions which have no relevance to *practical* criteria for the success of their explanations."[15] The 'practical' here has a pretty clear reference to the context of communication, and success in communication. We know these things to be irrelevant to the philosophy of science. It follows at least that Collins is introducing irrelevancies into his discussion. Scriven makes remarks similar to those of Collins about "understanding" and "intelligibility".[16] He points out that we can design tests, of a questionnaire sort, to determine the success of our efforts. These, he tells us,[17] are "objective". Such questionnaires can test whether the verbal stimulus evokes a certain response, namely, a dispositional response to the effect that if the person is presented with the questionnaire then certain answers are given. The presence of such a dispositional response is no more relevant than the presence of the simpler response, the disappearance of the quizzical look. One still wants to know, *Ought* he to have that response? And that is all one wants to know *qua* philosopher of science. Of course, if by "successfully produces knowledge" one means "knowledge of the sort which ought to have been produced" then one is back at the normative issue. That issue, however, as we know, can be discussed independently of the discussion of explanations$_B$ and of the responses explanations evoke. The introduction of the concept of explanation$_B$ by Scriven and Collins can at best obscure the crucial issue.

Scriven introduces a number of criteria in terms of which he proposes to evaluate explanations. What I want to do is examine these. This will show that he systematically confuses the normative question of the philosophy of science about explanations with the descriptive question of behavioural psychology about explanations$_B$.[17a]

Scriven calls "our evidence for the truth of the statements actually made"[18] the *truth-justifying* grounds of the explanation, and if " ... we have insufficient grounds for the assertions actually occurring in what we normally call an explanation ... we would certainly be justified in complaining that the explanation was ill-supported or − in the event of actual falsehood – inaccurate." He calls role-justifying grounds the grounds we have " ... for thinking that the statements made are adequate for a certain task – the task of explaining (in *some way*) whatever it is they are supposed to explain." Where such grounds are missing, though "it may be that the statements offered are well-supported and true," they "do not fully *explain* what they were supposed to explain; and here we might plausibly say that the explanation is incomplete or *inadequate*." Finally, "it may be that, through misunderstanding, the proposed explanation a not the *kind* required; then we would describe it as inappropriate or *irrelevant*." In this case the missing justification is a matter of the grounds which

support our interpretation of the practical requirements of the person or public to whom we address our explanation – for example, that they need an explanation of someone's behaviour in terms of his intentions rather than his muscular operations, or of the rocket's variation in apparent brightness in terms of properties of the rocket rather than the mechanisms of vision. The considerations which lead us to propose one rather than another type of explanation I shall call *type-justifying grounds*.

The bare statement of Scriven's distinctions make clear he is discussing success of the explainer in the transaction of explaining. *Our* discussion will therefore have to introduce such "practical considerations" of success, even though we know such considerations are in fact irrelevant to the philosophy of science. It is the analysis of Scriven which requires us to introduce these matters, not the analysis of explanation.

We may divide our discussion into *three steps*.

Step One: "Explanations successfully produce knowledge." We are told by Scriven and Collins that what is produced is "intelligibility" or "understanding", as we saw just above. These additional things we shall turn to directly, but "understanding" must presuppose knowledge, so at present we are discussing a necessary condition for the successful production of the "understanding", which production is, for Scriven and Collins, definitory of explanations. The use of a set of sentences to successfully produce knowledge in the explainee requires at least the following necessary conditions to obtain.

(1a) There must be a deduction of the event to be explained from premisses which include at least one law statement.

(1b) The premisses must be true.

We are in the extended process of justifying the inclusion of these two conditions for *all* explanations, i.e., all scientific explanations. Scriven does admit that for *some* explanations these are necessary conditions. In such cases he takes (1a) to be *not part* of the explanation but rather part of its justification. Specifically, the law statement is part of the role-justifying grounds.[19] We shall have to return to this point later. But what is important now is that we need not settle the issue. The question he raised was what were the necessary conditions for a set of sentences to be successfully used to explain. Scriven and ourselves are, for many cases, agreed that (1a) is a necessary condition. If the explainer offers, say, "*a* is *G*" as an explanation of why *a* is *F*, then a necessary condition that he be successful is that there be available in the context the law "$(x) (Gx \supset Fx)$" which establishes by deduction the relevance of *a*'s being *G* to *a*'s being *F*. On this Scriven and ourselves are to a large extent agreed. For the present discussion about such necessary conditions that agreement suffices. The further disagreement, whether the law is part of the explanation or of the justification of the explanation, is a matter which can be postponed until later.

It turns out, then, that (1a) is part of what Scriven calls role-justifying grounds. (1b) is clearly among the truth-justifying grounds.

We may now add these necessary conditions.

(1c) The person addressed must understand the language being used.[20]

(1d) The person addressed must be able to recognize that logical (evidential) connections obtain.

(1e) The person addressed must have evidence that the premises are true, and, in particular, *in the case of generalities, evidence that justifies their law-assertion*.

($1e_1$) What he has accepted to be available as evidence is a matter of psycho-socio-historical fact, and in that sense not objective.

($1e_2$) Whether the statements accepted as evidence constitute *good* evidence, i.e., testify to the truth of the premises, is a matter of logic, and objective. (1d) guarantees that if the explainee has good evidence, he will recognize its worth.

An explanation, or rather, an attempt at offering an explanation, will be justified in case these necessary conditions are all fulfilled. These necessary conditions thus all constitute *justifying* conditions, in Scriven's sense. What we want to do is to compare this list to that of Scriven. Clearly, (1a), (1d), and (1e) are all type-justifying grounds, upon Scriven's classification. That term thus gathers in several quite different things, which makes its use as misleading as Scriven believes it to be illuminating. Worse: ($1e_2$) is among the type-justifying grounds; it is also among the truth-justifying grounds. That is, Scriven's distinctions overlap. Clarity would surely demand the avoidance of such systems of classification. At least, that is what our elementary logic texts tell us. On the other hand, one might suggest Scriven misses the overlap because he notices only (1e$_1$), missing (1e$_2$) completely. That would mean Scriven does not distinguish between what is accepted as evidence and what is accepted being worthy, logically speaking, of acceptance as evidence. To put it in slightly different terms, it would mean Scriven does not distinguish explanations$_B$ and w-explanations$_B$. The failure of Scriven to notice his classification of grounds overlaps at (1e$_2$) suggests just such a diagnosis, to the effect that Scriven is more interested in the descriptive psychology of communicating than in the normative discipline which is the philosophy of science. This diagnosis is substantiated when one recognizes two kinds of "truth-justifying" conditions: those involved in (1b) and those involved in (1e), insofar as the latter involved (1e$_1$): the former sort of "truth-justifying" conditions are objective, the latter sort subjective. Indeed, it is precisely this distinction which enables one to draw the distinction between an explanation being objectively worthy and an explanation being subjectively worthy. The fact that these two sorts of "truth-justifying" grounds are grouped together means Scriven does not wish to distinguish what must be distinguished if one is to define a normative category of what is objectively worthy of acceptance as an explanation. Which is to say once again that he is more interested in descriptive psychology than in the philosophy of science.

Step Two: "Explanation successfully produces knowledge of a certain sort, namely, such knowledge as one wants or could reasonably want to have in order to understand." Here the understanding is, of course, scientific understanding.[21] Now, unfortunately both Scriven and Collins fail to tell us what they mean by "scientific intelligibility" or "scientific understanding". These concepts are left completely unanalyzed. In effect they leave it that a person understands, that explaining has been successful, when the quizzical look disappears. They do not raise the issue about the sort of knowledge which can reasonably be said to constitute scientific understanding. But we are not surprised about this, since it is simply another aspect of their failure to distinguish explanations$_B$ from w-explanations$_B$. That is, it is another aspect of their abdication of the normative enterprise of the philosophy of science. As for ourselves, we above examined scientific understanding in terms of process knowledge.[22] That permits us to supplement the Scriven-Collins sort of position.

In this present step two we have it that an attempt at offering an explanation will be justified in case the necessary condition is fulfilled that the explanation provide the kind of knowledge which is wanted. Our grounds for holding that the explanation provides such knowledge clearly come under

Scriven's general label of type-justifying grounds. The statement of this necessary condition must, however, be qualified in certain ways. Such qualification amounts to distinguishing various kinds of type-justifying grounds, that is, it amounts to making Scriven's rather crude classification somewhat more adequate.

Necessary conditions for success at explaining will include the following.

(2a) The explanation provides the knowledge the explainee wants and could reasonably expect to have.

Explanations may be available to satisfy his wants (and they would therefore be justified on these type-justifying grounds) whereas the explainee cannot reasonably expect to have them because any of necessary conditions (1e), (1d) or (1e) of step one are not satisfied (in which case different type-justifying considerations would exclude them). We see that various considerations relevant to the success of the transaction of explaining so to speak interact with each other in a way Scriven's classification obscures.

(2b) A necessary condition for success is that the explainee have no unreasonable desires for knowledge.

Again, this is a matter of type-justifying considerations.

Two cases come to mind. In the first place, the explainee may desire anthropomorphic Aristotelian explanations. Such a desire is unreasonable, since, in point of fact, such "explanations" do not provide knowledge, but the illusion of knowledge.[23] In the second place, the explainee may desire a scientific explanation of a sort which nature will not justify, the world simply not conforming to the level of explanation desired. For example, a process explanation may be desired for the case of a system involving a long sequence of tosses of an unbiased coin, but process knowledge is (known to be) not available for such a system: one must settle for imperfect statistical laws, that being the best nature permits. The explainee must scale his wants, with respect to knowledge, to a level which can be satisfied, given the way the world is. Thus, the type-justifying considerations might include a point to the effect that the explainee is aiming too high.

The explainee might also be aiming too high relative to the evidence in fact available. We discuss this in step three.

(2c) Given that the wants are reasonable, they are generically of a type and specifically of two sorts. What the explainee wants generically is a scientific explanation, an explanation which shows in terms of some other event (the initial conditions) why the explanandum must have occurred, in some pre-analytic sense of 'must', a sense which requires explication. The sort of explanation he wants specifically will be a function of which of two kinds wants he has. A reasonable interest may be of one of two sorts, either idle curiosity or some pragmatic interest. We justify our explaining by showing that the explanation will produce the kind of knowledge that is wanted. To satisfy his wants generically it must be a scientific explanation, according to the non-problematic sense the pre-analytic 'must' acquires upon explication. Specifically it must satisfy either an idle curiosity or some pragmatic interest. This already divides Scriven's role-justifying grounds into two cases, generic and specific, and further subdivides the latter according as whether the interest is idle curiosity or pragmatic. We see the further inadequacies of Scriven's classification.

Suppose we make the appropriate assumptions about (1a)–(1e) of step one. Suppose further we are supplying the explainee with what satisfies, or, at least, ought to satisfy, his wants generically. It is then a question of satisfying his wants specifically. We have two cases: ($2c_1$) both idle curiosity and pragmatic interest require satisfaction; ($2c_2$) only some pragmatic interest is present requiring satisfaction.

($2c_1$) Here the explanation must satisfy the necessary condition that either (i) the explanation produces process knowledge, if such knowledge be available, availability being a matter of (1e) of step one; or, if such knowledge is not available, the condition that (ii) the explanation provide imperfect knowledge as close to the ideal of process as the evidence ((1e) of step one) permits.

We have previously argued for these necessary conditions when we argued that process knowledge constitutes the ideal of scientific explanation. We do not need to repeat that justification. What it is important to note about it here is that the justification involves considerations of all three of the sorts Scriven mentions. Type-justifying conditions enter when one asks what one could reasonably one to know out of either idle curiosity or pragmatic interest. Truth-justifying conditions enter automatically with the considerations that led us to hold the premises of the explanation ought to be true. Role-justifying conditions enter in the following way, at least if we take seriously Scriven's definition of role-justifying conditions. Imperfect knowledge may be offered which satisfies the truth-justifying conditions but does not give the explainee all the knowledge he wants. In that case the explanation is imperfect and trivial, and the explainee is apt to characterize it perjoratively as simply constituting a "mere prediction technique" rather than a "real explanation" which "increases his understanding". That is, he rejects it is a *bad* explanation relative to what he could reasonably expect. This would seem to be a matter of inadequate role-justifying grounds.[24] Certainly, Scriven tells us that where role-justifying grounds are absent, the statements offered "do not fully *explain* what they are supposed to explain; and here we might plausibly say that the explanation is incomplete or inadequate."[25]

The reason considerations of all three sorts enter into the justification of the *one* necessary condition ($2c_1$) is that ($2c_1$) is inserted as a *normative* feature. The quizzical look may disappear without this necessary condition being fulfilled, but it ought not disappear. That is, this necessary condition ought to be fulfilled, and an explanation is successful normatively only if it is fulfilled. That fact that Scriven ignores the convergence of his conditions at this point is further evidence that he ignores the normative issue, and is concerned only with the descriptive psychological problem of explanations$_B$.

($2c_2$) Here the desire for knowledge is a matter of some pragmatic interest. In this case the necessary condition that must be fulfilled is that just sufficient knowledge ought to be presented and no more than enough knowledge to satisfy the pragmatic interest.[26] Idle curiosity requires one to introduce all relevant laws; pragmatic interests, as we know, cast their nets more narrowly. In this case we have what Scriven calls type-justifying considerations.

Step Three: In step one (1e) we considered evidence. In step two we considered what one might want to know. In step three we note that these can diverge, and look at the consequences of such divergence. (1e) of step one places limitations on what can be offered. The best that *can* be offered on this basis may not be enough to satisfy one's wants. *The unsatisfied wants that result are the motive behind the activity of scientific research.* Because one wants more than can be given owing to the limitations of available evidence, one goes out to deliberately gather more evidence.[27]

In this case the desires, wants, interests, with respect to knowledge, which enter into the type-justifying considerations reject as adequate any explanation permitted by available truth-justifying grounds. These unsatisfied interests initiate the search for more adequate explanations, that is, the search for evidence (truth-justification) such that if it points correctly then the explanation at which it points (role-justification) tells us everything we want to know (type-justification). And ultimately, of course, it is just this dissatisfication which what is currently available which pushes us towards the ideal of process knowledge, since idle curiosity, which in general is alone capable of leading us that far, is in fact among the carefully cultivated desires in the members of the scientific community.

To conclude this discussion of Scriven's proposed three-fold classification of criteria for evaluating explanations: (A) The criteria are for the success of explaining and not for the worth of explanations. (B) The criteria are concerned largely with explanations$_B$ and not with w-explanations$_B$. (C) Insofar as the criteria are about explanations$_B$ the list is not exhaustive; nor are the kinds exhaustive. (D) If the criteria are taken to be about w-explanations$_B$, the classes they are divided into overlap. (E) Criteria of all three kinds are relevant to justifying the worthiness of w-explanations$_B$. (F) Insofar as the criteria concern worthiness, they include items not specific to explaining but which apply generically to any sort of discourse at all. It follows that a discussion of explanation-justification which is adequate for the normative discipline of the philosophy of science, requires in one respect an enlargement of the set of criteria (the descriptive respect), in another respect a contraction of the set (the normative respect), and in any case a revision of the classification which, as it stands, is neither exhaustive, nor into mutually exclusive classes, and furthermore is based on differentia which put into separate classes what should come together so far as the normative classification of explanation is concerned. To put the inadequacies in a sentence, Scriven concentrates on explanations$_B$, rather than on explanations.

3.2. FORMALIST CRITICISMS OF THE DEDUCTIVE MODEL

Now that we have examined in some detail the idea of *explaining*, we are in a position to examine some formalist criticisms of the deductive model of explanation. These criticisms are directed against various sets of formal criteria that have been laid down in attempts to pick out all and only those deductive arguments that are explanations. Hempel and Oppenheim were the first to state such formal criteria for evaluating explanations purporting to be scientific. The formalist criticisms that these criteria are inadequate have taken three lines. One (adopted, e.g., by Davidson and by Kim) argues that, according to these criteria, any event can explain any other event. Another (adopted, e.g., by Eberle, Kaplan and Montague, and by Morgan) argues that there are patterns satisfying the Hempel-Oppenheim criteria which are nonetheless clearly not acceptable as scientific explanations. A third (adopted, e.g., by Omer) argues that when the criteria are tightened up to avoid the counter-examples the second set of critics introduce, then many acceptable scientific explanations do not satisfy the criteria. The upshot of these criticisms is that they are often taken (e.g., by Morgan and by Omer) to imply scepticism as to the viability of the deductive model of explanation. We must therefore examine these criticisms in detail. What we shall do is re-examine the point of the Hempel-Oppenheim discussion, evaluate the Hempel-Oppenheim criteria in the light of this, and then argue that none of the mentioned lines of criticism has touched the thesis that Hempel and Oppenheim were concerned to develop. It will turn out that the more significant of these sorts of criticism (e.g., that of Omer) are plausible only if one is less than careful about the explanation/explaining distinction.

In Part III of their classic study of scientific explanation,[28] Hempel and Oppenheim laid down certain formal criteria for sentences being a scientific explanation. These criteria were set forth as part of an extended argument describing the nature of scientific explanation and defending the idea that explanation so described is the correct account of explanation and understanding in such disciplines as biology and psychology.[29] There are still those who hold that, e.g., psychology aims not at prediction but at understanding, and that there are levels of emergence in biology across which prediction is not possible. As the discussions in Part I, Section 4, and in Part II of their paper make clear, Hempel and Oppenheim are especially concerned to argue against such anti-scientific views about nature of psychology and of biology. Since these views are still about, the Hempel-Oppenheim discussion is still of great importance, still worth defending. Which is what I propose here to do. Specifically, I propose to discuss some recent criticisms of the formal adequacy of the Hempel-Oppenheim criteria. My general line will be that once the Hempel-Oppenheim criteria are placed in the context of their discussion, the criticisms can easily be deflected.

Let me first state the Hempel-Oppenheim criteria.[30] These depend upon certain linguistic notions, with which we should therefore begin. The fundamental linguistic framework is that of a first-order language L with identity. A singular sentence is one in which no variable occurs. An atomic sentence is a singular sentence in which no connective occurs. A basic sentence is an atomic sentence or the negation of an atomic sentence. A generalized sentence consists of at least one quantifier followed by an expression containing no quantifiers. A universal sentence (existential sentence) is one the quantifiers of which are all universal (existential). A sentence is purely general (universal, existential) if it is generalized (universal, existential) and contains no individual constants. A sentence is essentially general (universal, existential) if it is general (universal, existential) and is not equivalent to a singular sentence. There are three criteria used to specify what it is for a sentence to be a "law in L":

(L_1) A sentence in L is a fundamental law in L if it is purely universal and true;

(L_2) A sentence in L is a derivative law in L if (1) it is essentially but not purely universal and (2) is derivable from some set of fundamental laws in L;

(L_3) A sentence is a law in L if it is either a fundamental or a derivative law in L.

Similar criteria are used to specify what it is to be a "theory in L":

(T_1) A sentence in L is a fundamental theory in L if it is purely generalized and true;

(T_2) A sentence in L is a derivative theory in L if (1) it is essentially but not purely generalized and (2) it is derivable from some set of fundamental theories in L;

(T_3) a sentence in L is a theory in L if it is either a fundamental or a derivative theory in L.

Then, a pair of sentences T and C is an explanans for the singular sentence E in L if and only if

(E_1) T is a theory;
(E_2) T is essentially generalized;
(E_3) C is singular and true;
(E_4) T and C jointly entail E;
(E_5) There is some class K of basic sentences of L
such that (a) K entails C,
(b) K does not entail not-T.

(Condition (E_5) may also be stated as saying that there is a T-compatible class of basic sentences K such that K verifies C but does not verify E.) Finally, we have it that a singular sentence E is explainable by T if and only if there is some C such that T, C and E jointly satisfy (E_1)–(E_5).

The criteria (E_1)–(E_5) are conditions for an argument being *objectively* worthy of being an explanation, to recall our discussion from earlier chapters. Hempel and Oppenheim specifically consider relaxing the condition of truth in (L_1) and (T_1) and replacing it by the condition that the theory T mentioned in (E_1) merely be required to be well-supported by available evidence. But, they cogently reason this

... does not appear to accord with sound common usage, which directs us to say that on the basis of the limited initial evidence, the truth of the explanans, and thus the soundness of the explanation, had been quite probable; but that ampler evidence now available made it highly probable that the account in question was not – and never had been – a correct explanation.[30a]

We used this point, it may be recalled, in Section 1.4, when we discussed certain criticisms raised by Feyerabend against the deductive-nomological model of explanation. However, even if it is truth rather than evidence that establishes the *objective* worthiness of an essentially generalized statement for use in an explanation, it is nonetheless equally the case that evidence is relevant to establishing the *subjective* worthiness of such a statement: it is the available evidence that establishes when a generalization is subjectively worthy of being law-asserted and therefore subjectively worthy of being used in explanatory arguments. And *only if an argument is subjectively worthy can it be used in explaining transactions*, only then can it be used *by* someone to explain something *to* someone. We saw this in the preceding section, in our discussion of Scriven. The crucial point is that, simply because it is truth and not evidence that constitutes the condition of objective worthiness, it does not follow that evidence plays no role at all in defining the conditions for the acceptability of an argument as explanatory, or, at least, for its being put to an explanatory use: As we develop our critique of the critics of the deductive-nomological model, this point will have to be kept in mind.

(E_1)–(E_5) have the consequence that an explanation will, roughly, be of the form:

(I)   (i) $(x)(Fx \supset Gx)$
       (ii) Fa explanans
       (iii) Ga explanandum

Assuming (i) and (ii) are true, then (i) is a fundamental law and also the fundamental theory and essentially generalized T, while (ii) is the singular C. (ii) itself is K; it therefore entails C; and neither entails E nor entails the falsehood of (i).

This claim that is implicit in taking (I) as the model of explanation is that we have in (I) an *analysis* of many ordinary causal explanations that are not *grammatically* of the form (I). Thus,

Ga because Fa

or

(II)   *Fa* caused *Ga*

would normally be explanatory in everyday contexts. On the Hempel-Oppenheim account, (I) provides an analysis of (II): in spite of its grammatical form, (II) is implicitly an assertion of two premises plus the claim that another sentences follows logically from these; in other words, (II) is a collapsed argument the structure of which (I) makes explicit.

Davidson has presented an argument from which it can be concluded that if in (II), *Fa* explains *Ga*, then *Fa* explains any other fact, say *Hb*.[31] This argument of Davidson's turns upon accepting two principles of substitution, that of logically equivalent expressions ($Subst_1$) and that of coextensive terms ($Subst_2$). Note first, that

(1)   *Ga*

is logically equivalent to the identity statement

(2) $\hat{x}(x = x \ \& \ Ga) = \hat{x}(x = x)$

Note, second, that

$$\hat{x}(x = x \ \& \ S')$$

is as a matter of fact coextensive with

$$\hat{x}(x = x \ \& \ S'')$$

just in case

$$S' \equiv S''$$

is in fact a true material equivalence. Thus, if we assume (1), then if

Hb

is true then so is

$$Ga \equiv Hb$$

and

(3) $\hat{x}(x = x \ \& \ Ga)$

is coextensive with

(4) $\hat{x}(x = x \ \& \ Hb)$

And, finally, note that

(5) Hb

is logically equivalent to

(6) $\hat{x}(x = x \ \& \ Hb) = \hat{x}(x = x)$

Now substitute in (II). We first replace (1) by (2); this justified by the principle ($Subst_1$). In this then replace (3) by (4); this is justified by ($Subst_2$). Finally, in this replace (6) by (5); this is justified by ($Subst_1$). The result is

(III) Fa caused Hb

Thus, using the innocuous ($Subst_1$) and ($Subst_2$), we seem able to explain any event given we can explain one event. The justification of ($Subst_1$) and ($Subst_2$) is that they preserve truth-value. It seems, therefore, that one must conclude contexts like that of 'Ga' in (II) are not extensional. This challenges the Hempel-Oppenheim assumption of the extensionality of the language L, into which the language of science, and therefore that of explanation, is translatable. But in fact their analysis provides a way in which this challenge can be met. What they propose is that (I) analyzes (II). In effect, then, the series of substitutions, (1) by (2), (3) by (4), (6) by (5), take place in (I). Thus the transition from (II) to (III) is *really* the transition from (I) to

(IV) $(x)(Fx \supset Gx)$
        Fa
        ―――――――
        Hb

The relevant point is that the premisses in (IV) do not entail the conclusion. It is therefore *not an explanation*. It remains true that the substitutions via ($Subst_1$) and ($Subst_2$) have proceeded *salva veritate*. But *validity* has not been preserved. (II) is not merely an assertion the truth of which is to be preserved; upon the Hempel-Oppenheim account, it is a collapsed argument, and hence validity also must be preserved. ($Subst_1$) will preserve validity as well as truth. But ($Subst_2$) preserves only truth. So long as coextensiveness is merely matter of fact, as it is in the case of (3) and (4), the substitution of coextensive singular terms will not preserve validity. We can thus resist the claim that either causal explanations are non-extensional or any event can be explained.

EXPLANATIONS AND EXPLAININGS 193

The logical crux of the reply is the simple one that even though

$$\frac{(x)\ (Fx)}{Fa}$$

is valid, the argument

$$\frac{(x)\ (Fx)}{F\ (\imath y)(Hy)}$$

is never valid. For, the definite description

$$(\imath y)(Hy)$$

might be unsuccessful. The relevant argument which *is* valid is

$$\frac{E!\ (\imath y)(Hy)}{(x)\ (Fx)}$$
$$\frac{}{F\ (\imath y)(Hy)}$$

Similarly, while

$$\frac{(x)\ (Fx \supset Gx)}{Fa}$$
$$\frac{}{Ga}$$

is valid, the argument

$$\frac{(x)\ (Fx \supset Gx)}{F\ (\imath y)(Hy)}$$
$$\frac{}{Ga}$$

is not valid; at least, not unless it is taken to be an *enthymematic form* of

$$a = (\imath y)(Hy)$$
$$(x)\ (Fx \supset Gx)$$
$$\frac{F\ (\imath y)(Hy)}{Ga}$$

CHAPTER 3

Thus, we can take (Subst$_2$) as justifying the move from

Fa caused Ga

to

$F(\imath y)(Hy)$ caused Ga

when

$a = (\imath y)(Hy)$

only if the second of these two causal statements is construed as implicitly of the form

$a = (\imath y)(Hy) \supset [F(\imath y)(Hy)$ caused $Ga]$

Nor is this implausible. Consider a non-causal example. Clearly, the statement *which is reasonably construed as an enthymeme* that

(*)   If Moby Dick is a whale then Moby Dick is a mammal

is justified, while

If the object hunted by Captain Queeg is a whale then Moby Dick is a mammal

is, one would think, unjustified, save as a more extended enthymeme, one whch has not only the suppressed premiss that

All whales are mammals

but also the additional suppressed premiss that

Moby Dick = the object hunted by Captain Queeg

But, one might challenge, is (*) an enthymeme? This would compare to challenging that the *argument* (I) analyzes the *statement* (II). One can go part of the way to meeting this challenge – which is really to translational adequacy – by noting that one would normally respond to

(**)   If Moby Dick is a whale then Moby Dick is not a fish

by pointing out the falsity that

(***)   All whales are fish

That is, one would respond by noting the *unsoundness of an argument*, not by noting the truth-values of antecedent and consequent. Still, it does not follow that (**) is *part of* what is asserted in the case of (**); one could be construed as indicating the inadequacy of the evidence used to back the assertion of (**). However, consider

(****)   If Moby Dick is not a whale, then Moby Dick is not a fish

In this case the appropriate response would seem to be to the effect that: granted that whales are not fish, it simply doesn't follow from not being a whale that Moby Dick is not a fish. Here one challenges the validity of an argument. This indicates (****), and therefore (***) and (*) are not unreasonably construed as implicit arguments, with premises of the sort (**) as part of what is asserted. Thus, what are apparently mere assertions are not unreasonably construed as implicit arguments. And what holds for assertions like (*), seems plausible also for assertions like (II). On the other hand, it is perfectly evident that one can seriously maintain (I) analyzes (II) only if (I) can be defended against the various formalistic objections we are in the process of investigating.

We have not yet escaped, however, from objections of the Davidsonian sort. These can be raised even if we stick to formulations of the sort (I), and avoid ordinary causal statements of the sort (II). The point can be made by quoting a version of Davidson's argument that has been presented by Kim.[32]

Let the law '$(x)\ (Fx \supset Gx)$' subsume two events described by 'c has F' and 'c has G' .... Then, if 'b has H' is any event-describing sentence, the law subsumes the event described by 'b has H' and the event 'c has G'; for the former event is also described by '$(\imath x)\ (x = b\ \&\ c\text{ has }F)$ has H', which, together with the law '$(x)\ (Fx \supset Gx)$', but not by itself, implies 'c has G'. In fact, ... any law that subsumes ... at least one pair of events subsumes every pair.[33]

The argument here is this. We begin with the Hempel-Oppenheim explanation

(V) $(x)\ (Fx \supset Gx)$
       Fc
       ───────────
       Gc

Then it is argued that if this is an explanation then we ought also to treat

(VI) $(x)\ (Fx \supset Gx)$
       $H\ (\imath x)\ (x = b\ \&\ Fc)$
       ───────────
       Gc

as an explanation. (VI) does not meet the Hempel-Oppenheim criteria since the second premiss is not singular: since it contains a definite description it is implicitly of mixed-quantificational structure. Nonetheless, the uniqueness condition for the definite description

$$E! \, (\imath x)(x = b \,\&\, Fc)$$

is logically equivalent to

$$Fc \,\&\, \{(\exists x)(x = b \,\&\, Fc) \vee (\exists x)(x = b \,\&\, (y)(y = b \supset y = x))\}$$

Under the assumption that by the rules of language 'b' names one and only one thing, the second disjunct of the second conjunct is a tautology, and therefore so is the second conjunct. So the uniqueness condition is logically equivalent to

$$Fc$$

*The second premi*ss of (VI) involves a definite description that succeeds in referring as does an ordinary singular term just in case the second premiss of (V) is true. It would seem, then, that if we accept (V) as an explanation we should be prepared to accept the second premiss of (VI) as representing what could be represented by a singular sentence, and that therefore we should accept (VI) as an explanation. But now, given that the definite description is successful, that is, given that 'Fc' is true, then

$$b = (\imath x)(x = b \,\&\, Fc)$$

We may therefore use (Subst$_2$) to move from (VI) to

(VII) $(x)(Fx \supset Gx)$
       $\underline{Hb }$
       $ Gc$

The idea seems to be that if we accept (VI) as an explanation, (Subst$_2$) justifies our accepting this as an explanation. But (VII) is no explanation; it does not satisfy the Hempel-Oppenheim criteria. All that follows here is, what we decided above, that in explanatory contexts (Subst$_2$) is not valid. Still, one wants to say, the fact represented by the second premiss of (VII) is the same fact as is represented by the second premiss of (VI) and that therefore we still have in the move to (VI) a way of arguing one fact can be explained by any fact.

But, *exactly how* does (VI) explain Gc? Clearly, the explanation of Gc should turn upon the premisses jointly but not individually entailing 'Gc'; this part of the point of the deductive-nomological model constructed by Hempel and Oppenheim, when it lays down we need both a C and a T. Now look more closely at the deduction in (VI). What we must do first, as in any argument involving a definite description, is eliminate that term by expansion according to Russell's rules. Thus, the second premiss becomes

$$(\exists x) \{x = b \,\&\, Fc \,\&\, (y)(y = b \,\&\, Fc: \supset y = x)\}$$

so that (VI) is logically equivalent to

(VI*) $(x)(Fx \supset Gx)$
$Fc \,\&\, (\exists x) \{x = b \,\&\, (y)(y = b \,\&\, Fc: \supset y = x)\}$
_____
Gc

In this argument, the second conjunct of the second premiss is irrelevant to the deduction of 'Gc' from the explanans. $(VI) = (VI^*)$ is thus not effectively different from (V). All that it adds to the explanation of Gc is redundant information which it inserts into its second premiss. We are not really explaining Gc in terms of any arbitrarily chosen fact, i.e., deducing it from the law plus information about that arbitrary fact; rather, we are deducing it from the same law and initial condition as in (V), and then adding some inessential (but obfuscating) information to the premisses of the deduction. We therefore need not accept the Kim conclusion that if we can explain one fact, that fact can be explained by any fact.

This rejection of that conclusion turns upon the principle that (VI) = (VI*) is not essentially different from (V), so far as concerns explanatory power. Thus stated, however, the principle is pretty informal. Nor is its rationalization – beyond that of avoiding Kim's absurd conclusion – obvious. What we must do is spell out the rationalization of such principles.

Now, Kim has another argument against the Hempel-Oppenheim criteria.

It is trivial to show that the notion of D-N argument ... cannot coincide with explanation for the following is easily shown: for any law L and a true event-description D^1, there is a true singular sentence D such that 'L, D, therefore D^1' is a D-N argument. Thus, one law would suffice to explain any event you please. As an example: you want to explain why an object b has property F, for any b and F you choose. So you construct the following D-N argument: 'Copper is an electric conductor, b is F or b is nonconducting copper, therefore b is F.'[34]

To explain why b is H for any b and H given the law that $(x)(Fx \supset Gx)$, we use the following sort of argument:

(VIII) $(x)(Fx \supset Gx)$
           $\underline{Hb \vee (Fb \ \& \sim Gb)}$

           Hb

But this will not do what Kim hopes: it does in fact violate the Hempel-Oppenheim criteria. If (E_5) is to be fulfilled we need a K such that

(E_5) (a) K entails $[Hb \vee (Fb \ \& \sim Gb)]$
(E_5) (b) K does not entail Hb
(E_5) (c) K does not entail $(Fb \ \& \sim Gb)$

But these, clearly, cannot simultaneously be satisfied. So (VIII) is not a counter-example to Hempel and Oppenheim.

What, however, really is wrong with (VIII)? Why does it fail to be an explanation? That is, what other than the mere fact that it fails to satisfy certain formal conditions makes it inadequate as an explanation? Or, equivalently, why does failing in just this way to meet the Hempel-Oppenheim criteria mean (VIII) fails to be an explanation? We can see the *point* of excluding (VIII) as an explanation, *and therefore the point of including formal criteria to rule out these cases*, by noting the odd way the deduction goes. If the law is true, then it entails that

$\sim (Fb \ \& \sim Gb)$

i.e., the falsity of the second disjunct of the second premiss. Thus, the second pemiss is true if and only if the conclusion is true. Now, it is this which is the crux of the matter. One cannot say that (VIII) *rationalizes Hb* in terms of *another* fact. And *for this reason* one must reject (VIII) as an explanation of the fact Hb. That is, *for this reason* Hempel and Oppenheim are justified in laying down conditions (E_5) to exclude cases such as these. *It is the notion of explanation that justifies laying down the Hempel-Oppenheim criteria, not conversely.*

An explanation of an individual fact rationalizes that fact in terms of another individual fact. Thus, (I) rationalizes Ga in terms of Fa. Here, 'rationalize' means, roughly, *provide a reason for*, or, in other words, if Fa rationalizes Ga then the occurrence of Fa is the reason why Ga *must have occurred*. But Fa and Ga are *different facts*. That is to say, Fa and Ga are *logically independent* of each other. If we assume the logical framework established by the Hempel-Oppenheim language L, then, if Fa and Ga are logically independent, then we can understand the "must" of the assertion that Ga *must* occur given Fa, only by the assumption that a generality obtains asserting a regular connection between exemplifications of the property F and exemplifications of the property G. *Arguments of the form* (I) *explicate the explanatory "must"*.

The same point can be put in terms of "prediction". Where one predicts, one moves from a statement of individual fact one has verified (in the basic instance) by observation to a statement of *different* fact one has not yet verified (directly by observation). The fact on the basis of which one predicts is different from, logically independent of, the fact predicted. Yet the former provides a reason for expecting the latter; we predict Ga given Fa because the occurrence of the latter means the former must occur. Again, the only reasonable explication of this 'must' is in terms of there being a regular connection between exemplifications of F and exemplifications of G. What follows from this is, of course, the famous Hempel-Oppenheim thesis that explanation and prediction are symmetric: where one explains one could have predicted, and where one predicts one also explains.

It is of the greatest importance to recognize the centrality of the idea of prediction to the concept of scientific explanation. I have already mentioned that in defending the idea of a science of man one defends the idea that psychology aims at prediction, not at empathetic understanding. And further, the notion of prediction cannot be separated from the notion of lawfulness. There may be controversey about how best to understand the notion of scientific law, but philosophers of science otherwise as far apart as Popper and Goodman have recognized it: Popper's notion of "test" presupposes the idea of prediction as does Goodman's idea of "entrenchment". As we have suggested, the notions of explanation, prediction, and lawfulness are tied together through out cognitive interests – idle curiosity and our pragmatic interests in empirical knowledge. These cognitive interests are interests in understanding natural (including social) processes. Whether our interest is idle curiosity or some pragmatic interest, what we want to know about a process is not only what *is* happening but also what *will* happen and what *would* happen *were* certain things now to be other than they are. When we have knowledge like this, then we have attained a *scientific understanding* of the process. Moreover, where we understand, there we have an *explanation*. But such understanding as just described clearly involves *prediction*, that is, *reasoned* prediction, prediction based on *laws* that connect the present state of the system to future states. Moreover, such understanding requires us to be able to make counterfactual assertions, and these, too, presuppose *laws*. Thus, prediction based on laws is inseparable from the idea of explanation, of scientific understanding. It is to defend this idea of science, in which prediction and explanation and lawfulness go hand in hand, that such criteria as (E_1)–(E_5) are laid down.

Now, I believe that this point, that such criteria are not laid down in a vacuum, is often forgotten. Let me illustrate.

(IX) $\quad \dfrac{(x)(Tx)}{Ta}$

does not satisfy the Hempel-Oppenheim criteria. However, it can easily be modified to

(X) $\quad \dfrac{\begin{array}{c}(x)(Tx)\\Tb\end{array}}{Ta}$

This does satisfy those criteria — nothing in the latter requires C to be essential to the deduction of E from T! — yet (X) is intuitively inadequate as an explanation as (IX). It could be ruled out by introducing a new principle, let us call it, following Ackermann,[35] the *principle of redundancy*: one first drops information irrelevant to the deduction, and the result is an explanation only if the formal criteria are then satisfied. Ackermann introduces this principle to rule out a slightly different case, namely,

(XI) $\quad \dfrac{\begin{array}{c}(x)(Tc)\\C\end{array}}{C}$

This is intuitively inadequate, but is in fact ruled out by the Hempel-Oppenheim criteria: (E_5) (a) and (E_5) (b) cannot be jointly satisfied. Ackermann was working with a slightly different set of formal criteria that did not so neatly exclude it. Another example he considers

(XII) $\quad \dfrac{\begin{array}{c}(x)(Mx)\\Ma \supset E\end{array}}{E}$

where 'E' is any sentence other than 'Ma' or 'Ma'. This is intuitively inadequate as an explanation. Ackermann proposes to rule it out by a *principle of triviality*.[36] The Hempel-Oppenheim criteria happen to rule this one out also: If (E_5) (a) is satisfied there is a K that entails C; K entails C if and only if K entails $\sim Ma$ or K entails E; if the former then (E_5) (c) is not satisfied, while if the latter then (E_5) (b) is not satisfied. But whatever the status of (XI) or (XII) on the Hempel-Oppenheim criteria, we still have Ackermann's principles available for our use. The principle of triviality can be used to exclude (VIII), which, in this context, is essentially the same as (XII) – though, of course, (VIII) has already been excluded by the Hempel-Oppenheim criteria themselves. More importantly for present purposes, we could appeal to the principle of redundancy to establish that the explanatory content of (VI)–(VI*) is just the same as that of (V).

Very good. It is indeed nice to have available principles that do the job so neatly of excluding the intuitively inadequate. Yet a sense of unease remains. *These principles seem to be introduced in an ad hoc fashion with the sole purpose of eliminating counter-examples; they seem to have no other rationalization.*[37]

That (IX)–(XII) are inadequate as explanations is, we may all agree, intuitively obvious. But what accounts for these intuitions? As was said in another context, intuitions without concepts are blind. We should rely not only upon our intuitions but upon the *concept of scientific explanation* that underlies those intuitions. Principles should *not* be introduced *simply* to plug up gaps in the Hempel-Oppenheim criteria. Rather, no principle should be introduced unless it can be *argued* that it is a *necessary component* in the concept of scientific explanation that we are proposing and *attempting to justify by reference to our cognitive interests in understanding natural (including social) processes. No principle concerning what is to count as an explanation should be admitted unless it can be justified by reference to the cognitive interests to which we are appealing in order to justify the concept of scientific explanation that we are defending.* Relative to these cognitive interests we can in fact see each of the Hempel-Oppenheim criteria as components in this concept of scientific explanation. *Clearly, what they attempt to do is come to grips with the idea of explanation being also possible prediction.* It is not so easy to see the principles of redundancy and triviality fitting into their places in the concept of scientific explanation. Something like these no doubt has its place. Certainly, they do have their place as rules governing *conversational transactions* in general and therefore *explainings* in particular. But, as we have seen, explanations are one thing, and explainings are another. What is not so easy to see is that the principles Ackermann proposes are rules that apply to explanations rather than explainings. As it stands, they in fact have little direct connection with the concept of explanation: their justification does not derive directly from the concept of scientific explanation itself; instead, their justification seems to be nothing other than that they succeed as *ad hoc* adjustments in excluding (IX)–(XII). And to show how precarious they in fact are, let me introduce another way of excluding arguments of this sort. Each of (IX) – (XII) contains as its theory *T* a generality of the sort

(XIII) $(x)(Mx)$

Generalities of the sort (XIII) do not permit scientific prediction. They do not enable one to predict the exemplification of *one* property *given* the exemplification of a *different* property. Or, what amounts to the same, they do not permit the explanation of *one* individual fact in terms of a *different* individual fact. Generalities of the sort (XIII) do not permit this because they mention *only one* property. This insight enables us to reject as non-explanatory those arguments Ackermann excluded by his principle of redundancy. Only we have done so not by inventing a rule simply to eliminate counter-examples, but by invoking something deriving directly from what always should be our guide here, the concept of scientific explanation itself.

There is but one problem. It is not, however, in the concept of scientific explanation we have invoked but in the Hempel-Oppenheim criteria. According to the criteria (L_1)–(L_3), (T_1)–(T_3), generalities of the sort (XIII) are acceptable as premisses in scientific explanations. That is just the problem. We wish to exclude them, or, at least, arguments involving them. The concern of Hempel and Oppenheim with the symmetry of explanation and prediction[38] establishes their acceptance of the *principle of predictability*, that no argument

$$\frac{T}{C}$$
$$E$$

is an explanation of *E* unless *C* is logically independent of *E* and unless *T* is such that *E* could have been predicted given *C* Moreover, as we said just above, the task of a scientific explanation (of individual facts) is to show why *one* fact *must* have occurred, given some *different* fact has occurred.

Accepting that individual facts are logically independent, this 'must' can be explicated only in terms of deduction from laws. Explanation, then, *is* deduction from laws. But that is what (reasoned) prediction is, too. *So the principle of predictability is not only a necessary but a sufficient condition for a set of statements to be a scientific explanation.* Statements satisfying this condition do everything we expect a scientific explanation *as such* to do. The qualification 'as such' is necessary, since within the class of scientific explanations, some may be better than others; but a scientific explanation that is weaker than another does not cease to be a scientific explanation. And it may be more appropriate in some explaining contexts to use one sort of scientific explanation rather than another; but a scientific explanation does not cease to be such simply because its use may be inappropriate.

The discussion in Hempel and Oppenheim proceeds on the basis of accepting the principle of predictability as definitory of what it is to be a scientific explanation. Or rather, this is how I interpret them. Actually, they lay down four conditions of adequacy which must be satisfied by any formal account of what it is to be an explanation. These conditions are the following.[39]

(A_1) The explanandum must be a logical consequence of the explanans.

The justification here is that, if there were no deduction then "the explanas would not constitute adequate grounds for the explanandum."[40] The sense of 'deduction' is that of formal logic, i.e., that sense of 'deduction' explicated by the formal structure of the language L.

(A_2) The explanans must contain general laws, and these must actually be used for the derivation of the explanandum.

If this is to be applied to the explanation of an individual event then the explanans must also contain a singular statement which is not a law.

(A_3) The explanans must have empirical content.

This actually follows from (A_1), but Hempel and Oppenheim draw attention to it because they were concerned to argue against those who offered – and those who still offer – in the areas of biology and the social sciences explanations that violate this condition. Specifically, they wished to exclude teleological explanations, or at least those teleological explanations that involve the introduction of such non-empirical entities as "entelechies".[41]

(A_4) The sentences constituting the explanans must be true.

Here it is again the idea of adequate grounds that is relevant: "That in a sound explanation, the statements constituting the explanans have to satisfy some condition of factual correctness is obvious."[42] Hempel and Oppenheim extract these adequacy conditions from a couple of examples of scientific explanation.[43] They do not say whether they take these conditions as sufficient. They do suggest, however, a more general condition that can be used as a guide to introducing such further necessary conditions if these turn out to be required. This is the principle of predictability:

... an explanation of a particular event is not fully adequate unless its explanans, if taken account of in time, could have served as a basis for predicting the event in question.[44]

Hempel has continued since the 1948 "Studies in the Logic of Explanation", to take this principle as definitory of scientific explanation. Thus, in 1962 he maintained that an "explanatory account [of a particular event] may be regarded as an argument to the effect that the event to be explained ... *was to be expected* by reason of certain explanatory facts."[45] However, in their 1948 paper Hempel and Oppenheim do not in fact rely any further upon this principle; they simply stick with (A_1)–(A_4). Nor need they have done more, given what they were about. In the context of arguing against those who hold the anti-scientific view that in certain areas there are types of explanation that do not involve laws, the rules (A_1)–(A_4) are quite sufficient. It may be, however, that for other purposes it will be necessary to fall back upon the more general principle of predictability.

After laying down these conditions of adequacy, and after showing their importance in the arguments against those with anti-scientific viewpoints, Hempel and Oppenheim proceed in Part III to lay down formal criteria that meet the conditions of adequacy and pick out all and only those arguments that are genuinely explanatory. Now, it must be most strongly emphasized that *the basic features of the deductive-nomological model of explanation follow directly from the principle of predictability*. It therefore follows that to show conditions (E_1)–(E_5) are inadequate to define the concept of scientific explanation is not to show the inadequacy of the deductive-nomological model. Even if (E_1)–(E_5) are repeatedly tinkered with to accommodate new counter-examples and only successive failures result, there will still be no tendency to show the inadequacy of the deductive-nomological model. Morgan, for one, does not see this.[46] All that follows is that we still do not have a set of formal criteria for picking out those arguments which, according to the principle of predictability, actually are explanations. Nor, one should also note, does the repeated failure to establish such criteria in any way tend to show that the deductive principles one uses in any deductive-nomological argument are other than those of formal logic, i.e., those that can be represented in a language of the sort L that Hempel and Oppenheim use to state their criteria. The deductive-nomological model is, if you wish, scientific commonsense, and failure to capture that commonsense in neat criteria doesn't serve seriously to challenge that commonsense.

In fact, not only should we not be upset by there being counter-examples to (E_1)–(E_5), we should even begin by suspecting these criteria will be violated. What we are proposing to do is apply the principle of predictability to particular arguments. We are going to use this principle as a criterion for determining which arguments can be counted as scientific explanatory: could they be used to make reasoned predictions, based on some generality? If yes, then the argument is explanatory; if no, then it is not. We must be cautious in its use, however. The principle of predictability places constraints upon which arguments can count as explanatory. Here we must note that thse constraints are not just on the generality involved, but on the *argument as a whole*. (E_1)–(E_5) spell out certain *formal constraints* that are implied by the principle of predictability. *But the principle of predictability does not say that the only constraints to be imposed upon arguments are purely formal*. To be explanatory, an argument must be a possible prediction. That is, it must be capable of being *used predictively*. Now, for an explanatory argument to be *actually used in explaining and predicting*, then certain *subjective conditions* must be met: explainor and explainee *must have evidence* that justifies asserting the generality T in the argument and in asserting the singular premiss C. And it may turn out that in one evidential context an argument may be put to a predictive use while in another evidential context it could not be put to a predictive use. Since explanation and (possible) prediction amount to the same (given the concept of scientific explanation that we are justifying by appeal to our cognitive interests in natural processes), in those contexts where the argument could be used predictively it must be counted as explanatory, while in those contexts where it could not be used predictively it must be counted as non-explanatory. *If this sort of case obtains, then the principle of predictability does not propose a formal criterion that categorically excludes or includes the argument as explanatory but rather imposes a contextually variable constraint that counts the argument as explanatory in one context and as non-explanatory in other contexts*. It is precisely because the principle of predictability does not lay down that all constraints must be purely formal that we should be prepared for the consequence that no set of formal criteria on the order of (E_1)–(E_5) will ever pick out all and only explanatory arguments. So far as concerns the principle of predictability, tinkering with such criteria may be doomed to inevitable failure. On the other hand, the principle of predictability *does* impose certain formal constraints of the sort (E_1)–(E_5). Most importantly, it requires that in any explanation there is *deduction from a generality*. The question is not whether *some* conditions can be caputred in a formalism but whether *all* of them can be captured. Whether or not all the conditions are formal we can find out only by examining particular cases, that is, examining the particular case in the light of the principle of predictability and the concept of explanation that it defines. Indeed, it is just this task of applying the principle to particular cases that we are now engaged in. However, no doubt the thrust of the argument is obvious: it is a mistake to suppose that only formal criteria will suffice. I want to emphasize once again, however, that even if such criteria do not suffice, it does not follow that explanations are not deductive, in the formal logical sense required by the L used by Hempel and Oppenheim.

But let us get down, once again, to the particular cases we have been looking at. It is obvious that once the principle of predictability is accepted, no generality of the sort (XIII) could function as the T of an explanation. The principle of predictability has precisely the virtue that Ackermann's principles lack: it is clearly grounded in the concept of scientific explanation that we are attempting to articulate and defend. Its exclusion of such generalities as (XIII) from scientific-explanatory status can now be embodied in an appropriate modification of the Hempel-Oppenheim criteria. For present purposes all we need add is that any scientific-explanatory generality "$(x)(\phi x)$" must be synthetic and such that the propositional function 'ϕx' is a truth-functional compound the parts of which mention (at least) two logically independent properties, the occurrence of each being essential. (ψ does not occur essentially in ϕ if ψ is a well-formed part of ϕ and ϕ is logically equivalent to some well-formed expression ϕ' and ψ is not a well-formed part of ϕ'.) The last qualification, about essentiality, is meant to exclude such cases as

$(x) \quad [Mx \ \& \ (Fx \lor \sim Fx)]$

where the second conjunct is not essential. This criterion can be stated in terms appropriate to L with no difficulty. But I do not intend here to make it any more precise. As it stands it is precise enough for our purposes, namely, the elimination of such generalities as (XIII). This I concieve to be an important point. Formalization is a tool. We should formalize only to the extent it is necessary. Where only a sketch is needed, one should not give more. If this is accepted then we can that Hempel and Oppenheim are not to be criticized for omitting this further criterion for an argument in L to be an explanation. For their purposes, the criteria (E_1)–(E_5) that they laid down were clearly sufficient. Too often when I read the subsequent literature I get the impression Hempel and Oppenheim are being criticized for not laying down formal criteria capable of doing more than they needed their criterion for. I believe this criticism is most unfair. More distressing, however, is the strategy one often sees employed in this literature. A counter-example is dreamed up. Then the formal criteria are tinkered with as to exclude the counter-example. Then another counter-example. Then more tinkering. But throughout mere tinkering, adjustments uninformed by a principled view of scientific explanation, the sort of principled view that so clearly informs the work of Hempel and Oppenheim themselves. I stress this point in the hope of repairing at least some of the embarrassment the profession has recently suffered from its not being recognized.[47] *Such counter-examples as (IX)–(XII) in no way demonstrate or even tend seriously to show the inadequacy of the deductive-nomological account of scientific explanation.* They merely show the criteria laid down by Hempel and Oppenheim are not sufficient to pick out the class of arguments in L which are scientific explanations.

This is one of the main points I wish to make in this section. But I wish also to argue that the idea of a set of purely formal criteria is more or less out of place. In asserting this I do not wish to assert the irrelevancy of the formal element. On the contrary, as we have repeatedly said: it is essential. Deduction from laws is essential to explanation and prediction. We indicated the justification of this above, when we discussed the use of 'must' in explanation and prediction. Since deduction is formal, the formal component cannot be omitted. Yet, I want to argue, neither tinkering with (E_1)–(E_5), nor informed adjustment will yield a set of formal criteria capable of eliminating all arguments which have struck philosophers as being apparent counter-examples to the claim that the deductive-nomological model captures the central idea of scientific explanation.

The main point can be made effectively by considering one particular example of a generality mentioning two, logically independent, properties, viz,

(*) All ravens are black.

To the extent that the properties are logically independent, the principle of predictability can be satisfied. But, in point of fact, it is difficult indeed to notice a raven without noticing its colour. That means that as a *matter of fact* it is difficult to find contexts in which (*) could be used to predict. If we take testing to be crucial to confirmation, then, since prediction is essential to a test, there are few contexts in which (*) could be said to be confirmed. (This is why it is a poor example to use in discussions of confirmation theory.) And, finally, if we accept the principle of predictability then this implies that there are few contexts in which (*) could be used to explain.

(*) cannot be used predictively due to the contingent fact that in general one cannot notice shape without noticing colour. It is not formal criteria but this contingent fact that serves to render non-explanatory in mot evidential contexts arguments that have (*) as the major premiss. This, I think, establishes as clearly as one could want that formal criteria cannot *suffice* to pick out explanatory arguments.

But to be fully accurate, we need here the distinction of the preceding section between *explanations* and *explainings*. Provided (*) is true, then an argument

(**)   All ravens are black
       This is a raven
       ―――――――――――
       This is black

with a true minor premiss meets all the *objective* conditions that must be fulfilled for an argument to be used to make a prediction, and therefore to be, objectively, an explanation. But before such an explanation can be *used* in an *explaining* situation, certain *subjective* conditions must be fulfilled: there must be *evidence* that justifies asserting the minor premiss and law-asserting the generality that is the major premiss. However, if the argument is to be used predictively, then one must not *already know* that the conclusion is true — otherwise one would not be *pre*dicting that the conclusion obtains. Hence, *the evidence for the premisses must be such that it does not justify asserting the conclusion independently of asserting the premisses*. The evidence justifying asserting the premisses of (**) will *ipso facto* justify asserting the conclusion, since the argument is deductively valid; the point is that the evidence justifying asserting the premisses cannot provide grounds *other than* those premisses for asserting the conclusion. Thus, for example, if (**) is to be used predictively, then the evidence justifying the law-assertion of the major premiss *cannot* include the knowledge that the instance

This is a raven and this is black.

is true. For, if one knows that this instance holds, then one *already knows* that the conclusion of (**) is true, and in that case the argument cannot be used to *pre*dict that the conclusion is true. Since an explanatory argument is one that could have been used predictively had one not antecedently known the conclusion to be true, it follows that a similar *epistemic* point holds for explanatory arguments: *when an argument is used in an explaining situation, the available evidence must be such that it justifies asserting the premisses independently of asserting the conclusion.*

The principle of predictability thus not only imposes objective constraints on explanatory arguments; but also places constraints upon the epistemic contexts in which such arguments may be USED *to explain and to predict.*

With this conclusion in mind let us return to considering the various sorts of counter-examples that have been opposed to the Hempel-Oppenheim criteria. In due course we shall recognize in one important case ((XIX) below) among others that evidential context is crucial in determining whether an argument can or cannot be used to explain.

We can begin by noting that using the principle of predictability we have so far excluded (IX)–(XII) but not Kim's counter-example (VI) = (VI*) and (VIII). In particular, even though (XII) is eliminated the seemingly parallel (VIII) is not. Ackermann's principles, however *ad hoc* they seemed to be, at least eliminated all these arguments. In a sense, then, what we want to do is come up with an analogous set of principles, but in our case a set that is informed with the idea of scientific explanation, so that we too will be able to reject these same examples but now on *truly* principled grounds. Our procedure will be to first find grounds for why we should expect (VIII) to be intuitively inadequate and then grounds why we should expect (VI) to be intuitively inadequate.

Consider a counter-example constructed by Eberle, Kaplan and Montague.[48] We suppose there are no mermaids and that the Eifel Tower is a good conductor of heat. Take E to be "The Eifel Tower is a good conductor of heat $[Ce]$" and T to be "All mermaids are good conductors of heat $[(x)(Mx \supset Cx)]$". Then E is explainable by T; for we may take C to be "If the Eifel Tower is not a mermaid then the Eifel Tower is a good conductor of heat $[\sim Me \supset Ce]$" and K to be "The Eifel Tower is a mermaid". The argument purporting to be explanatory is

(XIV) $(x)(Mx \supset Cx)$
     $\sim Me \supset Ce$
     ―――――――
     Ce

The criteria (E_1)–(E_5) are satisfied. Also satisfied is the formal condition the principle of predictability places on the theory T in (XIV). But it is obviously not an explanation. Why not? Here again we can appeal to the principle of predictability: (XIV) is not an argument we could use to yield a prediction that Ce. What we note in the first instance is that the basic problem is *epistemic*: the form of the argument (XIV) renders it epistemically impossible that it be used in order to predict. Both T and C are essential to the deduction. Both are true. The problem is with C. If (XIV) is to be used predictively, then we must know that C is true. This means that we must know either that

'Me' is false and 'Ce' is true

or that

'Me' is true

Since as a matter of fact there are no mermaids, it is the former. But in that case the truth of the minor premiss C is not ascertainable independently of the truth of E which is to be predicted and explained. The logical forms of T and C together with the facts of the case make it impossible that one know C independently of knowing E. The facts and the form of (XIV) mean that the principle of predictability cannot be satisfied. Or rather, *if we understand the principle of predictability to require certain epistemic as well as logical conditions be fulfilled before an argument can be used in order to explain, then, given the facts of the case, (XIV) cannot be used to give a scientific explanation of the fact Ce that appears as its valid and sound conclusion*. The principle of predictability, we have argued, is the basic principle governing the concept of scientific explanation. We have therefore just seen why, in terms of the very concept of scientific explanation, (XIV) cannot be used to give a scientific explanation of why the Eifel Tower is a good conductor of heat. The basic grounds for exclusion turn upon the unsatisfiability of the epistemic constraints imposed by the principle of predictability upon which arguments can be used to give explanations. Once conditions of this sort are introduced, we cannot expect purely formal criteria to suffice. This not to say that adjusting (E_1)–(E_5) cannot bring about an exclusion of (XIV). Indeed, in this case it can be done very easily: The epistemic condition cannot be met because, in (XIV), E entails C; we therefore simply lay on the additional formal condition that E not entail C. For the present counter-example, this patching will do. But it will not do in general: there are, as Morgan has shown,[49] other counter-examples that it fails to exclude. We will not enter into the game of further tinkering, at least not until we know where we are going, not until we are guided by some principle. The point that must be made is that such adjustments will be *ad hoc* unless they are guided by the principle of predictability, where this is understood as requiring epistemic as well as formal conditions for arguments to be usable as explanations.

But now consider the case where we have an argument like (XIV)

$$(x)(Mx \supset Cx)$$
$$\underline{\sim Me \supset Ce}$$
$$Ce$$

but differing from the Eberle, Kaplan and Montague example in that the grounds for asserting the minor premiss are that

'Me' is true

In this case we know C to be true independently of E, which is to be explained. And that means that we *do* predict. In *this* context, then, the principle of predictability *is* satisfied. And so it would seem that in *this* evidential context we *do* have an argument that could be used to give an explanation. At least, there is nothing in the concept of scientific explanation, as defined by the principle of predictability, that would exclude its being so used.

Nonetheless, though it *could* be used to predict and therefore to explain, it would still be very odd to use it. This is because, when we have evidence that

$$Me$$

is true, it would be *odd* to assert the *logically weaker* statement that

$$\sim Me \supset Ce$$

or, what is logically equivalent,

$$Me \vee Ce$$

That is, in the evidential context that we are now considering (and which is not that of the Eberle, Kaplan and Montague example), it would be *less odd* to use

$$(x)(Mx \supset Cx)$$
$$\underline{Me}$$
$$Ce$$

to explain and predict that *Ce* than it would be to use (XIV) for that purpose. *In this context (XIV) is excluded, not because it fails to be explanatory, but because in general we ought to assert as much as our evidence warrants, and therefore, where the evidence permits, the logically stronger over the logically weaker.*

Two conclusions can be drawn. ONE: *Arguments that can be used predictively and to give explanations in one evidential context may turn out to be not usable in this way in different evidential contexts.* We will pursue this theme further in connection with example (XIX) below. TWO: *An argument that satisfies the condition of being a scientific explanation may in certain contexts be excluded from use because there is a BETTER explanatory argument available for use in the same context.* This, too, is a point that we shall pursue in detail in what follows. *Note, however, that the "better" argument need not be better in the sense of having greater explanatory power: it may be that it is to be preferred on grounds other than explanatory power, e.g., by virtue of certain general conventions governing conversational transactions.*

What we have succeeded in doing is eliminate as counterexamples arguments such as (XIV) and (XII) that Ackermann excluded by his principle of triviality. Only, we have done it, once again, not by the *ad hoc* introduction of a rule in response to counter-examples, but by invoking a principle essential to the concept of scientific explanation.

We can now also see more clearly why the deduction in explanatory arguments must have a *law* as a premiss, why explanations have to be of the sort (I)

$$\frac{(x)(Fx \supset Gx)}{Ga}$$

rather than, say,

(I*) $$\frac{Fa \supset Ga}{Ga}$$

For, it might be objected that (I*) just as much as (I) establishes why *Ga must* occur given that *Fa* has occurred. (I*) will not do, since with it prediction is not, on epistemic grounds, possible: one cannot know that the first premiss is true, given the second premiss, unless we antecedently know that the conclusion is true.[49a] The additional premiss that permits us to infer 'Ga' from the logically independent 'Fa' must be such that the singular facts which are the (ultimate) evidence for its truth have to be obtainable prior to the confirmation of the truth of the conclusion. One needs, in other words, not a singular premiss but a generality. Confirming (but not, of course, conclusive) evidence can be obtained for the major premiss of (I) prior to observing that a is G. One will have such evidence just in case one has observed that all *previously noticed* Fs have been Gs.

Let us turn to another counter-example. Ackermann[50] considers the case of

(XV) $(y)(Ty) \vee (x)(Jx \supset Ex)$
$(Tb \vee \sim Ja) \supset Ea$
────────────────────────
Ea

which satisfies (E_1)–(E_5). In order to see the validity of (XV) we need only recognize the following proofs are valid

(XV′) $(y)(Ty)$
$(Tb \vee \sim Ja) \supset Ea$
────────────────────────

Tb	from Premiss 1
$Tb \vee \sim Ja$	Addition
Ea	Modus ponens

and

(XV″) $(x)(Jx \supset Ex)$
$(Tb \vee \sim Ja) \supset Ea$
────────────────────────

$Ja \supset Ea$	from Premiss 1
$(\sim Tb \vee Ea) \& (Ja \vee Ea)$	from Premiss 2
$\sim Ja \supset Ea$	simplification from Line 4
Ea	conjunction of Lines 3 & 5

Ackermann excludes (XV) because (XV′) violates his principle of triviality and (XV″) violates his principle of redundancy. We must deal with it in a different way. (XV) could be used predictively if either (XV′) or (XV″) could be so used. But (XV′) violates the formal condition that the principle of predictability places on the logical form of laws, so it cannot be used to

214   CHAPTER 3

predict nor therefore to explain. On the other hand, in some contexts (XV″) *can* be used to predict. The minor premiss of (XV″) is logically equivalent to

$$(\sim Tb \;\&\; Ja) \lor Ea$$

In contexts where we know this to be true because we know that '*Ea*' is true, then (XV″) could not be used predictively. But in evidential contexts in which we know the minor to be true because we know

$$\sim Tb \;\&\; Ja$$

to be true, then we *can* use (XV″) predictively. In such contexts, then, it would seem that the argument *can* satisfy the cognitive interests (idle curiosity, pragmatic interests) that determine what we want to know as we attempt to understand natural (including social) processes. As we have suggested, whether the interest is idle curiosity or some pragmatic interest, what we want to know about a process is what is happening and what will happen, what did happen previously and what would happen if certain things were to be different. These judgments about what will, did, and would happen all depend upon laws to justify the inferences that that is how things must be given the way that they are. What we expect of an argument, then, is that it be such that it could have been used predictively. *Scientific understanding demands nothing more of an argument.* It is evident that, *relative to this standard*, the argument (XV″) does everything demanded of an argument for it to be explanatory when it is used in an evidential context in which we know its minor premiss to be true because we know its first disjunct is true. Since, in those contexts, it does everything that our cognitive interests demand of an explanatory argument, it should therefore in those contexts be counted as explanatory. Thus, *given that the principle of predictability defines the concept of scientific explanation, we see that Ackermann's* ad hoc *principle of redundancy too hastily dismisses as non-explanatory what should reasonably be counted, in certain contexts, as explanatory arguments.*

Nonetheless, we must also note that in such contexts, the evidence justifies asserting

$$\sim Tb \;\&\; Ja$$

rather than the logically weaker

$$(\sim Tb \;\&\; Ja) \lor Ea$$

In such contexts, then, it would be *odd* to use (XV″) rather than

(XV′″) $(x)(Jx \supset Ex)$
$\underline{\sim Tb \ \& \ Ja}$
Ea

In addition, in such contexts the *topic* is specified to be *Ea*: it is *this* that one is concerned to predict to be explain. Relative to explaining or predicting this fact, the first conjunct '*Tb*' of the minor premiss is *logically irrelevant*. The argument (XV′″), while explanatory, introduces *redundant* information. Assuming that it is a general convention governing all conversational contexts that it is better than to avoid irrelevant redundancies, then in any context in which (XV′″) can be used to predict or explain there is a *better* argument available for use, namely,

(XV″″) $(x)(Jx \supset Ex)$
$\underline{Ja}$
Ea

This is *not* to say that (XV′″) is non-explanatory. Nor, be it noted, is it to say that (XV′″) has less explanatory power than (XV″″). In fact, in terms of predictive, and therefore explanatory, power, it is clear that (XV′″) and (XV″″) have equal explanatory power, that one is neither more nor less imperfect than the other, that *so far as concerns explanatory merit* the one can be used interchangeably with the other. *The difference is not one between an explanation and a non-explanation, nor one of explanatory power, but simply one of conversational niceties: one introduces irrelevancies that the other does not, one is more verbose than the other.*

Ackermann uses his "principle of redundancy" to categorize (XV″) as a non-explanation. Reflecting upon the concept of scientific explanation, we now see that, for certain contexts, this is a false categorization. Ackermann's "principle of redundancy" is therefore inadequate as a principle used to pick out the class of explanatory arguments. But this should not surprise us. Ackermann introduced his principle, not by reflection upon the concept of scientific explanation, but simply in an *ad hoc* way as a response to certain examples that satisfied the Hempel-Oppenheim criteria (E_1)–(E_5) but which, by our intuitions, failed somehow to be usable for explaining something to someone.

As for (XV) itself, it has the major premiss

$$(y)(Ty) \vee (x)(Jx \supset Ex)$$

To know this is true we must have evidence either that (a) '$(y)(Ty)$' is true; or that (b) '$(x)(Jx \supset Ex)$' is true, or, better, is law-assertible. If (a), then a *stronger premiss* is available in the context, viz, '$(y)(Ty)$' itself, and (XV') is to be preferred to (XV). But in any case, the evidence fails to justify the assertion of a generality that is of a logical form capable of law-assertion, so that, in such contexts, no argument is available that could be used to explain or predict. If the evidence available is (b), then again a *stronger premiss* is available in the context, viz, '$(x)(Jx \supset Ex)$' itself, and (XV″) is to be preferred to (XV). But still, in *this* context, (XV) *can* be used predictively, and is therefore explanatory. It is just that another argument, *while having no greater explanatory power*, is to be preferred on other grounds.

Let us now pass on to a slightly different counter-example to the Hempel-Oppenheim criteria that has been presented by Thorpe.[51] He takes '$(x)(Fx)$' to be a true theory and considers

(XVI) $\quad (x)(y)[Fx \vee (Gy \supset Hy)]$
$\quad\quad\quad \underline{(Fb \vee \sim Ga) \supset Ha}$

$\quad\quad\quad Ha$

Here, T is logically equivalent to

$$(x)(Fx) \vee (y)(Gy \supset Hy)$$

and so follows from a true theory just as Ackermann's T does. But the logical equivalence makes it evident that we are in fact back to Ackermann's example (XV), save fro some obvious details that can be handled in obvious ways.

Finally, let us recall Kim's example (VIII)

$\quad\quad\quad (x)(Fx \supset Gx)$
$\quad\quad\quad \underline{Hb \vee (Fb \ \& \sim Gb)}$

$\quad\quad\quad Hb$

In this, we know from the first premiss T that the second disjunct of C is false. So we can know that C is true only if we know that E is true. Again, it is epistemically impossible to use this argument to make a prediction. Thus, the principle of predictability shows us why this example is intuitively inadequate as an explanation.

Now consider this proposed counter-example:

(XVIII) $(x)(Ux \supset Wx)$
$Ua \,\&\, Ka$
───────
$Wa \,\&\, Ka$

To know the truth of C one must know the truth of 'Ua' and of 'Ka' and therefore the truth of one conjunct of E. Hence, with respect to that conjunct, (XVIII) has no explanatory force, while with respect to the other conjunct it does have explanatory force. This is why we are not wholly inclined to reject (XVIII). Yet, if we take it that the event to be explained is a being *both W and K*, then we must conclude that (XVIII) has *not* succeeded in explaining *that* event. Hence, under the assumption that conversations are expected to *stay on topic*, the principle of predictability shows why our intuitions tend to reject this argument as explanatory.

The next supposed counter-example illuminates further some principles introduced above. The example[52] is

(XIX) $(x)\,[(Gx \,\&\, Ix) \supset Fx]$
$(Fb \vee Gb) \,\&\, Ib$
───────
Fb

This example illustrates a point that we have, in effect, already made. As we suggested, (E_1)–(E_5) can be adjusted. No doubt such adjustments could easily exclude (XVIII), for example. What (XIX) illustrates is that such tinkering with the formal constraints cannot in general succeed. (XIX) is one of those argument forms, however, that can sometimes be used to give explanations and sometimes cannot, according to the principle of predictability. (XIX) therefore shows that any attempt to define the class of explanatory arguments by purely formal criteria will inevitably fail.

The validity of (XIX), and the essentiality of T for the deduction, can be easily seen once we note that C is logically equivalent to

$(Fb \,\&\, Ib) \vee (Gb \,\&\, Ib)$

There are two cases. *First*: If we know C to be true through knowing that Gb and that Ib then we can predict E via T. In this case, (XIX) *is explanatory*. *Second*: If we know C is true by Fb and Ib then there is no prediction of Fb and therefore no explanation. Thus, *whether or not (XIX) is an explanation depends upon the evidential context in which it is used*. The mode of verification picks out one of the two disjuncts of C as the one *actually* used in the

218   CHAPTER 3

deduction: the other then becomes irrelevant. In the *first* case, where (XIX) is explanatory, something in C is simply not relevant to the explanation of Fb. This irrelevancy makes it akin to Kim's counter-example (VI*), a point to which we shall return in a moment. But the present point is that criteria that eliminated (XIX) simply on the basis of logical features would eliminate too much; other criteria that included it would include too much. Its formal structure determines only that it can be both explanatory and not explanatory, though not at the same time but depending upon the evidential context in which it might be used. It is the principle of predictability whch makes clear why all this should be so.

We have now to deal with cases such as Kim's (VI*). Here the principle involved will be somewhat different, in a sense not a principle about explanation at all, or, at least, about explanation as such, but rather one about contexts in which explanations are offered, that is, one about explainings, which are transactions, rather than explanations, which are arguments, sequences of sentences. But we should not by now be surprised if contextual elements turned out to be relevant to accounting for our intuitions. For, we have already noted that the evidential context can be relevant to determining whether an argument is or is not an explanation.

Consider the simple case

(XX) $(x)(Ux \supset Wx)$
        Ua
        ─────────────
        $Wa \vee Ka$

Our intuitive response is that (XX) is not so much an explanation of the fact that $Wa \vee Ka$ as it is of the fact that Wa. This is not to say the former could not be explained. After all, that could be done by

(XXI) $(x)[Ux \supset (Wx \vee Kx)]$
        Ua
        ─────────────
        $Wa \vee Ka$

On the other hand, however, (XX) certainly establishes that if Ua occurs then $Wa \vee Ka$ must occur, and equally certainly it could be used to predict that $Wa \vee Ka$, given Ua. It therefore seems to be an explanation. Now, it is this last intuition that I think we should accept. Or we should if we take seriously the connection of explanation and prediction. This means, however, that we should seek an alternative account for our hesitations about (XX), hesitations absent in the case (XXI). It is not hard to find such grounds. For we have already noted that the premisses of (XX) also yield the explanation

(XXIII) $(x)(Ux \supset Wx)$
$\underline{Ua}$
Wa

The difference between (XX) and (XXII) is that the explanandum of the former is *weaker* than what the explanans can be used to justify. '*Wa* ∨ *Ka*' is weaker than '*Wa*' in the sense that its *informative content* is less. The notion of 'Content' can be understood as follows: (a) The (logical) content of a sentence S is the class of non-tautologous sentences entailed by S; (b) if S_1 entails S_2 but S_2 does not entail S_1 then the content of S_1 is greater than the content of S_2; and (c) if S_1 and S_2 are logically independent, if neither entails the other, then their contents are non-comparable. We could spend more time on formal details here, *à la* Carnap, but what has been said suffices for what we are about. The rule upon which our hesitations about (XX) are based seems to be to the effect that an explanation that explains a fact weaker than it could explain is inappropriate. But "inappropriate" to what? What principle lies behind this rule? Is it a principle concerning explanations as such? Or a principle concerning something else? Well, I have already suggested the latter.

Explanations are presented to persons in contexts, explaining contexts. Someone has wondered about an event E, why did E occur? Why was the world this way rather than another? An explanation is offered in order to show why E must have occurred, why the world could not have been otherwise — that is, could not have been otherwise given there obtained some other agreed upon and known state of affairs C. Now, in the case of (XX), this argument would be offered as explanatory presumably when the explainee has wondered about *Wa* ∨ *Ka*. But (XX) could also be used to explain *Wa*. This conclusion has greater content than '*Wa* ∨ *Ka*'; the premisses of (XX) entail a conclusion stronger than the conclusion needed to satisfy the questioner. The premisses thus supply information which is *redundant* relative to the interests of the explainee. The principle thus seems to be that redundant information should not be supplied. Call this the *principle of conversational content*. We may specify this a bit more carefully as follows, by considering its rationale. Other things being equal, if one aims at an end then one ought to use the most efficient means available. If one's aim is to impart knowledge one should therefore say neither too much nor too little.

If one aims to inform someone on some topic one should *stay on the topic*; one should not rehearse everything one knows. In particular, if one is engaged in explaining one should not say more than is necessary to explain whatever it is the explainee has wondered about. So, given that the explainee has wondered about $Wa \lor Ka$, (XX) contains too much information; if the explainer were to use it then more would be said than would be needed efficiently to attain the end of explaining. Given the topic, then, and the principle of conversational content, one should use (XXI) to explain that $Wa \lor Ka$ rather than (XX). Which is not to say, however, that (XX) ceases to be an explanation; all that follows is that it normally would be inappropriate to use (XX) in that explaining context. But there is another part to the principle of conversational content that must also be mentioned. It is that part that says one should not say too little. If one is explaining one is offering an argument that could have been used to predict the fact to be explained. In the statement of the argument one should therefore state exactly what the individual fact is that would have to be used if the explanandum were actually to be predicted. If one is supposed to be able predict E using the law that appears as the law-premiss T, then one should give a C strong enough to yield that conclusion; a weaker $C \lor C'$ will not do.

This principle of conversational content that we have now introduced, and seen the rationale of, is fairly powerful. It rules out redundancies in both explanans and explanandum. Thus, it automatically excludes cases of the sort (X)

$$\frac{\begin{array}{c} T \\ C \end{array}}{E}$$

where T alone entails E, and of the sort (XI)

$$\frac{\begin{array}{c} T \\ C \end{array}}{C}$$

(These turn out to be bad on a number of counts. But that was to be expected.) It also rules out arguments of the sort

(XXIII) $\dfrac{\begin{array}{c}(x)(Fx \supset Gx) \\ Fa \ \& \ Ka\end{array}}{Ga}$

since the second premiss contains information redundant to the conclusion presumably wondered about. But (XXIII) is precisely parallel to Kim's (VI*), so this principle leads to an informed rejection of the latter, not as an explanation, of course, but as an explanation appropriate to an inquiry about the state of affairs represented by its conclusion.

The principle also applies to other cases. Thus, another counter-example to the Hempel-Oppenheim criteria that has been proposed by Kim[53] is

(XXIV) $(x)(Ux \supset Wx)$
$Ua \,\&\, Ka$
───────────
$Wa \,\&\, (Ka \vee Ja)$

This violates both the epistemic condition established by the principle of predictability and also the content condition established by the principle of conversational content. Again, the argument (XIX) previously considered

$(x)\,[(Gx \,\&\, Ix) \supset Fx]$
$(Fb \vee Gb) \,\&\, Ib$
───────────
Fb

which may be explanatory of Fb or not, depending upon the evidential context, turns out, when it *is* explanatory, to violate the principle of conversational content, since it contains *less* information in its second premiss than it should relative to its conclusion. To put the point in more detail; (XIX) is valid, but the only way it could be both sound (i.e., with true premisses) and capable of being used predictively (i.e., capable of being an explanation) is in an evidential context in which the fact that Fb could be explained also by the argument

(XIX*) $(x)\,[(Gx \,\&\, Ix) \supset Fx]$
$Gb \,\&\, Ib$
───────────
Fb

The difficulty is that if (XIX) is used in those contexts where it *is* explanatory, then one is not citing that fact one would, in that context, need if one were to be predicting that Fb. If one uses (XIX) one is being less than candid about what one is using to explain the fact being wondered about. The principle of conversational content therefore directs one to use (XIX*) rather than (XIX). But once again, let it be emphasized that in such contexts (XIX) does not cease to be explanatory; it is only that there are arguments that are conversationally more appropriate to the task of explaining. Akin to (XIX) is a counter-example proposed by Morgan: [54]

(XXV) $(x)\,[(Gx\,\&\,Ix) \supset Fx]$
            $(Fb \vee Gb)\,\&\,Ib$
            ―――――――
            $Fb \vee Ta$

This is similar to (XIX) but violates the principle of conversational content at a second place.

Closely related to the principle of conversational content is another principle worth mentioning. Consider [55]

(XXVI) $(x)\,(Hx \supset Jx)$
            $(Ha \vee Ja)\,\&\,(Ha \vee {\sim}Ja)$
            ―――――――
            Ja

The second premiss is logically equivalent to 'Ja' so that this argument has exactly the content of

(XXVI*) $(x)\,(Hx \supset Jx)$
            Ha
            ―――――――
            Ja

It is therefore not excluded by the principle of conversational content. Nor, in spite of the fact that 'Ja' occurs in the second premiss, is any epistemic condition violated, for C can stilll be known to be true independently of knowing whether 'Ja' is true. This is, of course, because the occurrence of 'Ja' in the second premiss of (XXVI) is an inessential occurrence. In this case we can appeal to the *principle of conversational conciseness* to establish that where one is explaining the fact that Ja it is more appropriate to use (XXVI*) than (XXVI). The same principle justifies its being less appropriate to use [56]

(XXVII) $(x)\,(Px \supset Qx)$
            $(Pa \vee Qa)\,\&\,(Pa \vee {\sim}Qa)$
            ―――――――
            $(Pa \vee Qa)\,\&\,({\sim}Pa \vee Qa)$

than to use the logically equivalent argument

(XXVII*) $(x)(Px \supset Qx)$
Pa
―――――
Qa

Again, the occurrence of 'Qa' in the second premiss of (XXVII) is inessential, as in the occurrence of 'Pa' in its conclusion. In all these cases, the factual verification depends only upon observing the fact represented by the sentences occuring essentially. In terms of predicting one individual fact on the basis of another, (XXVI) is the same as (XXVI*) and (XXVII) is the same as (XXVII*). Which is the same as saying that the informational content, as we defined 'content' above, is the same. As explanations, then, the members of these pairs are not distinguishable, and where things are this way the principle of conversational conciseness bids us use the less verbose form. With this in mind we recognize that the comment of Hempel and Oppenheim about (XXVII*), that

... this reformulation shows that part of the content of the explanandum is contained in the context of the singular component of the explanans and is, in this sense, explained by itself.[57]

is less than felicitous. It confuses the notion of "informational content" with the notion of "contained as a truth-functional component". In terms of the former there is *no* self-explanation. Only the second notion of content creates the illusion of partial self-explanation. But it is an illusion, since the idea of prediction and therefore of explanation requires only the notion of logical content.

There is one important feature that the principle of conversational conciseness secures. The principle of predictability requires that an argument that is to be used in explaining or predicting contexts has to be a valid argument. Otherwise, one has not shown why one fact *must* occur given some other fact has occurred. Thus,

(I) $(x)(Fx \supset Gx)$
Fa
―――――
Ga

will do, where

(I**) Fa
―――
Ga

will not. The same principle of predictability imposes epistemic conditions that require that one premiss must be a generality. (I) will do, where (I*)

$$Fa \supset Ga$$
$$Fa$$
$$\overline{}$$
$$Ga$$

will not. But

(I***) $[Fa \ \& \ (x)(Fx \supset Gx) \supset Ga]$
   $(x)(Fx \supset Gx)$
   Fa
   $\overline{}$
   Ga

would seem to be as good as (I). Why, therefore, do we not use it? The answer is that it *is* explanatory. It is just that it is too verbose. If one explains *Ga* by using (I), then one has shown, using a synthetic generality as an essential premiss, that *Ga* is a necessary consequence of *Fa*. The same is true if one explains *Ga* using (I***). Except, obviously (I***) hardly does the job so expeditiously. The super-major of (I***) adds no new factual information that is not already contained in (I); In fact, being a tautology, the super-major adds *no* information at all. Moreover, the super-major of (I***) is also *logically redundant*, in the sense that once the law-premiss is added to (I**) to give us (I), then one has everything that is *logically* required if we are to *deduce C from E*. That is, we have all that is logically required to obtain the explanatory *must* connection between *C* and *E*. Hence, the principle of conversational conciseness therefore directs us to prefer (I) to (I***).[58]

The principle of conversational conciseness plays another role. It is this principle that justifies the use, in many conversational contexts, of *enthymemes*. An enthymeme is not merely permitted, but is suggested by this principle to be most appropriate when the relevant law-premiss is, in the explaining context, obvious: we need not, and should not, say what everyone knows. It is this principle, then, that justifies the use of (II)

   Fa caused *Ga*

which is the "collapsed" form of the explanatory argument (I)

   $(x)(Fx \supset Gx)$
   Fa
   $\overline{}$
   Ga

But the issues here are difficult. Thus, the omission of the "obvious" may, in spite of the best of intentions, lead to confusion. The law supplied by the explainer to (II) may not be the law supplied by the explainee. And even where such confusion does not occur, simply by virtue of not being articulated the law supplied by the context may be thought by both explainer and explainee to have more explanatory power than it does. This last is in fact one of the major defects of "causal language" in ordinary discourse, and is why science over-rides the principle of conversational conciseness and proceeds as Russell's programme with respect to "causal language" requires, namely, by replacing the language of causes by that of laws and making the lawful component explicit in all explanatory contexts.

We seem now to have handled a good number of questionable cases that have been offered as counter-examples to the Hempel-Oppenheim criteria. The principles we have employed are not of a piece, however. The principle of predictability derives from the concept of scientific explanation. But this is not true of the principle of conversational content. Now, I think that this last is a defect of our analysis, and shall argue so in due course. More specifically, I shall argue that the principle of conversational content should be supplemented by a principle of explanatory content. But I propose to proceed to that discussion in a rather roundabout way in order to make some other relevant points.

Before proceeding, however, we should emphasize once again what is the consequence of violating the principle of conversational content or rather what is *not* the consequence. An argument which satisfies such criteria as (E_1)–(E_5) for being an explanation does not cease to be an explanation simply because it violates the principle of conversational content; all that follows is that it would be inappropriate to use it in contexts of explaining. Thus, in evidential contexts in which (XIX) is an explanation, it does not cease to be explanatory because the principle of conversational content is violated. After all, it continues to do what an explanation is expected to do, namely, show why the explanandum must have occurred, given some other individual fact did occur. All the principle of conversational content establishes is that it is better to offer

(XXVIII) $(x) [(Gx \& Ix) \supset Fx]$
$Gb \& Ib$
___
Fb

as an explanation of Fb than it is to offer (XIX) as an explanation of that fact. By way of analogy, suppose one has evidence that P. In that evidential context, the principle of conversational content establishes it would be inappropriate to assert $P \vee Q$; rather, one should assert P. But that does not make it in any way false or wrong to claim that $P \vee Q$. If one comes in from outdoors and is asked how the weather is, it is inappropriate to say that either it is raining or the sun is shining, even though one has said nothing false. This feature — that it excludes arguments as inappropriate for use in explaining contexts rather than excluding them from the class of explanations — is a feature that the yet-to-be-introduced principle of explanatory power shares with the principle of conversational content. The contrast is, of course, to the principle of predictability which excludes arguments from the category of explanation. We are about to criticize certain accounts of explanation for declaring inappropriate explanations to be non-explanations. These criticisms will not be affected by our later introduction of another principle of content.

We have already commented on the approach to purported counter-examples that proceeds on an *ad hoc* basis to thinker with the Hempel-Oppenheim critieria (E_1)–(E_5) until the counter-examples are eliminated from the class of explanations. In fact, we have seen that this approach is severely defective. While it may succeed in eliminating certain counter-examples, it does so only at the cost of blurring the distinctions upon which we have been insisting.

It was the Eberle, Kaplan and Montague counter-example (XIV) that started off the process of tinkering. One response was that of Kim.[59] He requires us to transform the singular sentence C into the logically equivalent C' in which no atomic components occur inessentially. Here, we assume C is not a logical truth (otherwise no component is essential). Now put C' into conjunctive normal form, and let C_i, $i = 1, 2, \ldots$, be the resulting conjuncts. Kim now lays down criteria additional to (E_1)–(E_5). The first is:[60]

(K_1) For all i, E does not entail C_i

Where C is a logical truth, take any logically equivalent C', and all $\tilde{C}_i$ are logical truths. To cover the case of theoretical explanation Kim adds another condition. In this case no singular C is needed to obtain E, so we take C to be any logically true singular statement and instead of (K_1) use[61]

(K_2) For all i, if C_i is not tautologous then E does not entail C_i.

The Eberle, Kaplan and Montague counter-example (XIV)

$$(x)(Mx \supset Cx)$$
$$\underline{\sim Me \supset Ce}$$
$$Ce$$

violates (K_2). For, the conjunctive normal form of the second premiss consists of the single conjunct

$$Me \vee Ce$$

which is entailed by E, i.e., Ce. The criteria (K_1) and (K_2) also exclude the Thorpe counter-example (XVI)

$$(x)(y)[Fx \vee (Gy \supset Hy)]$$
$$\underline{(Fb \vee \sim Ga) \supset Ha}$$
$$Ha$$

and the equivalent Ackermann counter-example (XV)

$$(y)(Ty) \vee (x)(Jx \supset Ex)$$
$$\underline{(Tb \vee \sim Ja) \supset Ea}$$
$$Ea$$

But the conditions (K_1) and (K_2) are also violated by the example (XIX)

$$(x)[(Gx \& Ix) \supset Fx]$$
$$\underline{(Fb \vee Gb) \& Ib}$$
$$Fb$$

But here we see the confusion. (E_1)–(E_5), (K_1), (K_2) jointly eliminate this argument from the category of explanations. As a categorical judgment this is, as we argued, wrong. In some evidential contexts, (XIX) does do what can be expected of any predictive or explanatory argument. Criteria for being an explanation therefore should not judge it categorically as non-explanatory. Of course, even when the evidential context renders it explanatory there are other explanatory arguments which, according to the principle of conversational content, would be more appropriate. But, to be inappropriate and to be non-explanatory are two things and not one. Kim's conditions turn the former into the latter, and thereby wrongly judge (XIX) as non-explanatory in all contexts. Kim's *unprincipled* tinkering thus leads him to exclude from the category of explanation what are in fact explanations. As for his own counter-example (XXIV)

$$(x)\,(Ux \supset Wx)$$
$$\underline{Ua\ \&\ Ka}$$
$$Wa\ \&\ (Ka \vee Ja)$$

this still violates the criteria so far laid down. Kim suggests that (XXIV) is objectionable because one of the conjuncts of E follows from C alone. In order to eliminate such cases Kim proposes in addition to clauses (a), (b) and (c) of (E_5) the further clause (d). We first put E in conjunctive normal form, and eliminate all inessentially occurring components. Let E_i, $i = 1, 2, \ldots$ be the distinct conjuncts of this normal form. Then the additional clause is

(E_5) (d) For all i, K does not entail E_i

Kim has in fact put his finger on the crucial point. His intuition does not mislead him. But he shows no grasp of the concept or principle that renders his intuition insightful. The principle is that of predictability: insofar as there are components of E that are not logically independent of the components of C, then to that extent prediction is not possible. It is this same idea that underlies the intuitive unsatisfactoriness of (XIV). But precisely because he does *not* grasp this principle Kim constructs criteria which eliminate as explanatory not only these arguments but also what are clearly explanatory, the arguments (XVI) and (XV) in the appropriate evidential contexts.

Even with all of Kim's tinkering, however, there remain unexcluded counter-examples. For example, Morgan's counter-example (XXV)

$$(x)\,[(Gx\ \&\ Ix) \supset Fx]$$
$$\underline{(Fb \vee Gb)\ \&\ Ib}$$
$$Fb \vee Ta$$

remains uneliminated. It was partly to eliminate such examples that Thorpe proposed an alternate set of tinkerings.

It is not part of his (Thorpe's) basic idea to prevent the introduction of irrelevant predicates into either the premisses or the conclusion. It is not his aim to exclude cases such as (XXVII)

$$(x)\,(Px \supset Qx)$$
$$\underline{(Pa \vee Qa)\ \&\ (Pa \vee \sim Qa)}$$
$$(Pa \vee Qa)\ \&\ (\sim Pa \vee Qa)$$

in which, Hempel and Oppenheim claimed, the explandum is partially explained by itself, insofar as part of its "content" is contained in the explanans. Thorpe argues, as we argued, that examples like (XXVII) really are explanatory. Thus, Thorpe's criteria are not without their insights.[62] But the problem is still as with Kim's: it proceeds more on the basis of intuition than on the basis of reasoned principle.

Thorpe begins with a series of definitions.[63]

(a) A class of consistent basic sentences M is said to verify a singular sentence S if and only if M entails S;

(b) An MEC (Minimal Evidence Class) for the singular sentence S is a class of consistent, basic sentences such that this class verifies S and no proper subclass of it will verify S;

(c) $K(M)$ is the conjunction of all the elements in a class of basic sentences M;

(d) A singular sentence S is said to be an MEC-form if S is written as the disjunction of its $K(MEC)$s, and each $K(MEC)$ is called an MEC-disjunct.

A sentence may have more than one MEC. For example,

(α) $(Pa \,\&\, Qa) \supset Sa$

has three MECs

(β) $\{\sim Pa\}, \{\sim Qa\}, \{Sa\}$

Each of (β) entails (α), but no subset of these sets does. In contrast,

$\{\sim Pa, \sim Qa\}$

is not an MEC for (α) since both its non-empty subsets entail (α). Each singular sentence is logically equivalent to its MEC-form. Thus, (α) is logically equivalent to

$\sim Pa \vee \sim Qa \vee Sa$

Thorpe keeps (E_1)–(E_4) as proposed by Hempel and Oppenheim, but replaces (E_5) by

MEC (i): No T-consistent MEC for C can verify E.[64]

This does the same exclusionary job that is done by Kim's (K_2): it serves to eliminate the Eberle, Kaplan and Montague counter-example (XIV), the Ackermann counter-example (XV), and Thorpe's own counter-example

(XVI). It does not, however, eliminate (XIX), and this thus far is in Thorpe's favour. However, he does wish to eliminate such examples as (XX)

$$\frac{(x)(Ux \supset Wx)}{Ua}$$
$$Wa \vee Ka$$

and Morgan's counter-example (XXV), examples which involve "the introduction of irrelevant predicates in the explanandum ... ".[65] This is the wrong track. As we argued, (XX) is adequate as an explanation; it is just that in conversational contexts other explanatory arguments are more appropriate. Nonetheless Thorpe proposes to exclude these as non-explanatory. He proposes to do this by blocking the introduction of "irrelevant" predicates into the explanandum. Defining T_i to be a quantifier-free instantiation of the theory T, he does this[66] with the condition

> MEC (ii): Any MEC of E must be a subclass of the consistent union of some MEC for C with an MEC of some T_i

This rules out any case where the E-sentence has some MEC-disjuncts that are not deducible from the set consisting of T and C. But he has not ruled out such cases as (XVIII):

$$\frac{(x)(Ux \supset Wx)}{Ua \ \& \ Ka}$$
$$Wa \ \& \ Ka$$

or

(XXIX) $(x) [(Rx \ \& \ Sx) \supset (Sx \ \& \ Tx)]$
$\underline{Ra \ \& \ Sa}$

$Sa \ \& \ Ta$

Both these violate (K_2) but not MEC (ii). Thorpe proposes[67] to rule them out by excluding cases in which E and C have a common predicate:

> MEC (iii): If C_x is some MEC of C and E_i an MEC of E, then for any C_x and any E_i such that $K(C_x)$ and T jointly entail $K(E_i)$ it must be the case that C_x and E_i have no members in common.

This eliminates both the above cases. But unfortunately this condition also counts (XIX)

$$(x) \ [(Gx \ \& \ Ix) \supset Fx]$$
$$\underline{(Fb \lor Gb) \ \& \ Ib}$$
$$Fb$$

as non-explanatory, even though in certain evidential contexts this is, as we argued, reasonably understood to be explanatory. The problem here is that of Kim: unprincipled tinkering leads Thorpe to exclude from the class of explanations what are in fact explanations.

Nonetheless, as we also said, Thorpe's criteria are not without their insights. The virtues of his MEC "verificational" approach over the "syntactical" approach of Hempel and Oppenheim and of Kim "lies in the directness of its application, for it does not require the syntactic reconstruction of explanation forms for applying the criteria of acceptability as does the Kim revision."[67a] It is to be preferred, first, because the syntactic approach fails in intuitiveness through its reliance upon the reordering of sentences by syntactical manoeuvers that have no visible justification in terms of scientific explanation. Secondly, logical syntax is not a guide that can reveal the mechanics of D-N explanation and its power to serve as the logical design for hypothesis testing.[67b] The first point is not quite accurate: the MEC-approach is equally syntactical, turning as it does upon the apparatus of normal forms. It is just that the appeals are to different syntactical points. Nonetheless, the second remark is to the point, the approach through MECs does serve to focus attention upon the mechanism of verification. To this extent, it attends to the very points that are emphasized by the principle of pedictability. But instead of being guided by that principle, Thorpe lets himself be guided by counter-examples. As a consequence, he extends his criteria to exclude what should not be excluded.

3.3. EXPLANATIONS AND EXPLANATORY CONTENT

The only person who has approached the issues of the last section in a principled way is Omer. Even he is not wholly free of the "counter-example and tinker" spirit, but because he has reflected seriously upon principles his discussion turns out to be the most insightful. Nonetheless it is wrong. For Omer takes the wrong principle to be basic. What he begins with, and what informs his discussion throughout, is the principle of conversational content.

232   CHAPTER 3

The result is, as we shall see, both interesting and wrongheaded, sufficiently wrongheaded in fact that we shall propose replacing the principle upon which he relies with the somewhat different principle of explanatory content.

Omer organizes his discussion on the basis of two rules, both justified by the principle of conversational content. These are[68]

> (R_1):   In the explanans no sentence which has less informative content in the topic should be given when it is possible to give one which has more informative content,
>
> (R_2):   No sentence in the explanans should have less informative content than the explanandum,

where 'informative' content is to be understood as 'logical content' in the sense we above gave to that notion.[69] These rules are not yet quite correct. As stated, they exclude neither the case (XI)

$$\frac{\begin{array}{c}T\\C\end{array}}{C}$$

nor the case (X)

$$\frac{\begin{array}{c}T\\C\end{array}}{E}$$

where T alone entails E. (R_2) ensures sentences in the explanans cannot be weaker than E, but in the first case here neither C nor T is weaker than the explanandum, nor indeed is the explanans weaker than the explanandum. In order to exclude this case we should thus have in (R_2) both that E should not have less logical content than C and that it should have logical content equal to that of C. In order to exclude the second of these cases this should be further strengthened to apply to *any* sentence of the explanans, T as well as C. (R_2) must therefore hold that E should neither have less logical content than any sentence of the explanans nor have content equal to that of any sentence of the explanans. These emendations have the consequence that (R_2) now reads simply as:[70]

> (R'_2): E should be non-comparable (with respect to logical content) with any sentence of the explanans.

EXPLANATIONS AND EXPLAININGS 233

(R_1) and (R'_2) clearly handle several difficult cases. We have just seen they deal with two of these easily. For another consider (VIII):

$(x)(Fx \supset Gx)$
$Ha \vee (Fa \mathbin{\&} \sim Ga)$

$Ha \vee Ka$

(R_1) requires us to re-write the second premiss as the stronger

Ha

since the first premiss entails the falsity of the second disjunct. The result is

$(x)(Fx \supset Gx)$
Ha

$Ha \vee Ka$

and this is excluded by (R'_2) since E and C are now comparable. The exclusion of (VIII) is reasonable: it could not be used to predict that $Ha \vee Ka$. But, we must recall, Omer excludes it by the principle of content and not by the principle of predictability. This should not really surprise us, since very often it is entailment relations between C and E, i.e. their comparability with respect to content, that prevents an argument from being usable predictively. Yet we should also expect that in the long run the two principles will lead to different results, the one including as explanatory what the other precludes.

This soon becomes apparent. Quickly enough we discover Omer's rules, if used to define what it is to be explanatory, lead to the exclusion as nonexplanatory arguments that clearly, upon the principle of predictability, *are* explanatory. Thus, consider the case (XXIII):

$(x)(Fx \supset Gx)$
$Fa \mathbin{\&} Ka$

Ga

which includes the redundant 'Ka' in the premisses. By the principle of predictability, this argument does everything we might expect of an argument if it is to be explanatory of 'Ga'. Only, it contains the predictively redundant 'Ka'. Which is not to say it is not an explanation. All that follows – by the principle of conversational content – is that its use would be *inappropriate* in contexts of explaining E. Omer, however, judges it precisely in terms of

234   CHAPTER 3

this predictively irrelevant content in the explanans. Starting from the principle of conversational content, he requires this to be dropped. In order to exclude it, he interprets the phrase 'in the topic' in (R_1) to mean that in explaining situations one does not recite all the knowledge one has. Thus, (R_1) is to be understood as [71]

> (R'_1):   In the explanans no sentence not in the topic should be given nor should any sentence which has less informative content in the topic be given when it is possible to give one which has more informative content.

But Omer construes (R'_1) and (R'_2) as defining what it is to be an explanation. He therefore is excluding (XXIII) as non-explanatory contrary to what seems obvious, and in any case can be defended by appeal to the principle of predictability: that it *does* explain Ga. (R'_1) also serves to exclude (XIX)

$$(x)\ [(Gx\ \&\ Ix) \supset Fx]$$
$$(Fb \lor Gb)\ \&\ Ib$$
$$\overline{}$$
$$Fb$$

as non-explanatory in those evidential contexts in which, upon the principle of predictability, it is counted as explanatory. For, in those contexts one knows C is true by virtue of knowing

$$Gb\ \&\ Ib$$

is true. But the latter is more informative in the topic than the second premiss of (XIX). So in those contexts, the argument (XXVIII)

$$(x)\ [(Gx\ \&\ Ix) \supset Fx]$$
$$Gb\ \&\ Ib$$
$$\overline{}$$
$$Fb$$

is counted as explanatory while (XIX) is not. But again, while it may be *inappropriate* to use (XIX) where one could use (XXVIII), it would seem that (XIX), in the correct evidential contexts, *does* predict and explain the fact that b is F. Omer's reliance upon the principle of conversational content leads him to judge as non-explanatory what should not be to judged.

But even in terms of his own principle he does, it seems, go too far. Thus, consider the case where we have (XXVI*)

$$(x)(Hx \supset Jx)$$
$$Ha$$
$$\overline{Ja}$$

which has as a logically equivalent argument (XXVI)

$$(x)(Hx \supset Jx)$$
$$Ha \vee Ja$$
$$Ha \vee \sim Ja$$
$$\overline{Ja}$$

and also has as a logically equivalent argument

(XXX) $(x)(Hx \supset Jx)$
$Ha \vee Sa$
$Ha \vee \sim Sa$
$\overline{Ja}$

Both (XXVI) and (XXX) are, by (R_1), *non-explanatory*,[72] since '$Ha \vee Ja$', '$Ha \vee \sim Ja$', '$Ha \vee Sa$' and '$Ha \vee \sim Sa$' are all of less logical content than 'Ha'. As well, (R_2) excluded (XXVI) from being an explanation, since the second premiss of the explanans has less logical content than the explanandum. On the other hand the *logically equivalent argument* (XXVI*) remains classified as *explanatory*. But this is surely perverse. In order to verify the premisses C in either (XXVI) or (XXX) one needs only to observe exactly the same fact as one must observe to verify the C in (XXVI*). And the E is exactly the same in each case. So far as concerns capacity to be used predictively these arguments are equivalent. Each shows why Ja must occur given that one knows Ha has occurred. As explanations they are, surely, equal; they are surely, equally adequate at yielding that explanatory *must*. The only difference is that (XXVI*) says concisely what the other two say less concisely. They should therefore be classed, not as non-explanations, but as explanations for which there are more appropriate ways of expression. Kim also considers the case of (XXVI). He proposes to exclude it by insisting upon re-writing.[73] We conjoin all singular premisses and eliminate components that occur inessentially; we then put the result in conjunctive normal form, and count each conjunct one singular premiss. Only when this is done are the (E_1)–(E_5), (K_1), (K_2) conditions applied. Upon this basis, (XXVI) is excluded as explanatory while its logically equivalent argument (XXVI*) is reckoned to be explanatory. Again one wants to insist, against Kim as against Omer, that the predictive power of the former with respect to *Ja* is exactly the same as the latter, and that the explanatory power of the one is no less than the explanatory power of the other. Moreover, one wants also to say that the *C* both in (XXVI) and in (XXX) has *exactly the same logical content* as the *C* in (XXVI*). So if Omer's rules judge the latter to be an explanation they should judge the former to be explanatory also, if those rules really turn upon content *alone*. We may contrast these cases to such cases as (XXIII) or (XIX) where the rejection turns upon the explanandum being comparable in content with parts of the explanans. In effect, then, Omer so understands his rules as to eliminate cases which include redundancies in their expression as well as cases which include redundancies in their informative content. Like Hempel, Omer tends to confuse the notion of "informational content" with the notion of "contained as a truth-functional component". His slip here is due, I think, to his not thinking through his basic principle thoroughly enough, or, what is the same, to his falling into the mere tinkering approach.

In any case, Omer takes (R_1') and (R_2') as he takes them, and proceeds to lay down a set of formal conditions an explanation must fulfill. He begins with a notion introduced by Ackermann and Stenner, the notion of the set of *Tc* "ultimate conjuncts" of a set of sentences *T*.[74] Consider, first, the sequence of statements $W_1, \ldots, W_n$: this is a sequence of truth-functional components of *T* just in case *T* can be built up from the sequence by the formation rules of L. Then, a set of ultimate sentential conjuncts *Tc* of *T* is any set whose members are sentences of a longest sequence ($W_1, W_2, \ldots, W_n$) of truth-functional components of *T* such that *T* and the conjunction W_1 & W_2 & ... & W_n of the W_i are provably equivalent. If *T* is a theory then this definition holds with the stipulation that each of the W_i is also a theory. Further, if *T* is a set of sentences, the set of ultimate conjuncts *Tc* of *T* is the union of the set of ultimate sentential conjuncts of each member of *T*. Using this notion we can immediatley restate (R_2') as:[75] If the set *T* of sentences is an explanans for *E* then for any Tc_i ($Tc_i \in Tc$), Tc_i is non-comparable with *E*.

The criteria for being an explanation, as proposed by Omer, are these:[76] A set *T* of supposedly true sentences constitute an explanans for the supposedly true sentence *E* (where *E* is neither tautologous nor self-contradictory) if and only if:

(O_1): Tc entails E;
(O_2): Tc does not entail not-E;
(O_3): Some Tc_i ($Tc_i \in Tc$) is a universal law;
(O_4): No Tc_i ($Tc_i \in Tc$) is comparable with E.

(I shall comment below on the replacement of the Hempel-Oppenheim "true" by "supposedly true", but for the present we may ignore this point.) The condition (O_4) ensures that conjunctions cannot be used to derive E. It further ensures that at least two Tc_i must be used to derive E. Thus, it rules out such cases as (IX):

$$\frac{(x)(Tx)}{Ta}$$

and if E is singular then a singular premiss C must be used to derive it. But (O_1)–(O_4) do not ensure that the law is essential to the deduction of E, nor do they exclude what Omer wants to exclude, namely, cases where there is redundant information in the explanans. Omer therefore adds a further condition:[77]

(O_5): There is no proper subset Tc' of Tc such that Tc' entails E.

Ackermann also once used a condition identical to (O_5) when he laid down a set of criteria for being an explanation.[78]

As we would expect, these are too strong. Thus, (O_4) excludes as non-explanatory our old friend (XIX)

$$\frac{(x)\,[(Gx\ \&\ Ix) \supset Fx]}{(Fb \vee Gb)\ \&\ Ib}$$
$$Fb$$

which we have argued really is explanatory in certain evidential contexts. On the other hand, (O_4) does one thing which *is* required by the principle of predictability to hold of any explanation, "T, C, therefore E": C and E are logically independent. We would expect such confused conditions, however, from anyone who approached scientific explanation, not from the perspective of scientific explanation, i.e., the principle of predictability, but from the viewpoint of conversational niceties, i.e., the principle of conversational content. Even in its own terms, moreover, (O_4) does not do what it is supposed to do.[79] It is supposed to exclude such cases as

$$(x)(Wx \supset Fx)$$
$$\sim(Wa \supset Fa) \vee Gb$$
$$\overline{}$$
$$Gb$$

where we seem to be able to explain a ball's being green by citing the fickleness of women.[80] But (O_4) does not rule out the equally objectionable (cf. (VIII), above)

$$(x)(Wx \supset Fx)$$
$$\sim(Wa \supset Fa) \vee Gb$$
$$\overline{}$$
$$Gb \vee Rb$$

where the same generality explains the ball's being either green or red. But this should not surprise us. What is wrong with these examples, and the second in particular, is not that they present redundant information. Rather, like (XII), of which these are both special cases, and (XIV), what is wrong is that it is impossible for them to be used predictively: we can know the C of the premiss to be true only if we also know the conclusion to be true. That is, what is wrong is determined, not by the principle of conversational content, but by the principle of predictability.

It turns out, though, that (O_5) is the really interesting case. Like (O_4), it ensures that something required by the principle of predictability be obtained. For (O_5), this is its guarantee that a law be essential for the deduction of E. But it also excludes such arguments as (XXIII)

$$(x)(Fx \supset Gx)$$
$$Fa \ \& \ Ka$$
$$\overline{}$$
$$Ga$$

which should, by the principle of predictability, be counted as explanatory. It is again conversational niceties, not the idea of scientific explanation, that is used to justify such exclusions by (O_5).

But Omer pursues this last point further, and argues (O_5) is not yet strong enough. Some redundancy it does remove, but not all. It rules out *elements* in the explanans that carry redundant information. But it does not rule out redundant information *wherever* it occurs. So in the name of the principle of conversational content, he goes after this information also. The cases Omer has in mind are such as the following. If we are asked to explain

$$Wa \vee Ka$$

then we should not use (XX)

$$(x)(Ux \supset Wx)$$
$$\underline{Ua}$$
$$Wa \vee Ka$$

but rather (XXI)

$$(x)[Ux \supset (Wx \vee Kx)]$$
$$\underline{Ua}$$
$$Wa \vee Ka$$

The law premiss of the first argument entails but it not entailed by the law premiss of the second; the latter law is therefore weaker and less informative than the former. The principle of conversational content leads Omer to conclude that where we wish to explain the weaker conclusion we should use the less informative law. On the same grounds he argues [81] that we should use

(XXXI) $(x)(Fx \supset Hx)$
$\underline{Fa}$
Ha

rather than

(XXXII) $(x)(Fx \supset Gx)$
$(x)(Gx \supset Hx)$
$\underline{Fa}$
Ha

Guided by this line of thought Omer replaces his (O_5) by [82]

(O_5'): It is not possible to find sentences $S_1, \ldots, S_n$ ($n \geq 1$) such that for some Tc_i, say $Tc_1, \ldots, Tc_n$:
(a) $(Tc_1 \& \ldots \& Tc_n)$ entails $(S_1 \& \ldots S_n)$;
(b) $(S_1 \& \ldots \& S_n)$ does not entail $(Tc_1 \& \ldots \& Tc_n)$;
(c) on replacing $Tc_1, \ldots, Tc_n$ in Tc by $S_1, \ldots, S_n$, the result, Tc_s, entails E.

Omer recognizes that since Tc entails E but not conversely, the informational content of Tc will always be greater than that of E. His idea is to keep it to a minimum. This is the job of (O_5'). Since in any particular case we take the explanandum to state the topic, what is to be explained, what (O_5') does is guide us in selecting the law most appropriate to that topic. Specifically, it leads us to choose the weakest law possible. Morgan has shown that the result is far too strong: we have ended up with a set of conditions that are automatically violated. Thus, suppose we have an argument explaining E by means of Tc, and satisfying (O_1)–(O_4). Let Tc be $(Tc_1, \ldots, Tc_n)$. Take S_1 to be $Tc_1 \vee E$ and for each $i = 2, \ldots, n$ take $S_i = Tc_i$. Then Tc_1 entails $S_1 = Tc_1 \vee E$ but $S_1 = Tc_1 \vee E$ does not entail Tc_1 (for, if it did then E alone would entail Tc_1, and, contrary to our hypothesis, (O_4) would be violated). We now have a Tc_S; it is $(Tc_1 \vee E, Tc_2, \ldots, Tc_n)$. But if Tc_1 entails E then so does this Tc_S, and (O_5') is violated.[83] Morgan points out that it still will not do even if we make the obvious modification of (O_5') and require that the modified premiss Tc_S must, like Tc, satisfy (O_1)–(O_4). For, now too much is ruled out. Even

$$(x)(Px \supset Qx)$$
$$Pa$$
$$\overline{\quad Qa \quad}$$

turns out to be excluded. Take the law to be Tc_1 and replace it by $S_1 = Tc_1 \vee (x)(Qx)$ or by any $Tc_1 \vee E'$ where E' entails but is not entailed by E. The result is

$$(x)(Px \supset Qx) \vee (x)(Qx)$$
$$Pa$$
$$\overline{\quad Qa \quad}$$

which satisfies (O_1)–(O_4). But since this latter argument exists, we *can* find such sentences that (O_5') requires us to exclude the former argument. Which makes the proposed criteria far too strong.[84] Morgan suggests that the obvious move for Omer to make is to retreat to (O_5). But this is clearly wrongheaded. The obvious move for the tinkerer to make is to try to so redefine the Tcs as to prevent from appearing therein the disjunctions by means of which Morgan so cleverly constructs his counter-examples. What Morgan fails to see when he suggests the retreat to (O_5) is what would, I guess, be missed by anyone who is too bemused by formalism to recognize where the real philosophical issue lies. The important philosophical point is that *Omer's move from* (O_5) *to* (O_5') *was guided by a principle, and that principle stands whatever the formal inadequacies of* (O_5') *turn out to be*.

Omer argued that it follows from the principle of conversational content that we ought to prefer (XXI) to (XX) and (XXXI) to (XXXII). He argued that it follows from his basic principle that if we are to explain an event E then we should prefer that argument which employs as its law-premiss the weakest of possible laws. This derivative principle requires that, given the topic to be explained, an argument involving any but the weakest of laws ceases by definition to be an explanation.

Now, I believe this derivative principle to be wrong and therefore also the principle from which it follows. At least, that latter principle, the principle of conversational content, must be such that other principles can over-rule it. Which, indeed, is what we have been arguing in giving primacy to the principle of predictability. Hence, we may conclude that Omer is wrong in taking the principle of conversational content as basic to his defining of what it is for an argument to be a scientific explanation. What we are talking about here is serious philosophy, far more important an issue than whether Omer has succeeded in constructing a formal definition that picks out as explanatory all and only those arguments that his basic principle leads him to wish to include as explanatory. Not surprisingly, neither (R_1) nor (R_2) nor even the principle of conversational content is mentioned in Morgan's comments on Omer's discussion.

In order to get Omer's derivative principle in focus let us consider a use to which he puts it. It is a use to which it can be put in spite of Morgan's objections to (O_5'). It is in fact a really neat and interesting twist, and is, I think, the most important point to flow from Omer's valuable discussion. What Omer does is use the principle that we choose as explanatory only that argument which uses the weakest law possible, in order to *attack* the deductive-nomological model of explanation.

..., Professor Feyerabend has criticized deductibility as leading to the demand that all successful theories in a given domain must be mutually consistent. Professor Hempel's reply [85] was that one and the same phenomenon may be deductively subsumed under different and logically incompatible laws or theories. But if the fate of the D-N model is the minimal law, Professor Hempel cannot make this reply.[86]

Hempel's reply to Feyerabend consists of pointing out that Fa and Ga can be subsumed under either the theory

(XXXIII) $(x)(Hx \supset \sim Kx)$
$(x)(Fx \supset Hx)$
$(x)(Hx \supset Gx)$

or the theory

(XXXIV) $(x)(Hx \supset \sim Kx)$
$(x)(Fx \supset Kx)$
$(x)(Kx \supset Gx)$

Both these theories explain why Ga must have occurred given that Fa has occurred. But (assuming there are Fs) they are inconsistent. At most one is true, and therefore by the Hempel-Oppenheim condition (T_1), at most one could be a theory. It follows that at most one of the explanatory arguments could – objectively speaking – in fact be an explanation. In practice, however, we may not have evidence available to decide between (XXXIII) and (XXXIV). In practice, then, two inconsistent theories may very well be successful – relative to available evidence – in enabling us to predict and explain Gs on the basis of Fs. Such is Hempel's reply to Feyerabend. If Omer is correct, however, an argument with (XXXIII) as its law premisses does not explain being G in terms of being F. Nor does an argument with (XXXIV) explain being G in terms of being F. The only relevant explanation in this area is one which has

(XXXV) $(x)(Fx \supset Gx)$

as its only law premiss. The existence of inconsistent theories does not imply the existence of inconsistent explanatory arguments. So Hempel's reply is not successful.

Implicit in what we have already said is our defence of Hempel who is, on this point, it seems to me, absolutely correct. The principle of predictability justifies explaining an F's being G using either (XXXIII) or (XXXIV) or (XXXV). Omer excludes the former two using the principle of conversational content. But this latter is, contrary to what Omer believes, not a principle determining what arguments are non-explanatory, but a principle determining when certain explanations are inappropriate to use in explaining. Thus, Omer does not establish that the same facts cannot be explained by inconsistent theories; at best he shows that where the same facts can be explained by inconsistent theories then it is inappropriate to use either and more appropriate to use some weaker theory entailed by both. But *this* doesn't show that Feyerabend is correct.

EXPLANATIONS AND EXPLAININGS 243

Now, we have argued that Omer is wrong to take the principle of conversational content as his basic guide in deciding which arguments are to be counted as explanatory. Rather, it is the principle of predictability which is basic. But even if we accept this, it does not follow that his use of the principle of conversational content in this last case is wholly wrong. Indeed, I think he is correct. If we take him as arguing that certain explanations are more appropriate than others then he is undoubtedly correct. The principle of conversational content *does* imply that (XXI) is more appropriate than (XX), that (XXXI) is more appropriate than (XXXII), and that an explanation using (XXXV) is more appropriate than one using either (XXXIII) or (XXXIV).

This, however, is the real problem. For it seems to me that each of these last three judgments is just wrong. But if this is so, then there is something wrong here with the principle upon which Omer has over-relied but which is here applicable, a principle we have hitherto accepted, namely, the principle of conversational content.

We may begin with the case of (XXXI)

$$(x)(Fx \supset Hx)$$
$$Fa$$
$$\overline{Ha}$$

and (XXXII)

$$(x)(Fx \supset Gx)$$
$$(x)(Gx \supset Hx)$$
$$Fa$$
$$\overline{Ha}$$

We can make this concrete by taking F to be an environmental stimulus and H a behavioural response. Then G represents the internal state of the organism. It is true that (very often) we can predict the response, given the stimulus, without having to know anything about the stuffings of the organism. Yet those stuffings exist and, we also know, are relevant to the stimulus evoking the response it does: they *do* make a difference. Moreover, *it is a better explanation that cites these variables*. Thus, *where the laws mention more of the relevant variables there one has a better explanation.*[87] A set of laws that omits relevant variables is imperfect relative to a set of laws that does not omit those variables. And where an explanation uses less imperfect laws then its *explanatory content* is greater. Since the less imperfect is clearly preferable to the more imperfect when it comes to explanation, we can state the *principle of explanatory content*, that in explaining contexts, if other things are equal, explanations with greater explanatory content (i.e., those based on less imperfect laws) are to be preferred to explanations with less explanatory content. This does not mean weaker explanations are not explanations; it merely implies that where a stronger explanation is available then it is the latter which is appropriate to use in explaining. The reason for this is simply that when someone wonders "Why?" about something one may presume that, other things being equal they desire the *best* explanation available. Of course, sometimes other things aren't equal. Sometimes we do not want the best explanation; a weaker one may suffice for our purposes. Thus, if our interest is pragmatic, and the knowledge is for application, then, if we simply want to be able to manipulate the environment so as to make things H, then we will not wish to be bothered about any variable other than F. Relative to such a pragmatic interest, the stuffings of the organism turn out to be irrelevant, as, indeed, Skinner has said. Even so, we will want as much knowledge as is necessary to achieve what we want to achieve, and anything more imperfect than this will be inappropriate. But our interest is not always pragmatic, sometimes it is a disinterested interest, idle curiosity. Where a pragmatic interest might lead us to settle for an explanation with less explanatory content than some other one also available, idle curiosity will not allow us to do this. In either case, however, it is evident that the principle of explanatory content will direct us to use more contentful arguments where the principle of conversational content directs us to prefer the less contentful.

Previously we simply took over from ordinary contexts the principle of conversational content. Nor should we abandon its use completely. After all, it reckons as *inappropriate* such arguments as Kim's (VI) = (VI*) and such explanations as (XXIII):

$$(x)(Fx \supset Gx)$$
$$Fa \ \& \ Ka$$
$$\overline{Ga}$$

And for good reason: in explaining situations one does not want the explainor to recite every irrelevancy he happens to know. What we see now, however, is that there is a principle of content, that of explanatory content, that takes precedence in contexts of explaining over this principle of conversational content. In such contexts, the aim is to offer an explanation, and, moreover, not to offer just any explanation that is available but to offer the best explanation available, the one the laws of which are least imperfect, have the greatest explanatory content. It is because the principle of explanatory content is a guide about how best to achieve what is aimed at in explaining contexts that it takes precedence over the principle of conversational content. So we are directed to use (XXXII) rather than (XXXI). Only after we have satisfied principles specifically to the point do other considerations such as those of conversational niceties become operative.

We should not think that the principle of explanatory content applies only to the laws involved, as in (XXXI) and (XXXII). It also has consequences with respect to the explanandum. Thus, consider the case where a person wonders "why?" with respect to

(S_1) $Wa \vee Ka$

and we have a choice of explaining to him by using either (XX)

$$(x)(Ux \supset Wx)$$
$$Ua$$
$$\overline{}$$
$$Wa \vee Ka$$

or (XXI)

$$(x)[Ux \supset (Wx \vee Kx)]$$
$$Ua$$
$$\overline{}$$
$$Wa \vee Ka$$

As Omer argues, correctly I think, the principle of conversational content directs us to use the latter. But in fact neither seems to be the best. Rather, what one should use is (XXII)

$$(x)(Ux \supset Wx)$$
$$Ua$$
$$\overline{}$$
$$Wa$$

One should use (XXII) even though its explanandum is

(S_2) Wa

rather than the (S_1) that was originally wondered about. It is the principle of explanatory content that directs us to prefer the third of these explanations. How? Well, if one is wondering about (S_1) one is wondering with respect to a, why does it have a certain disjunctive property. Which is to say that one is wondering with respect to a, why does it have a certain *determinable* property. In contrast to (S_1), which asserts a determinable property holds of a, (S_2) asserts a *determinate* property holds of a. Now, if one predicts something has a determinate property one has thereby predicted it has the various determinable properties connected with that determinate property. The relationship between (S_2) and (S_1) makes this clear; the relationship between being red (determinate) and being coloured (determinable) does it just as well. To predict precisely where Neptune is to be found is to also predict it is to be found somewhere beyond Uranus. Suppose, then, that a person wonders whether there is a planet (call it "Neptune") somewhere beyond Uranus. He will be satisfied if we not only predict that determinable fact but also if we predict a more determinate fact by predicting precisely where Neptune is to be found. Equally, if we explain why Neptune is at some particular point beyond Uranus then we have explained why it is at some point or other beyond Uranus. If a person wonders about (S_1) he will be satisfied if we offer him (XXII) which predicts and explains the more determinate fact (S_2). But however this may be, it does not *follow* from these facts about what *does* satisfy someone that, when it is (S_2) he wonders about, what we *ought* to offer him is an explanation (XXII). In fact, if we take the principle of conversational content as our guide then we should *not* offer him (XXII). Nor (XX). Instead we should offer (XXI). But this consequence of that principle is surely wrong, however. For, the demands of the explanatory context require the principle of explanatory content to take precedence over the principle of conversational content. We ought to offer the wonderer, and he ought to welcome, other things being equal, the explanation which makes the most determinate prediction. The explanation offered should certainly remain *on the topic*, to use Omer's phrase, but *an explanation yielding a more determinate prediction is as much "on the topic" as an explanation yielding a less determinate prediction.* And, in addition, *an explanation yielding a more determinate prediction is better from the viewpoint of explanatory content than one yielding a less determinate prediction.* If the wonder prompting the "Why?" with respect to (S_1) is not simply blindly focused on (S_1) but is a genuine wonder about why things are the way they are *in the topic*, then one will have a better understanding of why things in the topic are as they are the more determinate is the fact in the topic that is predicted and explained. In explaining transactions we want to do the best we can for the explainee, so the principle of explanatory content directs us to prefer (XXII) to either (XX) or (XXI). The demands of the context of explaining require the principle of explanatory content to take precedence over that of conversational content.

We should also remember that explanations — arguments that can be used predictively — are employed in contexts other than that of those transactions we have called explainings. In particular, we should remember that they are used — predictively — in contexts in which they are deliberately put to the test. Suppose we wish to test

(S_3) $(x)(Ux \supset Wx)$

We then find an a which is U. On this basis we make a prediction. This prediction may be either

(S_4) Wa

via (XX) or

(S_5) $Wa \vee Ka$

via (XXI). Since we are testing the law, we are as interested in predictive failures as well as successes. The failure of (S_4) is

(S_6) $\sim Wa$

and that of (S_5) is

$\sim Wa \ \& \sim Ka$

Now, (S_6) clearly refutes (S_3). It is strong enough to do this. In contrast,

(S_7) $\sim Wa \vee Ka$

is not strong enough to refute (S_3). On the other hand, (S_5) is too strong. From the point of view of providing a test, (S_5) contains more content than is needed to refute the law (S_3) in which we are interested. It is harder to discover the facts that would refute a law that predicted Neptune to be somewhere or other beyond Uranus that it would be to discover the facts that would refute a law that predicted Neptune to be at some definite particular place. From the viewpoint of one setting out to test (S_3), the most reasonable choice of explanatory argument to be used to predict thus turns out to be (XXII), where the non-obtaining of what is predicted is neither too weak nor too strong *vis-à-vis* the refutation of the hypothesis that is its law-premiss. Thus, the demands of what it is to put an hypothesis to the test require us once again to prefer (XXII) to (XX). (Which, to repeat, is not to say the latter ceases thereby to be an explanation). Conversational niceties might require in general that we do not assert the weaker (S_5) when we could utter the stronger (S_4). But here it seems that it is not so much this principle at work, determining us to prefer (XXII) to (XX), but rather, again, the specific scientific demands of the context. Moreover, these same demands also exclude preferring (XXI) to either of the other two. For in the context of testing, the focus is upon the law being tested, and to substitute an explanation with a different generality as its law-premiss is simply *not to stay on the topic*.

We have just distinguished two contexts in which explanatory arguments are used. One is the context of explaining and predicting, where the aim is to impart knowledge presumably already at hand. The other is the context of testing where the aim is to acquire knowledge. It is in this latter context, that of research, that conflicting explanations are important, cases like our two mini-theories (XXXIII)

$(x)(Hx \supset {\sim}Kx)$
$(x)(Fx \supset Hx)$
$(x)(Hx \supset Gx)$

and (XXXIV)

$(x)(Hx \supset {\sim}Kx)$
$(x)(Fx \supset Kx)$
$(x)(Kx \supset Gx)$

In terms of explanatory content, ar explanation based on either (XXXI) or (XXXIV) would be more appropriate than one based on (XXXV)

$(x)(Fx \supset Gx)$

that is, more appropriate were the theory to be true. Recall that Hempel and Oppenheim in the conditions (L_1) and (T_1) laid it down that any law appearing in an explanation had to be true. This, clearly, is required by the principle of predictability. Objectively speaking, if some P is not Q then

$(x)(Px \supset Qx)$
Pa
─────────
$\qquad Qa$

gives no reason for supposing that Qa *must* occur given that Pa occurs. If the argument is unsound, with a false generality as its major premiss, then it does *not* guarantee the truth of its conclusion. And in that case it is incapable of being used to give an objectively justified prediction. Neither, therefore, does it explain. Hempel and Oppenheim recognized this when they laid down their principle (R_4). Of course, we may reasonably believe to be true generalities that are objectively false. In such cases we are subjectively, if you wish, justified in accepting as explanatory arguments which use these generalities as premisses. Yet objectively, we are mistaken in accepting such a generality to be a truth and mistaken in accepting an argument based on such a generality to be an explanation.

It is for reasons of this sort that Omer should not have substituted 'supposedly true' for Hempel and Oppenheim's true when he came to state his criterion for being an explanation.[88] But this substitution should not surprise us. Omer is guided by the principle of conversational content, the point of which derives from the niceties of conversation rather than from the objective content of what is to be conveyed. In an explaining transaction, the (honest) explainer does not, in practical terms, distinguish what is objectively true from what he accepts as true. If one's concern is simply to formulate a rule to guide the explainer in efficiently conveying what he accepts to the explainee, then one will come up with the principle of conversational content and in one's instructions the explainer one will tell him to convey in accordance with this rule that which is "supposedly", i.e., what he accepts as, true. But if one is concerned about the *worth* of the explanation — its *objective* worth — then one will insist upon truth, not just supposed truth. The taking of the principle of conversational content as basic derives from a concern about communication, a concern over-indulged at the expense of a concern about the objective worth of the items being communicated, that is, their objective worthiness for being accepted by either explainer or explainee. We are therefore not surprised that Omer has wrongly substituted 'supposedly true' for 'true' when he states the conditions an argument must satisfy in order to be an explanation. But this is to drift from our present concern, namely, the two mini-theories (XXXIII) and (XXXIV).

These two theories are (assuming there are Fs) incompatible. Hence, at most one can be an explanation. They could together be available for use only in a context in which the evidence tended to confirm both and decided between neither. Now, in explaining, while we *ought* to convey only what is objectively true, we *can* convey only what is supposedly true. But then, if the evidence tends to confirm both the contrary theories, it follows neither is worthy of acceptance as true. It would seem, therefore, that if we insist, as we should, that in explaining one attempts to explain on the basis of what one *knows*, then we cannot use either of the contrary theories but should use only what they have in common, namely, (XXXV). Note that this conclusion is not the same as Omer's; it does not tell us that an argument deriving '*Ga*' from '*Fa*' and (XXXIII) is *not* an explanation. For, that argument *is* an explanation *if* that theory is, as a matter of objective fact, true. It is just that even if it is true we do not know that it is and therefore should not use it. Rather, we should use only those arguments that are based on (XXXVI), which (if (XXXIII) is true) is of less explanatory content. However, this consequence holds only if we are thinking of contexts of explaining in the sense of conveying explanatory knowledge already at hand. It does not hold if the aim is to do something other or more than simply conveying such information. In particular, it does not hold *if we shift from the context of explaining to the context of research*. In this context we are still not permitted to offer arguments based on generalities known to be false, but we can offer arguments we do not know to be false. We can offer them provided they are offered as *mere* hypotheses, generalities that might be true but which we do not know to be true, generalities that have yet to be tested.[89] Provided they are taken in this way, we can even make predictions in the context of research that are based on incompatible hypotheses. Tentative explanations, conflicting explanations, are always possible in the context of research. That in fact is they very nature of the game there. Research proceeds on the basis of more or less shrewd guesses and attempts to narrow our range of ignorance by eliminating those guesses which are false until it leaves us with one hypothesis that we may accept as true. That is, accept as true to the extent that the Humean limits of induction permit us to be justified in accepting a generality as true. This last is important. For it means that in principle, each time we use an argument predictively we put its law-premiss to the test. We choose pragmatically where to draw the line between generalities we accept as true and those we are treating as true. Human fallibility makes it impossible to draw the line sharply, once and for all. Once we recognize the contingencies upon which these things depend, we shall not be surprised to find there is no once for all answer whether it is preferable to use in explanatory fashion an argument based upon (XXXVI) or an argument based upon one of the two conflicting theories (XXXIII) or (XXXIV).

Omer attends to matters concerning the linguistic context in which explanations are offered. That, after all, is why he took the principle of conversational content to be basic. But he does not, it now turns out, attend *sufficiently* to such contexts. Had he done so he could not have offered the defence of Feyerabend that he did. For, he would have had to recognize there are contexts, quite good scientific contexts, in which it is perfectly reasonable to advance explanations based on incompatible theories.

3.4. NARRATIVE AND INTEGRATING EXPLANATIONS

The relevance of the context of research/context of explaining distinction is of particular importance with respect to a sort of explanation which is sometimes suggested to be without appeal to laws. Dray has directed our attention to "narrative" explanations in history,[90] and Goudge has suggested that what he calls "narrative" and "integrating" explanations in biology do not conform to the deductive-nomological model.[91] We shall discuss these issues by focusing upon the discussion of Goudge. It will turn out that it is not so much that these explanations fail to conform to the deductive-nomological model as that they fail to conform to cases of this model commonly examined by philosophers of science. Specifically, we can say that what Goudge directs our attention to in narrative and integrating explanations is not the absence of laws but the presence of imperfect laws. But there will remain a feature of narrative explanations that still does not fit the deductive model. But this will not tell against the deductive model, for, we shall argue, this feature does not derive from the context of explanation but rather from the context of research.

Goudge once said of his book, *The Ascent of Life*, that it

> ... presents issues arising from the synthetic theory of biological evolution, the mid-twentieth century descendant of Darwin's theory of 1859. An investigation is made of a number of linguistic and conceptual shifts that have taken place since Darwin, and, along with this, some of the modes of explanation and some metaphysical implications of the theory are explored. It is argued that evolutionary explanations do not all conform to the deductive-nomological model as this is exemplified in the explanations of the physical sciences. There are, in addition, other types, such as "integrating" and "narrative" explanations, which do not involve laws or make positive predictions. It is further argued that among the metaphysical implications of the theory, the ideas of 'direction', 'novelty', 'progress', and 'purpose', suitably specified, have a place.[92]

The book he thus describes received the Governor-General's Non-Fiction Award in 1962. This award was appropriate. The book was both the first detailed exploration of the synthetic theory of biological evolution by a philosopher of science and also an important early critique of the *standard* logical empiricist accounts of science then current, a critique along the same lines as those associated with such names as Kuhn and Lakatos, but (one should add) a critique rather more sober in what it rejected and certainly more sober in its tone. More conversative also, in its willingness to venture into metaphysics. Earlier generations had been willing to explore the metaphysical implications of science – one thinks of Alexander, Russell, Broad – but almost everywhere positivism had triumphed and such explorations were thought to be explorations of the meaningless. Kuhn and Lakatos are hardly yet free of this positivist bias. Goudge's willingness to proceed into these areas now normally left to those fascinated by the pompous pseudo-religiosity of such thinkers (though that word comes to the lips only with difficulty here) as Teihard de Chardin is perhaps due to his being situated in Canada and more particularly in Toronto, where, due largely to the influence of G. S. Brett, the earlier British tradition of metaphysical realism and an interest in science had been swamped neither by positivism nor by "linguistic analysis". Goudge's *Ascent of Life*, then, is a major critique of some aspects at least of the positivist account of science, but, in its caution and sobriety is also a critique of the more radical anti-positivist positions of Kuhn and Lakatos.

In his description of *The Ascent of Life*, Goudge draws our attention to *narrative* and *integrating* explanations and contrasts these to the sort of explanations found in the physical sciences and which philosophers of science have called deductive-nomological and hypothetical-deductive explanations. Two models of explanation have, I believe, dominated the thinking of philosophers of science, and it is these models Goudge has in mind when he argues narrative and integrating explanations do not conform to the standard models. The right of these two models to claim to be standards of excellence need not be disputed even if one disputes their right to dominance, as does Goudge. What must be recognized is that even if there are explanations that do not conform to these models, the deductive model may well still be defensible. Both the standard models derive from classical mechanics. Both are rightly characterized as "deductive-nomological". The first model is the familiar one of *process*. This derives from Newton's explanation of the behaviour of the solar system in terms of the gravitational force-function. We know the system is closed. We have a complete set of relevant variables. And we have a law describing how these variables interact and how the system develops over time. From a knowledge of the values of all the variables at one time, we can deduce, thereby explaining and predicting, the values of the variables at any other time. And we can also deduce what would happen were values of the variables to change or be changed in any way. With such knowledge we know all the past and future behaviour of the system, given we know its present state, and we know all the things that would be occurring were it different from what it is. As we have argued, such knowledge provides us with a model of complete understanding, so far as individual processes are concerned. The other model of explanation deriving from Newton is that of explanation, not of individual facts and processes, but of laws. This model is that of *abstractive axiomatic theories*. In classical mechanics some sorts of systems are described by the gravitational force-function, others by Hooke's Law. Newton's Laws apply indifferently to these systems; they are generic, rather than specific. They apply to all systems of a genus, and succeed in doing this by abstracting generic features common to the laws describing the behaviour of the specific systems. The generic laws of classical mechanics are logically so inter-connected that they can be arranged into a deductive system with Newton's three laws, and the composition law of vector addition of forces, as axioms. These laws place limits upon the form which process laws take in specific systems. They must, for example, conform to the Law of Inertia. This means that when a scientist examines a specific sort of system not previously investigated, then the generic laws of the theory predict that there is a law, for the specific sort of system, there to be discovered, and that it will have such and such a form; the scientist is therefore guided in his research by the theory, and made reasonably certain by it that his search will be successful. Kuhn, in particular, has directed our attention to this feature of theories (or paradigms, as Kuhn calls them).[93] Abstractive axiomatic theories are epistemically desirable by virtue of their capacity to unify our knowledge. And they are desirable from the point of view of the researcher by virtue of their capacity to guide research.

When discussing explanation, philosophers of science, for better or for worse, but certainly at times with misleading emphasis, have tended to stress either laws for which, as in process laws, explanation and prediction amount to the same thing, i.e., the first model, or axiomatic theories, i.e., the second model. But, as we know, there do in fact exist intermediate and complicating cases. Consider a generic law applied to a system of some species within that genus. Unlike a process law, it would not make determinate predictions; it will place generic restrictions upon what will happen in the particular system, but will not predict specifically what will happen. The knowledge that results from applying a generic law to a particular system will be determinable and gappy, imperfect relative to what process knowledge yields. However desirable generic laws are from the view point of explaining laws, they are less desirable than process-type laws from the viewpoint of explaining particular facts. Nonetheless, if laws yielding determinate predictions are not available, one will have to explain particular occurrences as best one can, with gappy generic laws — though of course, scientists will aim to fill those gaps by discovering more specific laws. Another complicating sort of case can be seen when one recognizes that the use of process laws presupposes knowledge of what happens on the boundary, either that the system is closed, where nothing enters the system, or, more generally, the boundary conditions describing what crosses into the system from the outside. To explain what happens upon the boundary one may have to invoke laws describing generically quite different sorts of systems. Thus, in order to explain why heat starts to cross the boundary of a thermodynamic system at a certain time one may cite the behaviour of something other than a system merely thermodynamic, e.g., the behaviour of an experimenter. The explanation of what happens in the thermodynamic system will involve laws explained by two quite different theories. Moreover, explanations of particular facts in terms of laws from generically different areas may use only generic, i.e., gappy or imperfect, laws from those areas.

Through their focussing upon the two dominating models philosophers have tended to ignore explanations in terms of gappy laws and explanations in terms of laws part of quite different theories. It has been a tendency to concentrate on ideal cases at the expense of, if you wish, the empirical facts of explanation, that many explanations fall short of these ideals and are more complicated than the ideals suggest. It was Goudge's virtue to direct our attention to these empirical facts. The narrative and integrative explanations that he has located in the various explanations deriving from the theory of evolution are both types that embody the features we have mentioned that have been ignored in the concentration on the ideal cases.

Goudge draws our attention to the existence of gappy or inexact or quasi-laws,[94] and, in connection with this, the role of statistics.[95] He points out that such laws can have explanatory force, even though such explanations do not conform to any simplified deductive model of explanation,[96] one in which explanation and prediction are symmetric.[97] To see the sense of all this, consider a particular example.[98] Let 'F' denote a certain species of animals, let 'G' denote the property of being in the vicinity of a forest fire, let 'H' denote a degree of stamina above a certain amount, and let 'I' denote the property of outrunning the fire it is in the vicinity of. Then

(G_1) $(x) [(Fx \& Gx) \supset (Hx \equiv Ix)]$

represents that a member of that species will survive the fire if and only if

EXPLANATIONS AND EXPLAININGS 255

it has degree of stamina H. Degrees of stamina are, as 'degree' indicates, ordered, so that to have stamina below a certain degree is to exclude having above that amount. There is, therefore, *one and only one H* such that (G_1) holds. This permits us to form a definite description of H, and refer to it as "the degree of stamina sufficient to permit animals of species F to outrun endangering fires", which we may abbreviate as 'H^*'. Since, *as a matter of fact*,

(G_2) $H = H^*$

(G_1) yields

(G_3) $(x) [(Fx \& Gx) \supset (H^*x \equiv Ix)]$

which represents that an animal of the species F will outrun the fire just in case it has the stamina that is sufficient for outrunning it. (G_3), like (G_2) and (G_1), is also synthetic. Now, we may not know an identificatory hypothesis (G_2); it may be that all we know is

(G_4) $(\exists f)(\mathcal{H}f \& f = H^*)$

where $\mathcal{H}$ is a genus of which the various degrees of stamina are the species. (G_4) is the same as

(G_5) $(\exists ! f) [\mathcal{H}f \& (x) [(Fx \& Gx) \supset (fx \equiv Ix)]]$ [99]

If we know (G_5), then we know (G_3) obtains. But in the absence of an identificatory hypothesis (G_2), we have no knowledge of the sort (G_1). Law (G_3) compared to (G_1) is gappy. Moreover, it cannot be used to predict. Without knowing specifically what H^* is, i.e., without an identificatory hypothesis of the sort (G_2), we cannot determine whether H^* is present in an animal independently of that animal actually surviving a fire; for, in order to determine a is H^* we must deduce

(G_6) H^*a

from

(G_7) Fa
(G_8) Ga

and

(G_9) Ia

With (G_1), we can use the presence of F, G and H to predict I, but when our knowledge is limited to (G_3), we cannot predict I, that an animal will survive a fire. We can, however, use (G_3) to explain, even though we cannot use it to predict. Given (G_3), then (G_7), (G_8), and (G_9) permit us to deduce (G_6). We can now use (G_7), (G_8) and (G_6) as initial conditions, and the law (G_3) to deduce and thereby explain (G_9). Thus, with a gappy law like (G_3) we can sometimes explain *ex post facto* where we cannot predict.[100] There is a way of generating predictions from a law of the sort (G_3), however, by using statistics. If we know (G_3), and observe in a sample of things F & G that 30% of them are I, then we can infer 30% of them are H^*. From this we can generalize to the population, that 30% of things F & G and H^*. We can then predict that of Fs endangered by a forest fire, 30% of them, by virtue of their having sufficient stamina, will survive the fire. Explanation and prediction by means of gappy laws is thus not the neat sort of business it is with the process model. Nor as neat as that suggested by deductive-nomological model as it is usually presented, anyway.[101] One must note, however, that even with such *ex post facto* explanations from gappy laws as sketched above, there is deduction from laws, viz, gappy laws. As part of his general attack on the deductive model, Scriven has used examples similar to that analyzed above to conclude that since explanation is not always symmetrical with prediction, therefore such explanation does not involve deduction from laws.[102] As the above account shows, this argument of Scriven is invalid. All this is by now familiar to us. Like Scriven, Goudge appeals to such examples[103] to argue that the symmetry of explanation and prediction does not everywhere hold, but, unlike Scriven, Goudge does not draw the conclusion that there is explanation without deduction from laws.

Goudge introduces integrating explanations as follows[104] and cites an example:

The combined inquiries of paleontology and historical geology, supplemented by various other special disciplines such as taxonomy, permit the broad outlines of the history of life on the earth to be reconstructed. As a result, what has taken place can be summed up in a general historical statement an excellent example of which is the following:

(E) 'Living organisms are all related to each other and have arisen from a unified and simple ancestry by a long sequence of divergence, differentiation, and complication from that ancestry.'[105]

Two things should be noted about the explanation (E). First, it is a generality: it is a statement about all organisms and, therefore, about all species. Second, it involves mixed quantification: it asserts that for any organism, there are ancestors, and it places certain restrictions upon the nature of these ancestors, asserting that all individuals have a common sort of ancestor, that these earliest ancestors were simple, that the intermediate stages involve ancestors increasing in complexity over time, and that this increasing complexity is a matter of divergence, differentiation, and complication. We can think, very crudely, of this law as having the form

$$(x) \, [Gx \supset (\exists f, y) \, (\mathscr{F}f \, \& \, fy \, \& \, Ryx)]$$

Clearly, such a law will make determinable rather than determinate predictions. It predicts the *sort* of thing to expect, but not specifically what will be. Thus, one can, upon the basis of (E), expect homologies to exist. For example, the wing of the bat, the flipper of the whale, the leg of a horse and the arm of a man are all structurally similar (i.e., homologous) while being functionally quite different. (E) predicts that such homologies will exist but not that any one of these determinate homologies exist.[106] To that extent it predicts, though the predictions are not exact.[107] For that reason, explanations based on (E) are not the deductive-nomological or *process* type, in which determinate predictions are made. Moreover, *ex post facto* explanations are often possible using the law (E).[108] Goudge mentions[109] such biogeographic questions as, Why are tigers found in India but not Africa? Why are armadillos found mainly in South America? Using paleontology and historical geology, one can infer a great deal about the nature of and distributions of fauna in the past; and one can also infer the existence of such things as land bridges and deserts, the facilitators and barriers to migrations. If one now makes some further plausible hypotheses abut initial conditions, such as that of dispersal from a common centre, then one can use (E), the law of phylogenetic descent, to construct explanations answering questions of the mentioned sort. These explanations will turn upon the existence of chains of ancestors. This is the crucial fact (E) introduces. But since (E) makes only determinable statements, the specific details of the processes involved cannot be reconstructed. Here the contrast is to process explanations of the sort Newton provided for the solar system, where, if the present state is known the process law permits us to infer the whole state of the system at any previous time. Moreover, although (E) is generic, it is not here being used to explain laws but to explain individual facts. So the axiomatic model does not apply either.

Narrative explanations are resorted to when one attempts to explain such facts as that amphibians left their aquatic environment to become land-dwellers.[110] The explanation of the mentioned fact is in terms of the amphibians developing limbs and this development, and the consequent ability to live on land, "*seem, paradoxically*, to have been adaptations for remaining in the water"[111] during periods of drought, when it was a favourable adaptation to migrate from an evaporating pool to some other body of water. Gradually, adaptations to non-aquatic food (insects, plants, etc.) would result in land-dwelling fauna.[112] A narrative explanation attempts to answer the question, Why?, asked in the context of an assertion about some event E to the effect that E happened because s, and in particular where the assertion that E because s is in some way paradoxical[113] or at least implausible, judged to be so (and therefore judged as not reasonably to be accepted) on the basis of plausible and normally accepted assumptions.[114] The narrative explanation provides a *sketch* of a *fuller* explanation of how it could be that E happened because of s. It thereby removes the paradoxical aspect of the explanation,[115] showing that the assertion of E because s is, after all, not unreasonable, or, if you wish, intelligible. The narrative explanation does not rely upon any simple generality connecting E and s. This is, of course, precisely where explanations of this sort seem to conflict with the normal deductive-nomological model. The generality that would come to mind as connecting E and s when one asserts that E because s is, "Whenever an event of the same kind as s occurs then an event of the same kind as E occurs". But this generality is, normally, depending upon how one instantiates the free variable 'same kind', either false or tautological.[116] If the former, then it could not be explanatory, and if the latter then it is best construed as a principle of inference rather than a premiss,[117] and it again could not be explanatory. Rather than citing a generality connecting in some simple way the explaining and explained events, the narrative explanation consists of a story which links the two events by describing other intervening events and other relevant constraining or boundary conditions.[118] Now, we must recognize that Newton can do the same, with his process knowledge of the solar system. In the latter case, however, it is a matter of deduction from a law. In the narrative explanation, however, no such deduction occurs. Or, rather, the known laws impose determinable conditions upon the sorts of events that can be supposed to intervene, and, while the events interpolated by the narrative satisfy these determinable conditions, they are in fact given more determinate characterizations than can be deduced from known laws and initial conditions. As Goudge puts the point concerning these gappy or quasi-laws, "The proper conclusion seems to be that law-like statements in selectionist theory serve not as premises for deductions but as components in evidential systems which render *explicanda* intelligible or rationally credible."[119] Thus, narrative explanations make assertions about events that are more specific than those that can be deduced using the known laws. These assertions say more, in other words, or have greater factual content, than can be justified in terms of available lawful knowledge. So narrative explanations inevitably contain a hypothetical element.

> The sequence constructed is, of course, no more than a *possible* explanation at this stage; and it is worth noting how much use is made throughout of the possibility-expressing auxliaries, 'could', 'would', and 'might'.[119a] They indicate the conjectural nature of many of the conditions postulated by the pattern.[120]

In process explanations one can deduce all the specific links of the chain of causes and effects linking two events; in narrative explanations some of the links are literally missing, filled in only by guess-work, or hypothesizing.[121] One introduces narrative explanations in those cases where our knowledge is gappy and imperfect, determinable rather than determinate.[122] The aim of narrative explanations is to answer a question to the effect: how possibly could that have happened?[123] Such questions can, of course, be answered *best* by citing a process law and using it to deduce and explain the whole process. But most often process laws are not available; and certainly this is so in evolutionary theory. One mut answer the question as best one can, then, relying on the imperfect knowledge available. The result might well be a "likely story",[124] the generic framework of which is constituted by the imperfect laws one knows and the specific details of which are but plausible hypotheses.

> The aim is to make the sequence of events intelligible as a relatively independent whole ... the explanatory pattern ... forms a coherent or connected narrative which represents a number of possible events in an intelligible sequence. Hence the pattern is appropriately called a 'narrative explanation'.[125]

The 'intelligibility' here is, one must emphasize, *scientific* intelligibility, intelligibility in terms of laws.

We shall return to this last point directly, but a couple of other points should be made first. These, too, have to do with the inapplicability of the process and axiomatic models.

Narrative explanations are about processes. But in these explanations reference must be made to relevant boundary conditions. For example, in the case we examined, of the development of land-dwelling fauna, reference must be made to climatic conditions, for it is just such conditions which meant there was accessible to the amphibians an environmental niche less hazardous to them than remaining in water.[126] The information for these boundary conditions is derived, not from selectionist theory or from biology, but from other, contributory, sciences such as historical geology. This feature is one narrative explanations share with integrating explanations.[127] Since the sciences involved are generically different, the idea of their being unified axiomatically does not make much sense. For, that would involve more generic laws of which all these sciences would be specific cases.[128] The unity of science is indeed an ideal, but as yet we are far short of this ideal, and the facts Goudge is here directing our attention to make that perfectly clear.[129] Moreover, the relevant laws will often be imperfect. From them we will be able to deduce statements of individual fact, but the characterizations will not be specific but by means of definite descriptions (cf. 'H^*' above). Since the use of definite descriptions presupposes that these descriptions are successful, in Russell's sense, the explanations will presuppose laws in a way an explanation using only specific characterizations will not presuppose laws.[130] Thus, "an explanatory pattern of the sort was are examining is not so much a segment of a separate causal line stretching back into the past, as it is a portion of an intricate network having an enormous number of cross-connections."[131]

Another complicating feature is the historical nature of any evolutionary laws. A law is historical just in case that, in order to predict the future one needs to know not just the present state of the system but also the past states of the system.[132] (Mathematically, process laws of the sort Newton discovered are represented by differential equations, ordinary or (if the objects are fields) partial, while historical processes are represented by integro-differential equations of the sort first investigated by Volterra.)[133] Goudge points out that, at the most generic level of evolutionary theory, there are laws of the sort (E).[134] These laws unify more specific laws, and thereby explain them. (We must, of course, contrast using (E) to explain individual facts, and using (E) to explain laws.) It is at the more specific level that historical features enter.

> They [i.e., historical explanations] are not just incidental items which reflect the theory's 'undeveloped' nature. On the contrary, they are essential to it, and have their foundation in the fact that organisms are literally historical creatures. Their history is built into them. Hence no scientific account of organisms can be satisfactory if it abstracts them from their concrete history.[135]

This feature must not be confused with the fact that, so far as we know, evolution has occurred only once.[136] The laws of evolution have only one instantiation; but others are at least logically possible.[137] Neither historical uniqueness nor historicity precludes lawfulness. But both, and especially the second, considerably complicate the task of explaining, and this makes both the process and axiomatic models far distant ideals rather than useful accounts of the facts of evolutionary explanations.[138] Finally, we should note one further complication that separates biology from physics, where close approximations to the two ideals are in fact to be found. Physics deals with systems where the relevant objects and their relations to other objects are well understood; in other words, in physics we can often reasonably claim to have a complete set of relevant variables and a knowledge of relevant boundary conditions. In biology, this is not so. This is, of course, simply part of the fact that in this area our knowledge is imperfect. As a result of in particular not knowing boundary conditions, biological systems, either organisms or populations, must be treated as "open systems" rather than as "closed systems" of the sort the physicist has managed to discover.[139] Again, the process model, exemplified by Newton's explanation of the solar system, is inapplicable.[140] Goudge summarizes these points, concerning the historicity and openness of biological systems, in his remarks that

... an evolutionist has to deal with organisms and populations undergoing continuous, non-repetitive changes. These organisms and populations are open systems, ceaselessly interacting with the environment and with each other. Hence he can hardly hope to produce anything akin to astronomical forecasts.[141]

Again, however, none of this means that laws are irrelevant in explanation; to the contrary, they remain essential to such explanation. Goudge is not arguing an anti-scientific thesis, one that holds laws are not necessary for there to be explanations. Rather, all he is arguing is that explanations in biology, which *are* scientific, do not conform to the two models to which philosophers of science have fixed their attention.

Now, this last point might be challenged as follows, by reference to narrative explanations. I have said that Goudge is not denying explanations in evolutionary theory aim at scientific intelligibility, intelligibility in terms of laws. On the other hand, I have also pointed out that, according to Goudge, narrative explanations contain specific details which are not established fact but merely plausible hypotheses. These hypotheses are essential to the explanation; if they are deleted, the "likely story" that explains "how possibly" disappears, and with it *that* explanation. It would therefore seem that, for Goudge, the intelligibility he claims biology aims at is not simply scientific. For, if as Aristotle said,[142] poetry deals with the possible while history deals with the actual, then a narrative explanation, containing as it does an essentially hypothetical element, seems more to be poetry than history, at best a guess at an explanation rather than an explanation.[143] Partially, this is a quibble about 'explanation'. One starts to explain E by saying that E because s. This latter, however, turns out to be paradoxical or at least not reasonably acceptable by virtue of its apparently being highly improbable. The narrative explanation, in its hypothetical aspects, does not add further details of the explanation of E in terms of s but it *can* explain why the explanation of E because of s is not straightway to be rejected because of its apparent paradox or improbability. A mere story, based only on guesswork, may suffice to do the latter job of explaining away the apparent unacceptability of the explanation that E because s. Removing apparent paradox, rendering a proposed explanation plausible, can be an important aspect of marshalling evidence in its favour. Such moves may be particularly important in writing aimed at a non-professional audience: it is that sort of audience that would be most likely to bring to bear assumptions, which, until challenged, generate just that air of paradox the narrative explanation aims to remove. But narrative explanations can have a role with respect to a professional audience also, and once we see what this is, then one will not, I think, be any longer inclined to say that the existence of "likely stories" in narrative explanations in biology in any way argues that biology aims at an "intelligibility" of facts which is other than scientific intelligibility, intelligibility in terms of laws. Briefly put, the role of the "mere story" is to propose a research programme, one aimed at discovering laws, or, better, at removing imperfections; the aim of the story, then, is to bring about, in the long run, scientific intelligibility. But the point should be made in more detail.

A narrative explanation, one which renders "E because s" plausible by suggesting certain intermediate steps could possibly account for that change, has the merit of embodying a set of *detailed* suggestions about how the gappy explanation "E because s" could be improved. A narrative explanation contains a set of detailed hypotheses which, if subsequent research verifies them, will provide a less imperfect explanation of E. For the practicing scientist, then, a narrative explanation is both an explanation, but one which is gappy or imperfect, and a sketch of a research programme for eliminating the ackowledged gaps. This last point means, of course, that a narrative explanation, even in those parts where it is guess-work, is not mere poetry, not a mere story. It is, rather, the construction of an imagination subject to the discipline of science. The story-teller works within an established theoretical frame-work, viz, that constituted by the known generic or imperfect laws in the area. Within these guidelines he constructs his story. This story part of the narrative explanation is the proposal of a research programme, and as such has a legitimate place in scientific discourse.[144] For, after all, if our knowledge is gappy then at least idle curiosity and perhaps also our pragmatic interests will urge us to eliminate those gaps, and the means we use so to improve our knowledge is research: research is just that activity by which we reduce the imperfection of our explanations.[145] Typically, those who defend the deductive-nomological model concentrate on the product of research, viz, explanations in the sense of their model, while ignoring the process, and the role theories play in guiding that process.[146] Kuhn, in particular, has emphasized the interaction of theorizing and research practice;[147] essentially, this is what is involved in Kuhn's notion of paradigm.[148] Goudge is bringing out much the same point, with the idea of narrative explanations, that theoretical scientific discourse involves elements directed at explaining, deductively (the generic laws at least implicitly or contextually cited),[149] and elements directed at research (the same laws which limit the possible relevant hypotheses; and further hypotheses which constitute the story part of the explanation, hypotheses the ultimate acceptance or rejection of which will depend upon the outcome of the research).

The conclusion, therefore, is that narrative explanations, in spite of their distinctive features, do not violate the deductive model of explanation. Our thesis, that explanation is by deduction from laws, remains untouched.

3.5. ARE LAWS EVIDENCE FOR, OR PART OF, EXPLANATIONS?

What we must now turn to is the thesis of Scriven that, even where laws are available in explaining contexts, they should not be construed as part of the explanation. Such laws are, as he would put it, more properly thought of as being among the role-justifying conditions for an explanation than as part of the explanation itself.[150]

The basic idea behind the exclusion can be found if we examine in some more detail the descriptive-psychological approach to explanation.[151] We saw that such an approach to explanation is primarily scientific. What we have to do is to apply the typical theses of Scriven about explanation to the science (viz., psychology) which is used to support those theses.

In defining an explanation $_B$, one presupposes certain regularities obtain. These regularities connect stimuli (i.e., kinds of stimuli) to (kinds of) responses. A particular stimulus is an explanation $_B$ just in case it evokes a response of a certain kind, viz, the disappearance of the quizzical look. To say of a particular stimulus that it is an explanation$_B$ and to say that it brought about the disappearance of a quizzical look is to say tautologously the same thing. However, to say of a kind of stimulus that it is an explanation$_B$ is to assert a regularity, a regularity to the effect that stimuli of that kind in general or as a rule produce removals of quizzical looks. The 'in general' and 'as a rule' signal, of course, that the laws involved are imperfect. This becomes evident once we recognize that a stimulus of a given kind will produce that response in one context but not in another. For example, a verbal stimulus of the kind '*Fa*', i.e., a token of that type, may in one context succeed and in another fail in removing a quizzical look directed at '*Ga*'. In the one context, a sentence of the type '*Fa*' succeeds as an explanation$_B$ an in the other context, a sentence of that type fails. What would vary would be the explainee's knowledge. In the context in which sentences of that kind are explanations$_B$ the explainee knows that law that $(x) (Fx \supset Gx)$. In the other context, he does not. There is therefore no simple regularity connecting kinds of verbal stimuli with the kind of response definitory of being an explanation$_B$. On the other hand, there are certain kinds of stimuli which almost invariably remove the quizzical look. For example, if the explainee looks quizzical about why an ink-bottle spilled, one can almost invariably remove that look by producing a token of the sentence-type 'He gave it a sharp knock.' This is because the relevant knowledge (viz, that an ink-bottle will spill if knocked hard enough) is available to most everyone above a certain age and with the relevant cultural background. Where the law is a "truism" it will be available in most contexts. In any case, however, such knowledge is a relevant variable. It is only that under certain conditions it takes on the same value in the great majority of cases ("most everyone believes the truism that ... "). It may only then be safely assumed as constant. If we suppose the knowledge of the law $(x) (Fx \supset Gx)$ is such a case, where it is safe to assume such knowledge has been acquired, there will be a quite uniform regularity connecting verbal stimuli of the type '*Fa*'[152] with responses which are of the type "disappearance of a quizzical look with respect to '*Ga*' ", that is, in those circumstances, it will be a regularity that stimuli of the type '*Fa*' will be explanations$_B$ in respect of '*Ga*'. We have three relevant variables: the stimulus kind, the response kind and the knowledge kind. For the class of normally raised human beings of our culture group, these three variables will be related somewhat along the lines of (4.3) of Section 1.4:

$$(x) [Kx \supset (Cx \equiv (Fx \equiv Gx))]$$

One of these variables, the knowledge kind, we may safely assume takes on a constant value. This yields a regularity connecting the remaining two. This regularity will be of the form of (4.1) of Section 1.4:

$(x) [Kx \supset (Fx \equiv Gx)]$

The latter regularity will be imperfect, but for normal purposes of communication the imperfection will not be relevant. The normal circumstances in which we wish to communicate will be such that the law of the form (4.1) holds, and can be used to successfully predict. Our desire to enegage successfully in the transaction of explaining is a pragmatic interest. Pragmatic interest, we know from our discussion above,[153] in the search for knowledge casts its net less widely than idle curiosity. If the knowledge it secures is sufficient to reveal a means to the end defining the pragmatic interest then the pragmatic interest is satisfied. If the end be communication in normal circumstances, then the pragmatic interest will be satisfied with knowledge of regularities how to successfully provide an explanation$_B$ in normal circumstances. That means it will be satisfied with the imperfect knowledge schematized by (4.1) of Section 1.4. This omits one relevant variable, viz, knowledge, where that knowledge is a matter of "truisms". No need is felt to proceed to the law, which would be schematized by (4.3), in which this variable is explicitly taken into account. No need is felt because the pragmatic interest can be satisfied with less. (Compare our discussion of the poison-non-dying case in Section 2.4.) Thus, where our pragmatic interest has its defining end as the successful engaging in the transaction of explaining, and where the connection between initial conditions and explanandum is a "truism", then our pragmatic interest will be satisfied with knowing the regularity which connects an assertion of the initial conditions *qua* kind of verbal stimulus to the response kind which is the removal of the quizzical look in respect of the explanandum. Call this regularity for short simply the "explaining regularity", that is, the regularity concerning explaining and explanations$_B$. The point to be noticed is that this knowledge of the psychological regularity connecting stimulus kind to response kind, this knowledge of the "explaining regularity" is itself knowledge of a "truism".

Suppose we now ask for explanations of the behaviour of the explainee in the context of the transaction of explaining. We cite conditions (the verbal stimulus which is itself an assertion of initial conditions) and explanandum (the response, disappearance of the quizzical look), and in addition the imperfect law which we have called the "explaining regularity". Our interest in such an explanation of explaining will in general be pragmatic. Certainly, so far as Scriven and Collins are concerned, the interest is pragmatic.[154] That means we will be satisfied with the imperfect explanation based on the "explaining regularity". However, that regularity is a "truism", so it, too, need not be mentioned in order for this psychological explanation to produce a successful explanation $_B$ of some transaction of explaining.

Suppose we are considering explainee A and explainer B and the explaining transaction between them. In terms of schema (4.3) of Section 1.4,

$$(x) [Kx \supset (Cx \equiv (Fx \equiv Gx))]$$

A is of kind K (a standard human), and has appropriate knowledge and so is of kind C. B knows (4.3) obtains. He knows if he makes A to be of kind F then he will thereby have succeeded in making him of kind G. To be of kind G is to have a quizzical look with respect to an explanandum disappear. To be of kind F is to be present with a verbal stimulus consisting of a statement of initial conditions. B presents A with the stimulus, A thereby becoming F, and A becomes G. B has given A a successful explanation $_B$. If the appropriate knowledge which makes A a C is knowledge of a "truism" then B can make use of (4.1) rather than (4.3). Jones now wants to understand this psychological process which is the explaining transaction between A and B. Jones wants to understand, i.e., he wants an explanation. Jones is an explainee with a quizzical look about why A becomes G. That quizzical look is to be removed by Smith. Smith will explain the explaining to Jones by pointing out that A became an F. The connection, which we called the "explaining regularity", between A's being F and A's being G is itself a "truism" so Smith can take it for granted when explaining to Smith. Further, since Jones' interest is pragmatic, this imperfect explanation satisfies him. Thus, in order for Jones to adequately (relative to his interests!) understand the transaction between B and A as one in which an explanation $_B$ occurs, all that has to be pointed out to Jones is that A became an F. Jones is, of course, our philosopher — Scriven or Collins — who approaches explaining as a descriptive psychologist. Thus, where an explaining involves a "truism" they acquire adequate understanding of the transaction as an explaining when it is pointed out to them that B arranges that A becomes F. Which is to say that, B tells A the initial conditions. Thus, to be told that B tells A the initial conditions suffices for Scriven and Collins to come to an adequate understanding of the transaction as an explaining. They therefore define this factor and this factor alone to be "*the* explanation". And so it comes about that statements of initial conditions are, according to Scriven and Collins, fully fledged explanations.[155]

The defender of the deductive model takes a different line, of course. Such a one is not satisfied with a descriptive-psychological concept of understanding. That means he is not terribly worried about the laws in the communication context. Nor is he worried about communication. That means he has no pragmatic interest satisfaction of which at an early point will cut off further investigation. His interest is primarily normative: *ought* one's quizzical look to disappear. He argues as follows. The quizzical look ought to disappear only if the mentioned initial conditions necessitate the explanandum. The explication of the preanalytic notion of "necessitation" requires deduction from a law, which is therefore part of the explanation, if the explanation is to be adequate to its task of justifiably removing one's quizzical look. He can therefore grant everything Scriven and Collins say at the descriptive-psychological level and nonetheless insist on including, as part of the explanation, the law, even in the case where it is a "truism".

Scriven backs up his descriptive-psychological thesis with an argument.[156] Beginning with his descriptive-psychological notion of explanation (=explanation$_B$) he argues somewhat as follows. There is in the first place the verbal stimulus (statement of initial conditions). There is in the second place the relevant response (disappearance of the quizzical look). There is in the third place the context in which the former brings about the latter. Relevant parts of the context are such things as the explainee's state of knowledge, what he knows, the evidence he has, and so forth. In the case where the explanation involves a "truism", that "truism", *qua* known by the explainee, is part of the context. The defender of the deductive model insists on making it part of the explanation, that is, so to speaking transferring it from context to stimulus. Scriven then suggests that the context is infinitely complex, and that an infinite regress of sorts therefore results: we have no criterion in terms of which we can separate part of the infinitely complex context and assert that that part plus the stimulus constitutes "*the* explanation". "*The* explanation" will therefore become infinitely complex. But as that is absurd what we have to do is stop the sequence before it begins and include nothing of the context as part of the explanation.

If his interest were purely scientific, i.e., if he were interested in these behavioural-interaction situations which involve explainings simply as a matter of idle curiosity, then he could not make this sort of argument. From the viewpoint of idle curiosity there is no essential distinction between stimulus, response, and context. There are only relevant variables interacting in various ways. The stimulus/response/context distinctions enter only with pragmatic interests: context is constituted by the relevant variables values of which we take for granted ("normal conditions"); response is constituted by the relevant variables certain values of which we want to bring about; and stimulus is constituted by the relevant variables whose values we change as a means to our ends, viz, bringing about the appropriate response. (By the way, this is why the phrase "S-R theory" is doubly misleading. It is not a theory. And the motive suggested by the use of 'S-R' is that of a pragmatic interest, rather than idle curiosity.) That Scriven's investigations are guided by his pragmatic interest is thus further confirmed.

We see Scriven's argument breaks down. To suggest the context is infinitely complex is to suggest there is an infinite number of relevant variables. There is no reason to believe that this is true. On the other hand, the number of variables is undoubtedly quite large, and certainly includes more than the defender of the deductive model would want to include in the explanation. An argument along Scriven's lines can still be mounted, therefore.

In particular, it can be mounted, as Scriven sees,[157] in terms of evidence. If the explanation is to be acceptable, evidence must be available in the context to render it acceptable to the explainee. (Scriven calls such matters truth-justifying grounds, as we know.[158]) But that evidence itself requires further evidence. And so on, *ad infinitum*. I do not think the last follows, but certainly the chain of evidence may be quite long and complex. If, however, we include the first-level evidence, then there is no reason for not also including second-level evidence (the evidence for the first level evidence), and so on. We cut off the long, if not infinite regress, by insisting no part of the evidence is part of the explanation. As Scriven would put it, we distinguish an explanation from its truth-justifying grounds, and exclude any of the latter from the former in order to prevent a regress arising.

Scriven points out that the defender of the deductive model makes just this move which Scriven also makes.[159] He concludes the defender makes it for the same reason Scriven himself makes it. For Scriven the distinction explanation/(evidential) context is the same as the distinction stimulus/context which in turn is a function of Scriven's pragmatic interests in laws of communication as providing knowledge of means to ends. It is in order to keep that explanation/context distinction that Scriven separates the explanation from the truth-justifying grounds of the context. The defender of the deductive model does not (*qua* defender) have such interests. If he is to maintain Scriven's explanation/context distinction he must do so for different reasons. The defender draws a distinction between what makes an explanation objectively worthy of acceptance and what makes an explanation subjectively worthy of acceptance. An explanation is objectively worthy if its premisses are true (and entail the explanandum). The premisses are true, or false, independently of whatever the evidence may be that one has tending to confirm that truth or falsity. Whether an explanation is objectively worthy is in this sense — objective. And whether available evidence is good evidence is an objective matter. But *what* evidence is available is a psychological-sociological question, and therefore not objective. So whether an explanation is subjectively worthy is in this sense not objective— or rather, if you please, is subjective. It is the normative interests of the defender of the deductive model that lead him to make the explanation/(evidential) context distinction: the explanation is that in terms of which objective worthiness can be defined. To introduce evidence available into the explanation is to introduce a subjective element which would make it impossible to lay down a norm in terms of which the explanation could be evaluated as worthy or unworthy of acceptance for any time or place, irrespective of states of knowledge or ignorance. Introducing evidence into the explanation would enable one to define worthiness only relative to such as happened to have just that evidence. There would be no objective, timeless, norm of acceptability. In the interests of effectively defining such an objective norm, the defender of the deductive model separates the explanation from evidence which supports it. Thus, the defender of the deductive model agrees with Scriven on separating explanation and evidence, but not for the reasons Scriven suggests.

It is important to see this because it is by virtue of an analogy between the case of "truth-justifying" grounds and "role-justifying" grounds that Scriven proceeds to build his case that laws are not part of explanations.

> Hempel's notion [i.e., the deductive model] correctly avoids the ... temptation of saying that an explanation is incomplete when it does not include its *second-level* grounds [truth-justifying] (and then the third-level ...). But is it not the case that in requiring the explanation to include not merely the facts which are produced [by stating what we have called the initial conditions] but also the *first-level* role-justifying grounds [viz, the law statement], Hempel actually takes one step of just the same persuasive but illegitimate regress?[160]

Scriven argues that attempts to include grounds for explanations, of either of these two of his types, leads to the same regress. Hempel agrees that it is wrong in the one case. Therefore he should agree that it is wrong in the other case. So goes Scriven's argument. Now, this would be legitimate if the defender of the deductive model drew the explanation/evidence distinction in the way Scriven draws it, on the basis of the stimulus/context distinction, since the latter distinction serves equally, as we saw, to distinguish explanation as stimulus from knowledge of a law as part the context in the case where the law is one of Scriven's "truisms". But the defender of the deductive model does not draw the explanation/evidence distinction for the reasons Scriven draws it. That permits him to insist upon this distinction while also taking the first step of Scriven's regress in the second case, and including the "role-justifying" grounds as part of his explanation. Still, however, Scriven's regress might result in this case. The defender of the deductive model prevents it by arguing that there is a *reason* for including the law as part of the explanation. This reason again has nothing to do with Scriven's motives for the distinction. The defender thereby presents himself with a reason for taking the first step of Scriven's proposed regress without having to go on to take the rest of them.

We know there are two kinds of "role-justifying" grounds, that is, the grounds which establish the explanation satisfies the wants or interests of the explainee. What the explainee wants generically is a scientific explanation. Among the possibilities here he has a specific want. If it is idle curiosity, he wants the best scientific explanation available, viz, that which is closest to process knowledge. If his interest is pragmatic, he wants an explanation which yields a knowledge of some means sufficient to the end defining the pragmatic interest. What we are interested in now, however, is what he wants generically. Whatever he is provided with, if it is to be what he wants (whether he knows it is what he wants or not) — whatever he is provided with must be a scientific explanation. Now, it seems to me to be true of our pre-analytic notion of what is a scientific explanation of individual facts that such an explanation shows how one set of such facts (the initial conditions) necessitate the explanandum. But that notion of "necessitation" is is philosophically problematic, and requires explication. The only reasonable explication that has been proposed is in terms of deduction from a true statement of law. The defender of the deductive model includes the law as part of his definition of '(scientific) explanation' because it *is* part of that notion insofar as it can be reasonably explicated. The very idea of satisfying his generic interests, insofar as that idea can be explicated, requires us to include the law statement as part of the explanation. Having decided what anything must be if it is to satisfy him, we can then turn to the question of better or worse. However the latter may be answered, it is clear that the former can be answered only by selecting and grouping together things Scriven would separate into stimulus and context. Such grouping would not deny the latter separation. It would simply bypass it in the interests of indicating what must be present if the generic wants are to be satisfied (in an explicable way). And the point is, that this reason for including the law does not require one to take any further steps on any impending regress.

Scriven overlooks this line of reasoning because he asks a completely different question about explaining than does the defender of the deductive model. The latter asks, what must be present if the explainee is to have his generic interest satisfied? An answer to such a question can group stimulus and part of the context together, and no difficulties arise. Scriven asks the quite different question, how can I intervene in the situation in order to remove the explainee's quizzical look? But to intervene is to present the explainee with a stimulus! The answer to Scriven's question requires one to isolate the stimulus, and separate it from the context which is taken for granted. Scriven's question requires him to separate what the defender of the deductive model can put together. *Both* questions are legitimate, of course. But Scriven's is descriptive-psychological. Only the other question is one posed by the philosophy of science. Scriven's question simply indicates once again that his interest is a pragmatic interest in the behavioural psychology of communication. And it simply indicates once again how the psychological approach he adopts in order to satisfy his pragmatic interest prevents him from seeing how a philosophical defence might be given of the deductive model.

Behind Scriven's argument that laws are not part of explanations, speaking in historical terms at least, lies a discussion initiated by Ryle[161] and developed by Toulmin[162] to the effect that an inference

(5.1) Fa, so Ga

ought not be construed as an enthymeme, an incomplete argument, with

(5.2) $(x)(Fx \supset Gx)$

as its suppressed premiss and justified by virtue of

(5.3) $[(x)(Fx \supset Gx) \& Ga] \supset Ga$

being a tautology. Rather, Ryle and Toulmin argued, (5.1) is more properly construed as a complete argument, justified by the truth of (5.2). Just as the truth of (5.3) guarantees that the argument

(5.4) $(x)(Fx \supset Gx)$, Fa, so Ga

yields a true conclusion from true premisses, so the truth of (5.2) guarantees that the argument (5.1) yields a true conclusion from true premisses. If this case made by Ryle and Toulmin goes through then, clearly, the defender of the deductive model cannot hold that the law-statement has to be part of the explanation: (5.1) will itself be complete as an explanation.

Ryle's case is relatively simple. In effect he develops a version of a pattern of argument first put forward by Lewis Carroll.[163] The pattern goes as follows. The argument (5.1)

$$\frac{Fa}{\text{so, } Ga}$$

is justified by (5.2)

$$(x)(Fx \supset Gx)$$

The argument (5.4)

$$\frac{(x)(Fx \supset Gx)}{Fa}$$
$$\text{so, } Ga$$

just justified by (5.3)

$$[(x)(Fx \supset Gx) \& Fa] \supset Ga$$

The argument

(5.5) $\quad [(x)(Fx \supset Gx) \& Fa] \supset Ga$
$\quad\quad\quad (x)(Fx \supset Gx)$
$\quad\quad\quad \underline{Fa}$
$\quad\quad\quad \text{so, } Ga$

is justified by

(5.6) $\quad \{[[(x)(Fx \supset Gx) \& Fa] \supset Ga] \& (x)(Fx \supset Gx) \& Fa\} \supset Ga$

And so on. Now, (Ryle argues) if it is held that (5.1) is an enthymeme, and should be construed as including the justifying principle (5.4) among its premisses, then, equally, (5.4) should be held to be an enthymeme, more properly construed as (5.5), that is, as including the justifying principle (5.3) among its premisses. But then, again with equal justice, (5.5) should be construed as an enthymeme, with (5.6) as an additional premiss. However, the result will have a still different justifying principle, and this, too, will have to be added as an additional premiss. And so on, *ad infinitum*. Ryle suggests that if we make the first step, claiming (5.1) is incomplete, then we have equally to claim every successive argument in the series is incomplete. In that claim every successive argument in the series is incomplete. In that case, no argument is complete. But some argument must be complete if we are ever successfully to argue deductively. If this is so, then we should not take even the first step on the threatening vicious regress: (5.1) must not be construed as an enthymeme; it must be construed as complete in itself. Scriven accepts this argument as part of the general case he is making against the deductive model.[164]

The question is, then, can the defender of the deductive model (or, for that matter, of formal deductive logic) – can such a defender argue that there is reason for going as far as (5.4), holding that it is complete where (5.1) is not, while also stopping at (5.4), holding that it is not an incomplete version of (5.5)? What this question amounts to, obviously, is the question whether there any difference between the principle (5.2), on the one hand, and the remaining principles (5.3), (5.6), etc., that justify the arguments (as yielding true conclusions from true premisses) at all later stages in the regress? Once the question is posed the answer would seem to be obvious: (5.3), (5.6), etc., are all tautologies, while (5.2) is synthetic. To put it another way, (5.1) does not establish that the conclusion *must* obtain given the premiss obtains, whereas (5.4) does establish that the conclusion must obtain given the premisses obtain. Thus, since deductive arguments purport to give reasons why their conclusions *must* obtain, given their premisses, it follows that (5.1) is *deductively incomplete*, (5.4) *deductively complete*.[165]

Ryle's argument thus does not succeed. Toulmin developed a more sophisticated version of it. Ryle argues a general thesis about arguments like (5.1). Toulmin concedes some arguments of this sort are deductively incomplete; these are cases where the generality (5.2) is one of *natural history*. But where the generality is a *lawful generality* then it is more appropriate, logically, to construe it as a rule of inference justifying an argument complete in itself rather than as a suppressed premiss.[166] Where the generality is lawful, the argument (5.1) will be complete, and will be explanatory.[167]

Toulmin attempts to derive the distinction between generalizations of law and those of natural history in terms of a feature of the former absent from the latter. This is the distinction between *rule* and *scope of application*. Generalizations of law do not function as mere generalizations that can be falsified by a single counter-example, as, for example, "All swans are white" came to be known to be false when the first black swan was observed. To the contrary, counter-examples do not lead us to reject generalizations of law as false; instead, we retain the law or rule, and merely say that it is of less restricted application than we previously thought.[168]

One of Toulmin's examples is that of Snell's Law of refraction, that

(5.7) $\sin i/\sin r = c$

where c is a constant characteristic of the pairs of transparent substances. This law was well-confirmed. Yet counter-examples were found to it. For example, in calcite, we find an impinging ray splitting into two, and that Snell's Law holds for one of these rays and not the other. When this was discovered, scientists did not say something to the effect that "Snell's Law is only approximately true", nor did they give it up as false: Snell's Law is still found in text-books! All this is true. But it does not follow[169] that Snell's Law is a rule of inference rather than a generalization.

We may first notice that the same point about scope can be made even quite obviously *mere* generalizations. Consider[170] the generalization

Undergraduates dress informally

It would not be given up as false because one had a well-dressed Harvard undergraduate pointed out to one; rather one would say (perhaps) something to the effect that one was thinking primarily of schools in the mid-west, or in the Big Ten, or wherever. For Toulmin, the generalizations on the left would have to

(5.8)   *Generalizations*   *Scopes*

All professors dress formally   Everywhere
Undergraduates dress very informally   Midwest
Undergraduates take outside work   Not in liberal arts
   in term-time   colleges

be taken as rules of inference because one would not accept an instance as falsifying them but rather as requiring one to alter the scope. But surely one wants to say that if the generalizations on the left are thus understood, then they are *not* generalizations, but more akin to propositional functions; that is, one should take them as incomplete statements, which can be completed by either supplying a particular instantiation (e.g., 'on the Chicago campus') or else by quantifying (e.g., 'on all campuses', 'on some campuses', 'on most campuses', 'on all campuses in the Midwest'). Similarly, if we take (5.7) as stating Snell's Law, then Snell's Law is an incomplete generalization, a sort of propositional function requiring completion by some quantifier. What Toulmin calls "modifying the scope" is the replacement of one quantification by another. Which is to say, of course, that it *does* involve the rejection of one statement as false and its replacement by another – a point we made earlier when we dealt with a Popperian distinction between scientific laws and technical rules of calculation.[171]

But, as Alexander has pointed out,[172] the matter is more complicated than this in the case of science. The generalizations of (5.8) call for only one quantifier. The laws of science are usually multiply quantified. Thus, to take Snell's Law (5.7):

(i) it sums up relations between angles of incidence and the angles of refraction for *all* angles of incidence;

(ii) it tells us that for *all* the different specimens of the same two media the formula holds and the constant of proportionality is the same (it is a system-dependent constant);

(iii) it tells us that for *any* – or almost any – transparent medium *there is a unique* constant of proportionality; this constant turns out to be different for different pairs of media, and, in fact, specifically characteristic of those pairs;

(iv) it tells us that the formula holds for *all* times.

It is therefore falsified by

(a) finding the refractive index of substances varied systematically with time,

(b) finding a transparent medium for which the law did not hold;

(c) finding there were certain pieces of flint glass which, though identical in all other respects, had different refractive indices;

(c') finding there were certain species of flint glass, identical in all other respects with normal specimens but for which the relation was $\sin^2 i/\sin^2 r$ = constant;

(d) finding all previous observations had been wildly mistaken, and that the law which held for all transparent media was $\sin^2 i/\sin^2 r$ = constant.

276 CHAPTER 3

This makes clear that a shortcoming in one respect would not lead us to abandon Snell's Law altogether. Of the various sorts of falsification we may note that:

- type (c) does not occur in physical sciences – so far as we know – though they do occur in psychology;
- type (b) is not infrequent – in this case we do not "abandon" the law but note the restriction and speak of such things as "abnormal refraction";
- type (d) is wildly unlikely;
- type (a) has never yet been observed – though Whitehead thought it might be.

Finally, it is worth noting that in (iii) there is mixed quantification so that falsifications of type (c) and (d) involve verifying hypotheses *contrary to Snell's Law*.[173]

We may conclude, then, that merely introducing a "rule" vs. "scope of application" distinction does not establish that lawful generalizations are not logically of a form not essentially different – save, to be sure, in complexity – from generalizations of natural history. But this is only part of Toulmin's case. What he wants to focus upon is the apparently *non-falsifiable* character of lawful generalizations. This non-falsifiable character, testified to but not proved by the "rule"/"scope" distinction, is a result of their being *conceptual truths*.[174] Natural history takes concepts as given and attempts to discover matter of fact relations among the objects to which the concepts apply. Science, in contrast, provides a *re-conceptualization* of the data it studies.[175] Generalizations of natural history do not give meaning to the concepts used to state them; rather, they presuppose such meaning as antecedently given. In contrast, scientific laws give meaning to the concepts that appear in them: discovery of laws and reconceptualization go together because the laws are part of the very meaning of the new concepts the scientist uses to describe his data.[176]

This thesis, that laws of nature are somehow "conceptual truths", is defended by Toulmin by a number of different arguments.[177] We will look at four of these, and reject each as inconclusive. We may conclude, then, that Toulmin has made his case no more than did Ryle. But the final argument of Toulmin's we shall examine is of particular interest, for it raises a number of issues about the use of *models* in science that are worth commenting upon.

One argument Toulmin offers for the inseparability of laws and concepts is his claim that the concept of refractive index is meaningful only if Snell's Law holds.[178] To an extent this must be granted, but, even if it is, Toulmin's conclusion does not follow. The concept of refractive index is the "C" in (5.7). As (iii) makes clear, this concept is a *definite description*. What Snell's Law states is that this definite description is successful. So if Snell's Law is false, this definite description is not successful; it is therefore "meaningful" only if Snell's Law holds. The factual content here is embodied in (iii). From the fact that Snell's Law "determines" the "meaning" of 'refractive index', it does not follow that Snell's Law determines the meaning of the concepts appearing in the statement (iii). It therefore does not follow that Snell's Law determines the "meaning" of the concepts used to describe the data to which it applies.

A second argument of Toulmin's is this:

Newton's Laws of Motion are not generalization of the 'Rabbits are herbivorous' type; but they are not for this reason any the more tautological (cf. 'Rabbits are animals'); and this is because they do not set out by themselves to tell us about the actual motions of particular bodies, but rather provide a *form of description* to use in accounting for these motions.[179]

Laws of nature are not empty tautologies, but they are not empirical generalizations, either. Rather, they are substantive but *a priori* principles; or, as we have put it, "conceptual truths". This argument is also inconclusive. For, if one grants that Newton's Laws are abstractive generic hypotheses then one can grant they do not describe – i.e., describe determinately – the motions of actual bodies, but only give the form of those motions – i.e., a determinable or generic description of those motions. But, abstractive generic hypotheses are empirical generalizations not essentially different from generalizations such as 'Rabbits are herbivorous' – save, of course, for the occurrence of mixed quantification. And generic concepts and mixed quantification do not transform an hypothesis from being an "empirical truth" into being a "conceptual truth".

The third of Toulmin's arguments is quite different but, interestingly enough, turns on a similar point.

... what Snell discovered was ... the *form* of a regularity whose existence was already recognized. Ptolemy, Bacon and Kepler could not have studied refraction in the way they did unless they had been sure that there was some regularity to be discovered.... But though the existence of a regularity was clear to them, ... it remained to be found out what form the regularity took. This was what their experiments were designed to reveal, or rather, what they hoped to be able to spot from the results of their experiments.[180]

The final remark, about "spotting" the form of the regularity, ties in with Toulmin's idea that to discover a law is to re-conceptualize data. But, while we grant the substance of what Toulmin says in his passage, again his conclusion does not follow, that the concepts are theory-laden and that (therefore) the theory is a conceptual truth. This argument is a curious reversal of the second. In the latter, "form" referred to generic features. In the present case it refers to determinate features! What Ptolemy, and the others, believed was that *there is* a law. That is, they organized their experiments guided by this generic hypothesis – a piece of imperfect knowledge in this case, not a well-developed theory as in the previous case. To discover the form of the law is to discover what the *specific* function is that relates the variables they were interested in, viz, angle of incidence and angle of refraction. Now, it is true that the specific form of a law often so to speak "pops out of" the data; more often, it is only with great difficulty that one finds the pattern the data conform to; but in any case, it is not wrong to speak of the noticing of the pattern as a "conceptualizing of the data". That is, one way of describing Snell's discovery is to say he discovered how to *describe* the phenomenon of refraction in terms of the *concept* of refractive index as defined by (5.7). However, this description is a *lawful description*, since what it amounts to is a description of the facts in terms of the *empirical generalization* (iii). The appropriateness of describing a scientific discovery as one of "re-conceptualizing the data" does not entail that what is discovered is not an empirical generality.[181] We have made similar points previously, and they are equally valid as criticisms of this third of Toulmin's arguments.

What should be noticed is that the notion of 'form' for Toulmin hovers uneasily between a reference to a generic sense of form, as in the second argument, and a reference to a determinate sense of form, as in the third argument.[182] This is of considerable significance for Toulmin's discussion of concrete cases, such as classical mechanics, as I have argued elsewhere.[183] The present point is simply that the distinction between "rule" and "scope" alows Toulmin to obscure these two different notions of form.

Suppose a generality to the effect that As are B describes *specifically* how a sort of system behaves. This generality is the form, in the determinate sense, of the behaviour of these systems. Suppose that we now find exceptions to this rule. We do not know specifically what distinguishes these anomalous systems, but (let us suppose) that we have some determinable description of them. We can therefore say that there is a specific property of generic sort $\mathcal{G}$ such that, for all objects exemplifying this property, if they are As they are B. We have here a piece of imperfect *generic* knowledge. "As are B" is now part of a generic description of systems; we have shifted to the generic sense of form. As Toulmin would put it, however, we have shifted *scopes*, and it is the rule that gives the *form* of the regularity. So form moves from a context of specific or determinate descriptions to a context of generic descriptions.

What further obscures this is a feature of a good deal of scientific reasoning that Toulmin fastens upon as the substantial content of the fourth of his arguments we are considering. This feature is that of models. Now, these *do* have a substantial role to play in science. Toulmin, however, misunderstands that role. In particular, he fails to understand the role of modeling in the concrete example he picks to illustrate his point, namely, geometrical optics. What this misunderstanding yields is only spurious evidence, no solid case in support of his thesis that laws are rules of inference, conceptual truths.

Toulmin considers the case of calculating and explaining the length of a shadow of a rod illuminated by the sun, and correspondingly, calculating, i.e., explaining and predicting the height of the rod given the length of the shadow. The inference here presupposes (I am, for the moment, deliberately using the relatively vague 'presupposes') the Principle of the Rectilinear Propagation of Light. What Toulmin holds is that the argument by which the inference proceeds is not syllogistic.

The fact of the matter is that we are faced here with *a novel method of drawing physical inferences* — one which the writers of books on logic have not recognized for what it is. The new way of regarding optical phenomena brings with it a fresh way of drawing inferences about optical phenomena.[184]

Such a principle is important in physics, and embodies more than might at first be supposed:

... the importance for physics of such a principle as that of the Rectilinear Propagation of Light comes from the fact that, over a wide range of circumstances, it has been found that one may confidently represent optical phenomena in this sort of way. The man who comes to understand such a principle is not just presented with the bare form of words, for these we have already seen to be on a naive interpretation quite false [we return to this point in a moment]: he learns rather what to do when appealing to the principle — in what circumstances and in what manner to draw diagrams or perform calculations which will account for optical phenomena, what kind of diagram to draw, or calculation to perform, in any particular case, and how to read off from it the information he requires.[185]

The inference-rules are tied to diagrams, and, as the first of these quotes indicates, are part of the meaning of the descriptive vocabulary the scientist uses to describe the phenomena to be explained:

280   CHAPTER 3

... the very notions in terms of which we state the discovery, and thereafter talk about the phenomena, draw their life largely from the techniques we employ. The notion of a light-ray, for instance, has its roots as deeply in the diagrams which we use to represent optical phenomena as in the phenomena themselves: one might describe it as our device for reading the straight lines of our optical diagrams into the phenomena. We do not *find* light atomized into individual rays: we *represent* it as consisting of such rays.[186]

Toulmin's points are, basically, two: first, that the Principle of Rectilinear Propagation of Light is not an inductive generalization; and second, that the diagram plays so significant and distinctive a role in the inferences that the latter are not reasonably construable as purely formal and syllogistic. We shall look at these in reverse order, first getting a good perspective on the role of the diagrams, and then turning to the question of the logical status of the Principle of Rectilinear Propagation.

Geometrical optics[187] examines the *conditions of illumination*. It divides objects, first, into those that are self-luminous, like the sun, and those that are non-self-luminous, like Toulmin's stick and the earth upon which the shadow falls. The latter become illuminated, under certain conditions, in the presence of self-luminous objects. The non-self-luminous objects can be divided into the transparent objects (or media) and the opaque. If a self-luminous object illuminates one which is non-self-luminous and an opaque object is placed between them, then the illumination of non-self-luminous object ceases (though, of course, the interposed opaque object comes to be illuminated). On the other hand, if a transparent object is interposed the illumination of the second object does not cease. Opaque objects can be further distinguished into those that reflect illumination and those that do not. And so on. In the simple case of Toulmin's stick and its shadow, we have an opaque object, the stick, interposed between the self-luminous object, the sun, and the non-self-luminous object, the earth. With the interposition of the stick a portion of the earth ceases to the illuminated; this area is called the shadow of the stick. No object that "falls within the shadow" is illuminated. Geometrical optics aims to describe the areas and volumes in which objects in the vicinity of opaque objects can be illuminated by self-luminous objects. This is what is meant by saying that it examines the conditions of illumination. It turns out that these areas and volumes can as a matter of fact be fairly easily described in geometrical terms. It is these *spatial facts* about illumination that are represented in the diagrams drawn by the geometric opticians. The diagram is a *projection* of a set of three-dimensional geometrical relationships onto a two-dimensional space, accompanied by a reduction in size. The laws of projective geometry enable one to deduce from the diagram the facts one wishes to know about the actual areas and volumes of illumination. *The diagram is thus not essential to the inference.* As for the *laws* that are involved, they are generalities to the following effect:

Whenever a self-luminous and several non-self-luminous objects are in such and such spatial relations then any non-self-luminous object will be illuminated if and only if it is in such and such further relations to these objects.

Clearly, *there is no reason why laws of this sort should not be construed as general premisses in deductive explanations.* No doubt these deductive arguments will not be simple cases, syllogisms *in Barbara*, but they do not cease thereby to be deductive in the strictly formal sense. And the latter is all we are trying to defend against Toulmin. If all the latter means to point out is that not all scientific are *in Barbara*, then we need not disagree. But presumably he is trying to defend an *interesting* thesis, that explanations in geometrical optics are not formally deductive, and this thesis, we now see, he has given us no reason to think plausible, let alone acceptable.

In geometrical terms, volumes and areas can be conceived of as bundles of straight lines. For certain areas, one or more of the bounding edges are straight lines. For certain volumes, one or more of the edges are straight lines. As the diagrams of geometrical optics make more than evident, certain important boundary lines for the areas and volumes of illumination and non-illumination are straight and these edges are segments of straight lines extending to the self-luminous object which is the source of illumination. The importance of these boundary lines, together with the geometrical possibility of describing volumes as "bundles" of straight lines, permits one to *conceptualize* the areas of illumination as consisting of bundless of straight lines which, if the self-luminous object is sufficiently distant, all converge towards to the point where the self-luminous object is located. Light thus comes to be conceptualized as consisting of bundles of rays. Toulmin is therefore correct when he asserts that "We do not *find* light atomized into individual rays: we represent it as consisting of such rays";[188] but the failure of the description-explanation distinction for geometrical objects does not follow from this, as he believes. The laws (5.9) of optics and those of geometry make such representation possible. But the explanation of, e.g., the length of the shadow, is in terms of laws like (5.9). From the *fact* these laws make such representation possible, it does not follow these laws define the meaning of the concepts they contain. Such concepts as that of being self-luminous or of being opaque would be the same concepts even if the areas and volumes of illumination were bounded and edged by curved rather than straight lines. So Toulmin's claim, that the explanatory laws of geometrical optics determine the meanings of the concepts used in that discipline, does not hold. The theory-ladenness of the concepts, and, what is the same, the non-formal nature of the inference rules has thus not been established.

Besides the laws (5.9) there are other laws in geometrical optics. Thus, one comes to recognize differences among various transparent objects or media. The illumination proceeds, so to speak, through the transparent media; that is, opaque objects behind or within these are illuminated. Let two such transparent objects be brought into contact, and suppose the boundary between is flat (i.e., in the two-dimensional diagram which represents the situation, a straight line). If we now set up a stick so only a portion of the boundary is illuminated, then we discover that there is a discontinuity in the straight line which is the boundary of the area of illumination, and that this discontinuity occurs at the boundary between the two transparent objects. The nature of this discontinuity of the boundary of the area of illumination is in fact regular. The determinate form of this regularity was of course, discovered by Snell: it is given by Snell's Law (5.7), $\sin i/\sin r = c$, where c is a system-dependent constant, characteristic of the chemical kinds of the transparent media involved. Clearly, the simple sine formula is but part of the whole that Snell's Law actually is; the full statement would involve reference, of the sort sketched in (5.9), to self-luminous objects, to transparent objects and media, and to areas and volumes of illumination. This brings out well the point illustrated by the generalizations (5.8): such formulae as the sine fomula of (5.7) ought to be considered, not as the full statement of the scientific law, but as propositional functions with the appropriate quantifiers to be supplied by the context.

Laws (5.7) and (5.9) share an element of logical form. This is the feature that certain important edges of areas (and volumes) of illumination are straight (or flat), and, if the self-luminous object is sufficiently distant, then these lines, with occasional discontinuities, converge on the point source of illumination. It is this feature of shared logical form that is called the Principle of Rectilinear Propagation of Light. Thus, *the Principle of the Rectilinear Propagation of Light is an abstractive generic hypothesis for the theory we refer to as "geometrical optics"*. Its role here is similar to that of the Law of Inertia in classical mechanics. Toulmin perhaps grasps something of this when he asserts that

By itself, the principle tells us no additional facts over and above the phenomena it is introduced to explain ... its acceptance marks the introduction of the explanatory techniques which go to make up geometrical optics....[189]

This is not inept as a description of the logical relation between the abstractive laws of the theory in an area to the more specific laws applying to the specific sorts of systems the theory deals with only generically. But of course, as we saw, Toulmin blurs generic and specific — this is a consequence of his rule/scope distinction — and so his insight remains quite undeveloped. Be that as it may, however, what it is important to recognize is that the Principle of Rectilinear Propagation of Light does not cease to be empirical for its being an abstractive generic law.

Toulmin, of course, wants to argue the Principle is not straight-forwardly empirical. What is empirical is the determination of its scope.[190] The principle itself is a conceptual truth, part of the inference for, rather than a premiss of, explanatory arguments. What he offers in support of this is the claim the Principle is not an inductive generalization. The inappropriateness of such a characterization can be recognized when we recognize the absurdity of the induction

Light always has travelled in straight lines; what we have here is light; so what we have here will almost certainly travel in a straight line.[191]

He concludes

To say "Light travels in straight lines" is, therefore, not just to sum up compactly the observed facts about shadows and lamps: it is to put forward a new way of looking at the phenomena, with the help of which we can make sense of the observed facts about lamps and shadows.[192]

But the way Toulmin states the induction is very misleading. Light is, of course, *not* a bundle of rays. That is indeed only the way we *represent* it. The induction could not, therefore, be in any simple-minded way about light travelling in straight lines, as Toulmin suggests. It is much more appropriate to say that the induction is about lamps (self-luminous objects) and shadows (opaque non-self-luminous objects). But — to repeat — the Principle of Rectilinear Propagation does not say *just* that: it also adds a method of representation. And it is essential to add a further point. Whatever the exact form of the inductive inference involved, it is not one of simple enumeration: "All observed As have been Bs; this is an A; so almost certainly this will be a B" or "All observed As have been Bs; so All As are Bs." It cannot be this way since the Principle of Rectilinear Propagation of Light is an abstractive generic hypothesis: it involves mixed quantification. In this sense it is indeed quite inappropriate to think of it as "summing up" a set of observations.

This is important because it prevents the Principle from being falsified in the simple minded way in which 'All A are B' can be falsified by an A which is non-B. One can continue to hold the Principle even though one knows phenomena to which it applies but for which one cannot give a specific geometrical description of the straight lines one knows on the basis of the Principle to be discoverable. Thus, Kepler could hold the Principle, maintain it as not falsified, even though he could not discover the specific description the Principle asserts to be there for the case of refraction. It was Snell who discovered the *specific law* for these phenomena, a discovery which confirmed the *generic law* of the Rectilinear Propagation of Light. Thus, if one's paradigm for the conclusion of an inductive inference is a law falsifiable in the simple way "All A are B" is falsifiable — as it seems to be the paradigm for Toulmin — then the Principle of Rectilinear Propagation of Light is not a conclusion of an inductive argument. But it does not follow from this, as Toulmin wrongly supposes it does, that the Principle is not empirical law, and that it is not, in some more reasonable sense of 'inductive', the conclusion of an inductive inference. As part of his misunderstanding of the generic nature of the Principle of Rectilinear Propagation of Light, Toulmin speaks of the discovery of Snell's Law as introducing "a rule for extending the inferring techniques of geometrical optics to cover refractive phenomena."[193] This is satisfactory as metaphor, but misses the point that the generic Principle *already* applied in the area of refraction. It laid down *determinable* conditions on the facts. The discovery consisted in locating the *determinate* form of the law. With this discovery the diagram techniques of the stick-and-shadow case could, with suitable modifications, be applied in the refraction case. In this latter sense only did an *extension* of techniques occurs. Toulmin's forcing everything into the rule/scope distinction blurs the crucial distinctions.

Of course, the Principle has come to be falsified. The Principle could not account for the phenomenon of diffraction. This did not by itself falsify it. But a contrary theory, the Wave Theory of Light, *could* account for diffraction; those phenomena therefore confirmed this theory; and this theory entailed the falsity of the theory of geometrical optics. This is the usual way in which abstractive theories, containing laws involving mixed quantification, come to be rejected as false.[194] And here one must add, "false in all generality". For the Wave Theory entails that the Principle of Rectilinear Propagation holds "under certain conditions", and states precisely the limits of its validity.[195] We have been over this pattern before, in, for example, our discussion of (4.1) etc., in Chapter 1. Toulmin of course recognizes the limited validity of the Principle.[196] But he uses it (wrongly) to argue for the rule/scope distinction. What it really evidences, we now see, is the status of the Principle of Rectilinear Propagation as a piece of imperfect knowledge.

There remains one point that we have so far studiously avoided. Toulmin makes much of the metaphor that *light travels*: and such associated notions as *propagation, interception*, and so on.[197] There is an *analogy* or *model* being appealed to here, and we cannot ignore it, since it is so obviously significant in the discourse of physicists. We can say, first off, however, that *the analogy or model plays no role in the explanation of the facts geometrical optics accounts for*. Nothing in Snell's Law, (5.7), nor in (5.9) requires attention to these metaphors; the deductions, the explanations and predictions concerning "lamps and shadows" make no use of them. Once one recognizes this one realizes that Toulmin can draw from the existence of such metaphors no conclusions about the inadequacy of the deductive model of explanation. Yet the metaphors *do* play a role in science, as Toulmin realizes, and we must see what this is.

The model is that of some object travelling down a straight-line orbit, and light is taken to travel similarly down the straight lines that are its rays. There are positive aspects to the analogy — there are straight lines in both cases — and negative aspects or disanalogies — the lines trains travel on are separated by finite distances while the rays of light form a continuum. Other things one might not know whether they are positive or negative. Thus, the travelling metaphor suggests a "speed of travel".[198] The model thus suggests an hypothesis about light that may or may not hold. The task, then, is to test it. This locates the role models of the sort we are considering play in science. Their role is not located in the context of explanation but in the context of research: their role is to generate hypotheses that are then subjected to test.[199] None of this tells against the deductive model of explanation, however.

Another who has argued for such a role of models or analogies in science has been Goudge. We can perhaps get Toulmin's insights into a more accurate perspective by glancing at this latter account. Goudge develops his ideas about models and analogies with particular reference to the hypotheses which are part of narrative explanations. Recall that in the latter there are two components. One is deductive-nomological and explanatory in the stricter sense of the deductivist thesis; the explanation here is imperfect. The other part consists of hypotheses and guesses, more or less informed, that fill in the gaps in the imperfect explanation. This second part is not so much in the context of explanation as in the context of research. It is models and analogies which, at least sometimes, are the source of these hypotheses. And for biology, one model in particular is of importance.

Goudge argues in detail that inferences about whether a factor or character is an adaptation make implicit use of the concept of "purpose" or "proper function".[200] There is, however, no reason to suppose either "(i) that the internal structures and processes of organisms are [literally] purposive, [or] (ii) that the overt, macroscopic behaviour of organisms, other than human beings, is purposive".[201] Rather, "What we are entitled to affirm is that these structures, processes and behaviour show an *ostensible* design or plan".[202] Goudge argues in detail[203] that to thus speak of purpose has none of the metaphysical implications that such philosophers as Bergson have claimed to find in the fact of evolution, nor any of the objectionable features of vitalism and finalism.[204] By 'ostensible design', Goudge has in mind nothing more than something along the lines of Broad's definition of "teleological".[205] Suppose that we have a sort or kind of system. Systems of this kind have parts which are arranged, act and interact in certain ways. Systems of such a kind are teleological, exhibit ostensible purpose, if (a) the parts and their actions are such as might have been expected *if* the system had been constructed by an intelligent being to fulfill a certain purpose he had in mind, and if (b) when systems of this sort are further investigated under guidance of this hypothesis, hitherto unnoticed parts, arrangements among parts and processes of interaction are discovered, and these are found to accord with the hypothesis. Broad asserts that

... the most superficial knowledge of organisms does make it look as if they were very complex systems designed to preserve themselves in the fact of varying and threatening external conditions and to reproduce their kind. And, on the whole, the more fully we investigate a living organism in detail the more fully does what we discover fit in with this hypothesis.[206]

If we consider the narrative explanation we looked at previously[207] we can see where the purposive models play their role. That narrative explanation concerned the adaptation of amphibians to land by the development of limbs. The narrative consisted of a hypothetical account of how this adaptation proceeded. It is considerations based upon the hypothesis of ostensible purpose that (among other things) enable the narrator to construct a plausible story, that is (as we saw) a set of hypotheses worthy of further research. But the use here of 'hypothesis' in 'hypothesis of ostensible purpose' is misleading. One is not so much reasoning on the basis of an hypothesis but rather in terms of an *analogy*; natural selection and human intelligence are both *problem-solving* capacities[208] — the former solves problems involved with the maintenance of individuals and the perpetuation of their kinds, the latter problems of technology — and thinking of the situation faced by biological systems we can form hypotheses about how natural selection might solve the survival problems confronting those systems. Human problem-solving capacities provide a *model* for the problem-solving capacities of natural selection. As the quote from Broad correctly indicates, the model has been a fruitful source of hypotheses, though, as Goudge points out,[209] maladaptations as well as adaptations occur among the products of evolution, so that it is easy to overstate the case for ostensible design. However, this does not vitiate the utility of the purposive model, since the disanalogies between the human and evolutionary cases serve to explain the existence of the maladaptations. In particular, there is nothing in the human case that corresponds to the mutations of selectionist theory: the mutations always occur randomly relative to the needs of organisms, which means maladaptations are reasonably to be expected. When using the model or analogy to suggest hypotheses, the disanalogies must be taken into account. But this is, as Sellars has pointed out,[210] something which, while often neglected, is one of the more important features of the use of models in scientific theorizing.

One source, then, of the hypotheses appearing in narrative explanations is, according to Goudge, models and analogies. Goudge is neither the only philosopher nor the first to insist upon the role of models and analogies. We have of course, seen that Toulmin also insists upon the importance of analogies. N. R. Campbell was perhaps the first so to insist.[211] Some, such as Nagel,[212] point out that such models are not essential to explanations. In this, they are certainly correct; at least, so we argued for the case of Toulmin's and Goudge's examples anyway. From this Nagel concludes that such models are not essential to theories, that they are "merely heuristic". But if one thinks of research as a feature of science that should equally be accounted for as is explanation, then one cannot ignore those things that systematically generate hypotheses that research comes to verify. For this, the heuristics of science — theories and models — are essential. In drawing our attention to this point, Toulmin and Goudge are pursuing a theme that has been taken up of late by such other philosophers of science as Lakatos.[213] Hesse,[214] too, has emphasized the role of models in providing reasons for hypothetically extending a theory in one direction rather than another, and using these hypotheses to guide research. Her point is close indeed to Toulmin's. Since an extension of a theory to include an as yet untested new hypothesis may have no basis in the evidence supporting the original theory, the choice of one extension over another must, Hesse concludes,[215] be made for reasons internal to the theory, from which it would follow that if the use

288   CHAPTER 3

of models to select such an extension is the appropriate way to additional empirical content, then the model must be an integral part of the theory extended. Here one must qualify: all that follows is that the model is an integral part of the research tool, the hypothesis-generator, that also includes the theory. It does not follow the model is essential to the lawful and theoretical structure which is the product of the research; to this extent Nagel is surely correct; as we have said and argued. Moreover, Hesse's claim that models are essential is too strong; not all valuable extensions flow from analogue models. Other heuristics are possible,[216] which, in fact, may be essential to the theory (as the product of a research process) in the way in which a model is not. For example, the composition law of classical mechanics directed one to look for Neptune when the perturbations of the orbit of Uranus did not turn out as had been predicted.[217] Still, as Hesse – and Goudge – and Toulmin – point out, models are *often* important to research. On the other hand, we should also note that they are probably just as often a hindrance to research. Thermodynamics could only develop within certain limits under the treatment as caloric as a fluid. And (as Toulmin and Goudge explicitly point out) models can contain disanalogies as well as analogies, in other words, features that would yield false rather than fruitful hypotheses if they were to be used in hypothesis-generating. This being so, then what we need is a general theory of heuristics; a theory of when it is reasonable to rely upon theories and models as generators of hypotheses to be relied upon as research guides, and when it is reasonable to reject one such set of heuristics (or part of such a set) and replace it with another. Neither Toulmin nor Goudge sketch a general account of heuristics, however. Nor do they sketch a theory of when the use of a model to generate research-guiding hypotheses is more or less reasonable. But in this last, they are not without company. Kuhn, alone, has sketched such an account (it is in terms of a paradigm's "puzzle-solving" capacities),[218] and even that is a sketch.[219] Still, Goudge at least does develop a view of mind, the evolutionary view, that can provide a framework for accounting for the development and change of science, and in particular for the procedures of evaluation that determine the acceptability and unacceptability of models and paradigms.[220] As we have noted above, certain insights of Peirce that Goudge has discussed show how such an account might proceed.[221]

What emerges out of Goudge's discussion of modern selectionist theory of evolution is an account of science which is in many ways similar to that of Kuhn. Except that Goudge leaves out those more baroque twists about which one is afraid it is those features rather than the truths that have secured for Kuhn's theories the popularity they now enjoy.[222] Thus, in Goudge there is none of that equivocal language one finds in Kuhn about such things as the theory-ladenness of concepts — dangerously equivocal language since it makes it appear (wrongly, Kuhn protests) that Kuhn shares the irrationalism of such philosophers as Feyerabend or Toulmin. For, when Goudge attacks the standard accounts of explanation, does not draw the anti-scientific conclusion drawn by such philosophers as W. Dray. Thus, when Goudge discusses narrative and "how possibly" explanations, he is at pains to insist upon the contextual importance of laws.[223] The "intelligibility" such explanations aim at is,[224] Goudge argues, *scientific* intelligibility. Dray, in contrast, insists that laws are essentially absent. When one attempts to show an event in history is intelligible by offering a narrative explanation then the intelligibility is a matter of the behaviour being in accordance with rules, not in its being subsumable under laws.[225] Dray's case has been sufficiently demolished by others;[226] we need not go into it here. Suffice it to say that Dray's anti-scientific account of man was never part of Goudge's views. Goudge would, it goes without saying, defend the acceptability of narrative explanations in history. I suppose the significant difference would lie in the source of the hypotheses constituting the story the narrative relates. In biology the source of hypotheses lies in such analogies as that of the purposive model. In history it is presumably the imaginative capacity for understanding others that we have come to call *verstehen*.[227]

One might note, however, that if Goudge is correct, then, in biology, narrative explanations propose research rather than terminate it; in history, in contrast, narrative explanations satisfy curiosity, and thereby terminate research. But we can take this as evidence that historical narratives, unlike evolutionary narratives, are written (as Horace says of poetry)[228] to amuse and to edify, and are accepted upon those terms. To that extent, history is not scientific. But that is no problem, for then history is not explanatory either, but poetry. In any case, however, for Goudge there is no essential break between science and history, between man and nature: the same types of explanation apply in both. Indeed, for Goudge, man is so much a part of nature that the basic framework for understanding man is an evolutionary view of mind.[229] This is not to say Goudge is a metaphysical behaviourist (he rejects Watson's position on this point).[230] Rather, if we take his approval of Mead's social behaviourism[231] as a guide, then he presumably would opt for some form of psycho-physiological parallelism.[232] Upon this view, mind would be reckoned emergent, but such a view is, of course, not anti-scientific.[233] And the idea that the mental is emergent is one which Goudge seems to find congenial.[234]

But however these details go, the point remains that Goudge, like Toulmin, insists upon the role of models and analogies in science. Only, he is clear, where Toulmin is not, that this role is one in the context of research, not in the context to which deductive-nomological explanations are appropriate. In general, then, I conclude Toulmin has made no case that the laws or principles of science are somehow non-empirical, and that therefore the deductivist thesis is mistaken. On the contrary, the latter thesis emerges unscathed. Laws need not be construed as rules of inference. And if the latter is not successfully defended by Toulmin's arguments, then we may conclude we have been given no reason for being statements of law are not part of the explanation. We may conclude, Ryle, Toulmin, *et al.*, notwithstanding, that laws are indeed an essential part of the explanations we have been discussing.

3.6. CAN WE KNOW CAUSES WITHOUT KNOWING LAWS?

One last gambit used by those who attack the deductive model remains to be discussed. This one has been adopted by Scriven[235] and also by Collins.[236] It is, in effect, an attempt to refute the argument given above for including a law statement as part of the explanation. It is argued that besides "necessitation" based on laws there is also "causal" necessitation. We can know, it is argued, that causal connections obtain, even when we do not know any laws to obtain. Hence we have explanations which adequately satisfy our pre-analytic notion of a scientific explanation and yet which do not involve laws. So much the worse for the deductive model, it is concluded. Let us see.

Consider the following thesis:

(T)   If it is correct to say that (event) *A* caused (event) *B* in circumstances *C* then it is also correct to say that the law obtains that whenever an *A*-like event occurs in *C*-like circumstances then a *B*-like event follows.

We may express (T) by saying that whenever it is correct to make a causal assertion when it is correct to make the corresponding assertion of law. We must distinguish (T) from

(T*)   Whenever one correctly makes a causal assertion one is *thereby* also making an assertion of the corresponding law.

Collins argues that the deductive model has a certain amount of plausibility but that a close examination reveals the spurious nature of this plausibility.

The first step of his discussion involves a claim that certain explanations based on causal assertions are fully adequate as explanations. With respect to the explanation

(CE)   Why did B happen? A caused B.

Collins comments that

To the question, "Perhaps, but what is the rest of the explanation?" the answer should be, "There is no more to it."[237]

Let us call this thesis the thesis of the "adequacy of CE (= causal explanations)". Collins suggests that

Law deduction theorists contend that acceptable explanations that are actually given in the form "A caused B" or "A happened because B happened" or "B was a result of A" leave unstated the greater part of the explanation. [A] What they suppose to be left implicit are laws and conditions which, taken together with A, suffice for the deduction of B. [B] I wish to argue, on the contrary, that in so far as this other material is relevant to the explanations with the forms cited, it is not as the completion of a half-stated explanation, [C] but as the defence of a fully-stated explanation. [D] [238]

[I insert the letters to facilitate reference.] The laws which the defender of the deductive model insist are part of the explanation (cf. [A], [B]) are rather part of the defence of an explanation (cf. [D]); in Scriven's terms, those laws constitute role-justifying conditions for explanations rather than parts of explanations. The reason they are part of such a defence is that we have complete explanations without them (cf. [C]). Collins' argument thus turns on arguing for the thesis of the adequacy of CE. What he argues is that the law-deduction theory gains its plausibility from (T). But he argues that (T) comes to be mistaken for (T*),[239] that the law deduction theory holds only if (T*) is true;[240] that (T*) is false;[241] that (T) is probably true;[242] and that (T) is a statement about how to defend causal assertions.[243] He argues as follows that (T*) is false. He distinguishes

(t_1)   If A causes B then there is a law associating A-like events with B-like events.

from

(t_2)   If one knows A causes B then one knows a law associating A-like events with B-like events.

(t_1) assures that when a causal relation obtains then there is a law to be discovered. The truth of (t_1) guarantees the truth of (T). However, (t_2) is false.[244] From the falsity of (t_2) the falsity of (T*) follows. The plausibility of the law-deduction theory turns on the confusion of (T) with (T*). As for the use of causal assertions in explanations, the adequacy of such explanations, that is, the thesis of the adequacy of CE, is argued in terms of examples of things which are accepted as explanations.[245]

This last step of his case is simply wrong. From the fact that some things are accepted as explanations it does not follow that they ought to be accepted. Collins' defence of the thesis of the adequacy of CE is plausible only because he confuses a descriptive-psychological matter with a normative-philosophical matter. Which is to say he confuses explanations$_B$ with explanations. In short, he does not establish that the citing of causes is ever worthy of being accepted, i.e., that citing causes can ever be an explanation in the normative sense. It is open to the defender of the deductive model to hold that no causal assertion ever provided an explanation.

Now, I think it true that ever so many causal assertions do *not* explain. Think merely of the claims of astrologers. Ever so many causal assertions are accepted as explanatory when they in fact are not explanatory. Their acceptance is a matter of *superstition*. On the other hand, it would seem that the claim that no causal assertion ever provides an explanation is too sweeping: some causal assertions *do* provide adequate explanations. Collins would perhaps agree with this in its essence. He would say that some causal assertions, those which do *not* provide adequate explanations, are not correct in the sense that they cannot be defended in terms of laws, as (T) requires. And those which are adequate are those which can be so defended. The law appears as part of the defence. I wish to hold that, to the contrary, those causal assertions which *do* explain are those which do in fact involve the assertion of a law. Thus, the best way to defend the deductive model would seem to be not by rejecting the thesis of the adequacy of CE, but rather by accepting (T*), and defending (t_2). What we shall therefore argue is that the laws that are involved in correct causal assertions are imperfect. In particular, these laws will in general assert that there are certain relevant factors without making explicit what these factors are. That is, the laws will be of the sort we mentioned in the ink-bottle and the poison and non-dying examples which we discussed in Chapter Two. (These are the sorts of examples to which Collins appeals to make his case.) The laws which Collins has in mind in [A] and therefore in (T) are, clearly, the less imperfect laws which can provide explanations of the imperfect laws which we assert when we make causal assertions. These less imperfect laws are not asserted when we make the causal assertions; to this extent, [C] is correct. Further, the less imperfect can be used to explain the more imperfect of the causal assertions; we can use this to provide a reasonable sense to [D], though this sense is not what Collins has in mind, since he rejects (T*), the truth of which is presupposed by the just suggested reasonable sense. At any rate, we can suggest that it is because Collins identifies all laws with perfect laws that he fails to see the truth of (T*), that causal assertions *involve* the assertion of laws.

As was just mentioned, the examples which Collins uses[246] to make his case plausible are examples such as the ink-bottle or poison and non-dying examples. We saw in Chapter Two that such examples cannot be used as counter-examples to the deductive model. There is nothing about the examples which cannot be construed by the defender of the deductive model as involving the assertion of an imperfect law. To put the matter in terms of the schematized example (CE), there is nothing to prevent the defender of the deductive model from construing the explanation

(CE_1) A caused B (which is, save for tense, the same as "A causes B")

as the explanation

(CE_2) Whenever an A-like event occurs then a B-like event occurs;
and that is an A-like event
───────────────────────────
so, this is a B-like event

where (CE_2), on the one hand, makes explicit what is implicit in (CE_1) and, on the other hand, conforms to the deductive model, involving deduction from an imperfect law. Collins and Scriven use examples of the sort (CE_1). Since these can be construed via (CE_2) they do not refute the deductive model. It does not follow that they *ought* to be so construed. What the defender of the deductive model must do is present a positive case for construing causal assertions as essentially involving assertions of laws.

We may take (CE_1) to be read as

(a)   This event, which is A-like, causes this other event, which is B-like

or as

(b)   (This is A-like) causes (that is B-like)

This transformation of (CE_1) reveals the logical features of the 'causes' in (CE_1). It is, in the *first* place, a connective, the nomological connective we may say. It connects atomic sentences, i.e., statements of events or of individual fact, into complex sentences. In the *second* place, it is a non-truth-functional connective. The truth-values of the two atomic sentences in (a) and (b), or, what is the same, (CE_1), do not by themselves determine the truth-value of the complex sentence. The truth will depend not only on whether the simpler sentences have certain truth-values but also on whether th events or facts those sentences mention are of the appropriate kinds, which is the *third* feature to be noticed: the complex sentence will be true only if the atomic sentences contain predicates mentioning *kinds* which are causally related to each other. In the *fourth* place, the sort of causal relatedness of these kinds will vary from case to case, from context to context. To take a simple example, a change in pressure may cause a change of volume or a change of temperature. The causal relatedness in the former case is given by Boyle's (imperfect) Law, $p \propto v$, and in the latter case by Charles' (imperfect) Law, $p \propto 1/T$. In the former case, it is direct variation; in the latter case, it is inverse variation. If we now assert "the pressure of this system being changed caused its volume to change," on the pattern of (CE_1), then the truth-value of this sentence will depend on whether the assertion is taken to involve a causal relatedness of the two kinds of events of the direct variation sort or is taken to involve a causal relatedness of the inverse variation sort. Of course, the sort of causal relatedness on which the truth-value of the assertion depends may not be explicit in the assertion itself. It may only be implicit, something to be gathered from the context of the conversation. For all that, it is relevant to the truth-value of assertions which involve the nomological connective. Such dependence of truth-value on what is not explicit is common among non-truth-functional connectives. Compare the various meanings 'if-then', e.g., the definitional (as in "if he is a bachelor then he is unmarried") and the causal (as in "if it rains, then the streets will be wet"). The truth-value of 'if p then q' intended as the definitional 'if-then' will depend on whether 'p' and 'q' are definitionally related or not. If they are, say, causally rather than definitionally related, then the original assertion will be false. Whether the 'if-then' is to be understood causally or definitionally will not be explicit, and will have to be gathered from the conversational context. Exactly the same holds for the nomological connective: the sort of causal connectedness relevant to the truth-value of the sentence will have to be obtained from the context. We now have what we may call the *four crucial features* of the nomological connective.

Taken by itself and in its own terms, the nomological connective is philosophically problematic. It is a non-truth-functional connective. Since the usual explication of the analytic/synthetic distinction [247] presupposes that the only primitive connectives are the truth-functional connectives, it follows that that distinction is rendered inexplicable if the nomological connective is taken as primitive. Further, if it is taken as primitive then the problem arises about how we know when it obtains. This is, of course, the problem to which Hume addressed himself.[248] Call these the *two philoosphical problems* generated by taking the nomological connective as primitive.

The defender of the deductive model has to present a positive case that in asserting (CE_1) one is asserting (CE_2). Such a case can be made by showing (CE_2) provides an *analysis* of (CE_1), that is, the only reasonable explication of (CE_1). An argument to that effect is made by showing, *firstly*, that (CE_2) captures the four crucial features of the nomological connective. It is clear that the definition or analysis which (CE_2) proposes for 'causes' does capture those crucial features. It is perhaps worth noting that it is not 'causes' which is analyzed but 'causes' *qua* in a particular sentential context or environment. The third crucial feature has the consequence that such contexts are relevant. The specific feature of the context which is important is the mentioned kinds of events. The use of 'causes' in any particular context presupposes the causal relatedness of these kinds. (CE_2) makes explicit in its analysis just the specific sort of causal relatedness of the kinds which is presupposed by the use of 'causes' in that particular context. It makes explicit the specific sort of causal relatedness which we know to be present by the fourth crucial feature of the nomological connective. We do not, of course, give any general analysis of 'causes'. This is impossible because what is relevant to the truth-value of sentences involving the nomological connective is context-dependent. We can only give the schematic rule that wherever there is a correct causal assertion then there is involved in that assertion the assertion of *some* imperfect *law or other*, and then exhort one to look for the specific imperfect law which is present in any particular case where 'causes' is used. As the ink-bottle and the poison and non-dying cases make evident, it is not difficult to find the appropriate imperfect law. This leaves open the mere possibility of a counter-example, but one runs such a risk whenever one proposes an analysis of a concept whose meaning is context-dependent.[249] At any rate, within those limits of risk which context-dependence imposes upon us, it is clear that the proposed analysis captures the four crucial features of the nomological connections. 'Causes' turns out to be a connective − defined in the way (CE_2) indicates. It is non-truth-functional, since the truth of (a) or

(b) will depend not merely on the truth-values of the simple components, i.e., the minor premiss and the conclusion of (CE_2) and *also* on the truth-value of the (imperfect) law which is the major premiss. The kinds are relevant, as the major premiss of (CE_2) indicates. And, the specific sort of causal relatedness of these kinds is made explicit in the major premiss. Showing all this is the first step in defending the proposed analysis of (CE_1) in terms of (CE_2). The argument defending the proposal shows, *secondly*, that the analysis avoids the two philosophical problems. Since 'causes' is not a primitive connective, and since the analysis using (CE_2) involves no non-truth-functional connectives, it follows that the analysis is compatible with the usual explication of the analytic/synthetic distinction. Nor is there any problem about our knowledge when a causal relation obtains. Our knowledge of causal connections is the same thing as our knowledge of matter-of-fact generalities. Since the latter is philosophically non-problematic, so is the former. Showing these two things is the second step of defence.

This, then, is the positive case which the defender of the deductive model makes for his claim that the law is part of the explanation. It is part of the explanation (CE_1) because it is part of the analysis of (CE_1). In effect, this argument amounts to an argument that causal necessitation, upon which both Collins and Scriven depend, is philosophically problematic and is to be explicated in terms of logical necessity, in terms of logical deduction from laws. The law enters because its presence is required as a part of the analysis of the notion of causal necessitation.

The challenge to this analysis would enter in respect of the fourth crucial feature of the nomological connective. Collins and Scriven would argue that although the sort of causal relatedness may be relevant to the truth-value of the causal assertion (CE_1) = (a) = (b) it is not *part of* such an assertion. It is relevant in the sense of being a "role-justifying ground" for (an explanation based on) the causal assertion (CE_1). In this way they deny that (CE_2) can constitute an analysis of (CE_1). It follows that the *meaning* of the nomological connective must be independent of the law which appears in (CE_2). Neither Collins nor Scriven propose any analysis of the nomological connective. They must therefore take it to be a *primitive* notion.

Davidson has moved toward a conclusion similar to that of Scriven and Collins. He argues that (CE_2) does not analyze (CE_1), that, in other words, statements of the sort "A causes B" are not collapsed arguments. But he also argues that it is wrong to take 'causes' to be a connective; rather, he suggests, it is best construed as an ordinary two-place relation of ordinary first-order logic. This latter position has clear advantages over the view that takes 'causes' as a nomological connective. For, if 'causes' is an ordinary relation in an ordinary first-order language, there is no challenge whatsoever to the usual explication of the analytic/synthetic distinction. That first-order languages capture the logic of ordinary discourse is a major theme of Davidson's work.[250] Davidson's proposal does not, however, avoid the second of the philosophical problems confronting those who hold (CE_2) does not analyze 'causes', namely, how do we *know* when the causal relation (or connective) obtains? Davidson, unfortunately, asserts we *do* know, at times, that the relation obtains,[251] and lets it go at that. Here Davidson's position is similar to that of Collins, who also just asserts. Scriven is rather better than this, as we shall see. These positions of Collins and Scriven we shall turn to shortly, and the discussion of Collins will suffice as comment upon Davidson also. The other difficulty with Davidson's position is what we above called the "third feature" of the nomological connective, that the complex causal statement will be true only if the atomic sentences mention *kinds* that are causally related. Davidson obscures his failure to account for this feature by a spurious distinction between the events that are causally related and descriptions of those events. Kim has more recently attempted to remedy this defect in Davidson's account. The result is a position which either is more or less the same as that which we are defending or involves a bloated Meinongian ontology in which to all intents and purposes 'causes' is a special non-truth-functional sort of connective.

Clearly, the positions of Davidson and Kim are worth looking at. We shall first look at Davidson's defence of his thesis. It will turn out that whatever force it has in showing 'causes' is not a non-truth-functional connective, it is inconclusive against the deductivist thesis that (CE_2) analyzes (CE_1). We shall then turn to Kim's attempt to fix up Davidson's position so as to capture the relevance of kinds to the obtaining of causal connections. We shall conclude, as we just said, that either Kim is defending a position much like ours or his is a Meinongian version of the idea that 'causes' is a special sort of connective. This last will bring us back to the views of Collins and Scriven that we have just now been focusing upon.

Davidson's thesis is clear:

So far as [the logical] form [of causal statements] is concerned ... the relation of causality between events can be expected (no matter how "strong" or "weak" it is) by an ordinary two-place predicate in an ordinary, extensional first-order language.[252]

He argues for this by (a) arguing against the view that causal statements are collapsed arguments (i.e., the view we are defending), (b) arguing against the view it is a connective, and (c) arguing that his own view solves various philosophical problems. Our strategy should be clear: we shall argue that both (a) and (b) are unsound, and that with respect to (c), it leaves as many problems unsolved as it solves, and in particular that the problems it solves are equally solved by our position, that (CE_2) analyzes (CE_1). Diagnostically, we shall suggest Davidson's position is plausible only because he fails, as do Scriven and Collins, to recognize there are imperfect laws and that these, as well as determinate and process laws, can be used in explanatory contexts.

Davidson considers[253] the sentence

(f_1)   The short circuit caused the fire

and raises the question whether its form might not be given by

(f_2)   *The fact that* there was a short circuit *causes it to be the case that* there was a fire.

He raises several objections to treating (f_2) as involving a connective with the properties of a conditional. The first point he makes is that it cannot be a material conditional to the effect that "If there was a short circuit, then there was a fire":

No doubt (f_2) entails this, but not conversely, since (f_2) entails something stronger, namely, the conjunction 'There was a short circuit *and* there was a fire'.[254]

He uses this to argue against a proposal of Arthur Pap (among others) to treat the connective in (f_2) as a conditional. Pap writes that

The distinctive property of causal implication as compared with material implication is just that the falsity of the antecedent is no ground for inferring the truth of the causal implication.[255]

Davidson comments that

If the connective Pap had in mind were that of (f_2), this remark would be strange, for it is a property of the connective in (f_2) that the falsity of either the "antecedent" or the "consequent" is a ground for inferring the falsity of (f_2).[256]

This comment is fair enough. It does establish (f_2) is not a conditional.

This argument of Davidson does not affect the case we are making for (CE_2) providing an analysis of (CE_1). What we are proposing is that

(A_1)   Being A causes this to be B

is a telescoped argument

(A$_2$) Whenever something is A-like then it is B-like

This is A-like

———————————

This is B-like

To assert (A$_1$) is thus to law-assert the major premiss of (A$_2$), to assert the minor premiss, to assert the premisses entail the conclusion, and (therefore) to assert the conclusion. Clearly, Davidson's point is not successful against this analysis.

We should, of course, distinguish (A$_1$) from

(A$_3$) If this is A then [causally] this is B

which is a conditional of sorts. Upon our acccount this recieves the same analysis (A$_2$) as (A$_1$), save for what is asserted. For (A$_1$) and (A$_3$), the major premiss of (A$_2$) is law-asserted. But for (A$_1$) the minor is asserted while for (A$_3$) it is not asserted, merely treated hypothetically. The falsity of the minor of (A$_2$) is, as Davidson says, sufficient for the non-assertibility of (A$_1$). But not for the non-assertibility of (A$_3$).

Burks holds that (A$_3$) is a conditional with at least one of logical properties of ordinary conditionals, namely, transposition: "p is causally sufficient for q is logically equivalent to $\sim q$ is causally sufficient for $\sim p$".[257] Davidson comments that

... this shows not only that Burks' connective is not that of (f$_2$), but also that it is not the subjunctive conditional 'would cause'. My tickling Jones would cause him to laugh, but his not laughing would not cause it to be the case that I didn't tickle him.[258]

This is less felicitous. And, certainly, it is not a position we ourselves can endorse, since transposition holds for (A$_3$) if the latter is understood as (A$_2$). Thus, (A$_3$) transposes to

(A$_4$) If this is not B then [causally] this is not A

which we would analyze as

(A$_5$) Whenever something is A-like then it is B-like

This is not B-like

———————————

This is not A-like

300   CHAPTER 3

This argument is as valid as (A_2); and the major premiss of the one is law-assertible if and only if that of the other is also. But we may dispute the clarity of Davidson's example. We have

(*)   My tickling Jones causes him to laugh

and

(**) Jones' not laughing causes me not tickle him.

In (*) the 'causes', since it connects an action to its intended effect, carries that suggestion of asymmetry introduced by considerations of strategy. Since the 'causes' in (**) does not connect an action to its upshot, the suggestion of strategical asymmetry would be inappropriate. Here is one reason for hesitating to infer (**) from (*). In (*), something causes *me* to act or fail to act; the suggestion here is that of affording me a motive to act. This suggestion is quite inappropriate in the context of (**). Here is a second reason for hesitating to infer (**) from (*). In conclusion, therefore, Davidson has not shown transposition fails to hold for the causal tie.

But there are other properties of the ordinary conditional, i.e., material implication, that certainly do fail to hold of the causal tie. Consider[259]

(g_1)   It is false that if the Liberals win the election there will be an increase in welfare programmes.

This could easily be used in the course of a political debate. Normally, its speaker would not be understood to be either expecting or predicting a Liberal victory. But if we translate (g_1) as

(g_2)   ∼(The Liberals will win the election ⊃ there will be an increase in welfare programmes)

then we obtain a sentence logically equivalent to

(g_3)   The Liberals will win the election and there will be no increase in welfare programmes

since '$[\sim(p \supset q) \equiv (p \ \& \sim q)]$' is a tautology. However, no one asserting (g_1) would wish to assert (g_3); (g_2) therefore cannot translate (g_1). Upon the view we are defending, this is, of course, true. The conditional in (g_1), namely,

(g_4) If the Liberals win the election then [causally] there will be an increase in welfare programmes

should not be translated as in (g_2) but as

(g_5) (Normally) whenever Liberals win an election there is an increase in welfare programmes

The liberals will win this election
***
There will be an increase in welfare programmes

where the minor is neither asserted nor denied, but taken hypothetically, and the major is law-asserted. (g_4) is thus translated just as (A_2) translates (A_3). (g_1) is the negation or denial of (g_4); that is to say, it is the denial of (g_5). The latter, however, is not a single sentence but an argument. Denial therefore cannot consist simply of negating a sentence. A denial of (g_5) = (g_4) may be either (i) a denial of the validity of the argument, or (ii) a denial of the truth of the minor premiss, or (iii) a denial of the truth of the major premiss, or (iv) a denial, not of the truth, but of the law-assertibility of the major premiss. None of these involves asserting that the Liberals will win the election. That the causal connective, when negated, does not behave like a conditional is not at all surprising upon our point of view.

Another example of the same point is this.[260] In an appropriate context, someone might well assert that

(g_6) If Smith was convicted only if he was guilty, then justice was served

but he would not normally feel himself committed thereby to the assertion of

(g_7) If Smith was acquitted then justice was served.

But if (g_6) is translated as

(g_8) (Smith was convicted $\supset$ he was guilty) $\supset$ justice was served

we have a sentence which, assuming non-conviction is the same as acquittal, is logically equivalent to

(g_9) (Smith was acquited $\supset$ justice was served) & (Smith was guilty $\supset$ justice was served)

the English translation of which would be

(g_{10}) If Smith was convicted then justice was served, and if Smith was guilty then justice was served

Since (g_6) can be asserted without any committment to either (g_7) or (g_{10}), it follows (g_7) cannot translate (g_6). The causal tie thus differs in this way also from the ordinary conditional, that, for the ordinary conditional, sentences like (g_8) entail sentences like (g_9), while, for the causal tie, sentences like (g_6) do not entail sentences like (g_{10}). Again, our view, that causal statements are collapsed arguments, makes this something to be expected. But what exactly shall we say about a sentence like (g_6)? The antecedent of (g_6) is

(g_{11}) Smith was convicted only if he was guilty

which, upon our account, is the collapsed argument

(g_{12}) Whenever a person is convicted then he is guilty
Smith was convicted.
─────────────────────────
Smith was guilty

with the major premiss law-asserted and the minor taken hypothetically. It is this *argument* (g_{12}) which is the "antecedent" of (g_7). But, now, what connects this "antecedent" with the consequent of (g_7)? In the case of (g_{11}), the statements which are the antecedent and the consequent appear in (g_{12}) as themselves, the minor premiss and the conclusion; and the property mentioned in the minor is, by the major premiss, lawfully connected with the property mentioned in the conclusion. The connection between antecedent and consequent in (g_{11}) is provided by the major premiss of (g_{12}). In (g_7), however, the "antecedent" is not a statement attributing a property to an individual. The "antecedent" is itself an argument, and it makes no direct sense to try to supply a law-asserted premiss enabling us to deduce

(g_{13}) Justice was served

from (g_{12}). We must look for a more complicated account. What should be said can be brought out by noting the suggestion that (g_6) makes to the effect that *sometimes* being guilty is *not* causally necessary to conviction. Justice is served in Smith's case, says (g_6), if it is true that in Smith's case guilt was causally relevant. To say that sometimes guilt is not causally necessary to conviction is to say that the major premiss of (g_{12}) is literally false. Nonetheless is not wholly wrong: in spite of the exceptions it does get at something true. Which is to say, the law-assertion of the major in (g_{12}) should be taken to be the law-assertion of the piece of imperfect knowledge, something to the effect of

(g_{14})   Whenever a person is convicted then he is guilty unless he is Q

where Q specifies one or more exceptive conditions. Smith's being guilty will be causally necessary to his being convicted just in case Smith is not Q. What is crucial, then, for justice to have been served in Smith's case is that these exceptive conditions Q do not hold for Smith. This permits us to attribute the following complex structure to (g_6):

(g_{15})
- (p_1) Whenever a person is convicted then he is guilty unless he is Q, and if he is both convicted and not Q then, with respect to that person, justice is served.
- (p_2) Smith is convicted
- (p_3) Smith is not Q
- (c_1) Smith is guilty
- (c_2) With respect to Smith, justice is served

Here (p_1) is law-asserted, while (p_2) and (p_3) are taken hypothetically. (p_1) entails (g_{14}). The "antecedent" of (g_6) consists of drawing the conclusion (c_1) from the law premiss (g_{14}) that is entailed by (p_1), and the initial conditions (p_2) and (p_3); while the consequent of (g_6) consists of the further conclusion (c_2) drawn from the same initial conditions and the law (p_1) which stands to the (g_{14}) required in the "antecedent" as the less to the more imperfect.

Our analysis of these examples shows we are safe in accepting the general conclusion that Davidson draws, that the 'causes' of (f_2) is not, in any straightforward sense, a conditional: the analysis of the above examples shows that this is exactly what we would expect upon the account we are defending, that laws are part of what is asserted when a statement like (f_2) is asserted and that a statement like (f_2) is most appropriately understood as a collapsed argument.

Davidson, of course, does not accept our analysis. He offers a series of points designed to establish the inappropriateness of construing (f_2) as a collapsed argument.

This approach [that of construing (f_1) as (f_2)] no doubt receives support from the idea that causal laws are universal conditionals, and singular causal statements ought to be instances of them .... We might try treating (f_2) as the conjunction of the appropriate law and 'There was a short circuit and there was a fire' ... but then (f_2) would no longer be an instance of the law. And aside from the inherent implausibility of this suggestion as giving the logical form of (f_2) (in contrast, say, to giving the grounds on which it might be asserted) there is also the oddity that an inference from the fact that there was a short circuit and there was a fire, and the law, to (f_2) would turn out to be no more than a conjoining of the premisses.[261]

But since singular causal statements are obviously not conditionals, as we have seen, the idea that they ought to be instances of lawful universal conditionals is not plausible at all. Whatever support the construing of (f_1) as (f_2) receives, it does not derive from this quarter. The most to be said is that the assertion of singular causal statements should *involve* the assertion of laws. We can therefore safely ignore the criticism that if we construe (f_2) as a collapsed argument then it is no longer an instance of the law. As for the point that inference from the two singular statements and the law to (f_2) is nothing more than conjunction, we must grant that this is indeed the *logical form* of the inference. But we can also point out that a substantial change in *grammatical form* occurs in the shift to (f_2). Once we see this we can see why it *is* odd, as Davidson says, to hold the move to (f_2) is one of *mere* conjunction, while we also hold that, logically speaking, that really is all that is involved. Finally, the point that the law is more appropriately construed as grounds for, rather than as part of, the assertion (f_2), is one that simply begs the question at issue.

These arguments of Davidson therefore fail to establish that (f_2) cannot be analyzed as a collapsed argument. But this is not his main artillery. Davidson also argues that the 'causes' of (f_1) should not be thought of as a connective at all. That is, he attacks the move from (f_1) to (f_2). We do not take the connective of (f_2) to be unanalyzable, as do Scriven and Collins, but we do accept the move to (f_2). It is against this that Davidson's main argument is directed.

This argument attempts to establish that the connective in (f_2) both is and is not truth-functional. As for the latter, this is evident, since (f_2) may change from true to false if the contained sentences are switched.[262] On the other hand, Davidson argues, substitution of materially equivalent expressions *does* preserve truth, and therefore the connective of (f_2) must be truth-functional.

... (f_2) retains its truth if for 'there was a fire' we subtitute the logically equivalent '$\hat{x}(x = x$ & there was a fire$) = \hat{x}(x = x)$'; retains it still if for the left side of this identity we write the coextensive singular term '$\hat{x}(x = x$ & Nero fiddled)'; and still retains it if we replace '$\hat{x}(x = x$ & Nero fiddled) = $\hat{x}(x = x)$' by the logically equivalent 'Nero fiddled'. Since the only aspect of 'there was a fire' and 'Nero fiddled' that matters to this chain of reasoning is the fact of their material equivalence, it appears that our assumed principles have led to the conclusion that the main connective of (f_2) is, contrary to what we supposed, truth-functional.[263]

We have seen this sort of argument previously, in Section 3.2 above. The "assumed principles" Davidson mentions are ($Subst_1$), the principle of substituting logically equivalent expressions, and ($Subst_2$), the principle of substituting coextensive singular terms. When we discussed these in Section 3.2, we granted that the first applied to the contexts of (f_2). We did not grant that ($Subst_2$) applied, however. Davidson argues as follows:

If Smith's death was caused by the fall from the ladder and Smith was the first man to land on the moon, then the fall from the ladder was the cause of the death of the first man to land on the moon. And if the fact that there was a fire in Jones' house caused it to be the case that the pig was roasted, and Jones' house is the oldest building on Elm street, then the fact that there was a fire in the oldest building on Elm street caused it to be the case that the pig was roasted. We must accept the principle of extensional substitution, then.[264]

Recall, here, that we are arguing that

(h_1) Fa caused Ga

can be asserted just in case

(h_2) $(x)(Fx \supset Gx)$
           Fa
           ———————
           Ga

is sound, the major premiss can be law-asserted, and the minor can be asserted. ($Subst_2$) leads us from (h_1) to

(h_3) $F(\imath y)(Hy)$ caused Ga

which, upon our analysis is

(h_4) $(x)(Fx \supset Gx)$
           $F(\imath y)(Hy)$
           ———————
           Ga

But (h_4) is invalid, and so (h_3) is not assertible. (Subst$_2$) preserves truth, but not the assertibility of causal statements, insofar as the latter are construed as collapsed arguments, since it [Subst$_2$] does not preserve validity. Davidson's argument is to the effect that the move from (h_1) to (h_3) *is* satisfactory: if

(h_5) $a = (\imath y)(Hy)$

holds then (h_1) is, according to Davidson, assertible if and only if (h_3) is assertible. What *we* are committed to is the idea that (h_1) is assertible if and only if

(h_6) $a = (\imath y)(Hy) \supset [F(\imath y)(Hy) \text{ caused } Ga]$

is assertible. We arrive at something like Davidson's point if we can hold that in most normal contexts a sentence like (h_3) is *implicitly* of the form (h_6). Davidson holds

(h_7)   Smith's death was caused by the fall from the ladder

is assertible if and only if

(h_8)   The death of the first man to land on the moon was caused by the fall from the ladder

is assertible, given that

(h_9)   Smith was the first man to land on the moon

and where the grammatical form of (h_8) is also its logical form. In contrast, we hold that (h_1) is assertible if and only if the logical form of (h_8), as distinct from its grammatical form, is given by

(h_{10})   (Smith was the first man to land on the moon) $\supset$ [the death of the first man to land on the moon was caused by the fall from the ladder]

But as soon as we see this, we also see that Davidson has given us no adequate grounds to prefer his position to ours. He has thus *not* established that (Subst$_2$) holds for contexts of the sort (f_2).

Davidson's argument was to the effect that (f_1) could not be transformed into (f_2) because the connective in (f_2) would have to be both non-truth-functional and truth-functional. But he has not established the connective **is truth-functional.** We can therefore continue to construe (f_1) as (f_2), and

EXPLANATIONS AND EXPLAININGS

to construe 'causes' as a non-truth-functional connective. However, it must be emphasized that we hold, contrary to, say, Scriven and Collins, that this non-truth-functional connective is not primitive, that it is analyzable, and, if you wish, analyzable away, in terms of laws.

Here, however, Davidson attacks the analyzability thesis. He holds that we need not "be able to dredge up a law if we know a singular causal statement to be true"; all we need to know is that "there must be a covering law."[265] If we can know and reasonably assert singular causal statements without knowing a law then the thesis we are defending is false. His argument is this:

> ... very often, I think, our justification for accepting a singular causal statement is that we have reason to believe an appropriate causal law exists, though we do not know what it is. Generalizations like 'If you strike a well-made match hard enough against a properly prepared surface, then, other conditions being favorable, it will light' [call this (G)] owe their importance not to the fact that we can hope to render them untendentious and exceptionless, but rather to the fact that they summarize much of our evidence for believing that full-fledge causal laws exist covering events we wish to explain.[266]

The reply to this is that the laws that are part of singular causal judgments need not be causal laws of the sort mentioned in the final quoted sentence and which are contrasted to laws of the sort (G). Davidson assumes the causal laws of singular causal judgments have to be determinate and specific; (G) will not do. But in many cases all we have are laws of the sort (G). Hence, Davidson concludes, when we assert a singular causal judgment we may know there is a causal law but not know what, specifically, that law is. The reply, we know, to this argument is to deny the premiss that laws of the sort (G) will not do. Davidson, apparently, confuses the need for laws with the need for determinate, or even process laws. Which is to say that Davidson does not recognize that *imperfect laws* may be used when making singular causal assertions.

Davidson states his position on the relation of singular causal statements to laws as follows:[267] "if 'a caused b' is true, then there are descriptions of a and b such that the result of substituting them for 'a' and 'b' is entailed by true premisses" which include a determinate causal law. This brings out one of the more subtle points of Davidson's overall argument: when Davidson uses 'description' he means *definite description* rather than indefinite description. We have thus far rather calmly translated

(f_1)   The short circut caused the fire

into such schemata as

(i_1)   *Fa* caused *Ga*

where Davidson wants them translated into schemata of the sort

(i_2)   The F caused the G

How do we handle schemata such as (i_2)? Do they yield problems sufficient to force us to give up our analysis?

Now, there is no reason why definite descriptions cannot occur in causal contexts. Thus, we might have, instead of (i_1), the singular causal assertion

(i_3) $F(\imath x)(Hx)$ caused $G(\imath x)(Hx)$

Upon our account, (i_3) would expand into

(i_4) $(x)(Fx \supset Gx)$
         $F(\imath x)(Hx)$
         ———————
         $G(\imath x)(Hx)$

which is acceptable provided that it is valid (which it is) and provided that the premisses are true, which means, among other things, that the second premiss is true, that the definite description is successful. That is, this argument is acceptable as an explanation provided that

(i_5) $(\exists! x)(Hx)$

obtains. Normally one would not use a definite description unless (one had good reason to believe) it was successful, so (i_5) would normally be supplied by the context, implicit in it. Now, if 'the F' and 'the G' were successful, if

(i_6) $(\exists! x)(Fx)$
         $(\exists! x)(Gx)$

both held, then we might even have

(i_7) $F(\imath x)(Fx)$ caused $G(\imath x)(Gx)$

which, in English, is the tedious

(i_{7*})   The F which is F caused the G which is G

But (i_6) entails

(i_8) $F(\imath x)(Fx)$
         $G(\imath x)(Gx)$

The context would supply (i_6) if (i_7) were to be used in a normal way, just as the context supplies (i_5) for (i_3). But if this is so, then the obvious (i_8) would also be supplied by the context and the tedious (i_7) = (i_{7*}) could be shortened to

(i_9) The F caused the G

Let us now make the assumption that

(i_{10}) The F = the G

We need not make this assumption. Instead, we could take the relevant law to be of the form

$$(x)(y)[Fx \& R(x,y): \supset Gy]\ ^{268}$$

or even

$$(x)(y)[Fx \& R(x,y): \supset : (z)(R(x,z) \supset z = y) \& Gy]$$

which says that if an F Rs something then it Rs at most one thing and that thing is G. We could even add a further clause that the G is R-ed by at most one F. We could also have

$$(x)[Fx \supset (\exists y)(R(x,y) \& Gy)]$$

which states that for any F there is a G that F Rs, or even the stronger

$$(x)[Fx \supset (\exists! y)(R(x,y) \& Gy)]$$

If there is exactly one F then we could assert that "the F caused the G R-ed by the F", or, if we let '$G'x$' abbtreviate '$(\exists! y)(R(x,y) \& Gy)$' then we could assert ($i_9$)

the F caused the G'

So if the law were of the form last considered we could handle (i_9) without the special assumption (i_{10}). All these complications, including the last, are possible. But the crucial points are more perspicuous if we stick with

$$(x)(Fx \supset Gx)$$

310   CHAPTER 3

as our law, which we can do only if we make the assumption (i_{10}). Let us therefore stick with that (often unrealistic) assumption. In that case, we obtain the following account of (i_9) as a collapsed argument. *First*, (i_9) is to be expanded into (i_7), using context-supplied information. Then, *second*, (i_7) is to be expanded into

(i_{11}) $(x)(Fx \supset Gx)$
            $\underline{F(\imath x)(Fx)}$
            $G(\imath x)(Gx)$

where, *third*, the context supplies (i_6) and (i_{10}) as the further additional premisses needed to secure acceptability. *What is being explained here is why the G is a G.* What we have explained is that, since the G is an F the G is also a G. Just as (i_4) explains why the H is a G, so (i_{11}) explains why the G is a G. In the case of the H, the explanation would, presumably, be offered in a context in which it was wondered

(i_{12})   Why was the H a G?

(i_{12}) is a complex question, presupposing

(i_{13})   Does the H exist?

has been answered affirmatively, i.e., presupposing (i_5)

        $(\exists ! x)(Hx)$

holds. In the case of the G, the explanation would be offered in a context in which it was wondered

(i_{14})   Why was the G a G?

In this case, the presupposition would be

(i_{15})   Does the G exist?

an affirmative answer to which is available only if (i_6)

        $(\exists ! x)(Gx)$

holds. But the presupposition entails (i_8)

        the G is a G

so instead of asking (i_{14}) we might well ask,

(i_{16})   Why did the G occur?

The point is, (i_{16}) is a question that might reasonably be answered by (i_{11}). But (i_{16}) could be intended as a request for a quite different sort of explanation. It could be a question to the effect that

(i_{17})   Why did *the* G and *no other* G occur?

Similar questions could be asked about the *H*:

(i_{18})   Why was the *H* a *G* and why was there no other *G*?

This question requests an explanation for the two facts

(i_{19}) $G(\imath x)(Hx) \& (y)[Gy \supset y = (\imath x)(Hx)]$

In the same way, (i_{17}) requests an explanation for

(i_{20}) $G(\imath x)(Gx) \& (y)[Gy \supset y = (\imath x)(Gx)]$

which is a complicated way of asking for an explanation for

(i_{21}) $(\exists x)[Gx \& (y)(Gy \supset y = x)]$

(i_{11}) could not explain this fact. Thus, whether (i_{11}) answers (i_{15}) is going to depend upon the context, upon how (i_{15}) is understood.

I go through this because Davidson succeeds in obfuscating issues by exploiting this ambiguity.

> If what we are to explain is why an avalanche fell on the village last week, we need to show that conditions were present adequate to produce *an* avalanche .... We might instead have asked for an explanation of why *the* avalanche fell on the village last week. This is, of course, a harder task, for we are now asking not only why there was at least one avalanche, but also why there was not more than one. In a perfectly good sense the second can be said to explain a particular event; the first cannot.[269]

The last is just wrong. The first explanation *also* explains a particular event: it explains why, *at a particular place and time*, an avalanche occurred. The second explains that same particular event *and also* why no other events of the same kind occurred. But suppose we speak as Davidson does, and also hold that singular causal judgments do explain why particular events occur, then it will be rare indeed when a law will be known from which the facts to be explained can be deduced.[270] It will then be more than plausible that causal laws will, contrary to what we are arguing, not be part of singular causal judgments.[271] But Davidson has given us no reason to suppose that the question (i_{16}) normally asks more than (i_{14}), and therefore that it cannot normally be answered by an explanation like (i_{11}), or, indeed, one in which the law is more imperfect.[272]

As we saw, Davidson holds we should say the following about singular causal judgments: "if '*a* caused *b*' is true, then there are descriptions of *a* and *b* such that the result of substituting them for '*a*' and '*b*' in '*a* caused *b*' is entailed by true premisses" which include a determinate causal law.[273] What he should say, we now see, is this: "*a* caused *b*" holds if there are indefinite descriptions of *a* and *b*, say *F* and *G*, such that '*b* is *G*' can be deduced from '*a* is *F*' and a law. The law need not be a determinate causal law; it might well be imperfect. Furthermore, the properties *F* and *G* need not be such that 'the *F*' and 'the *G*' are successful definite descriptions, nor, therefore, such that *a* = the *F* and *b* = the *G*. And even if 'the *G*' is successful, and *b* = the *G*, "*a* caused the *G*" need not be taken as offering an explanation of anything more than why the *G* is *a G*. Finally, when definite descriptions are used to pick out the explaining and explained events, the causally relevant properties, those mentioned in the explaining law, need not be mentioned at all. Thus, if

>*a* is *F*

is

>The event described on p. *N* of the *Times* of date *d*

while

>*b* is *G*

is

>The event described on p. *M* of the *Times* of date *d'*

then we may quite correctly make the singular causal assertion

>The event on p. *N* of the *Times* of date *d* caused the event described on p. *M* of the *Times* of date *d'*

based on the law

>$(x)(Fx \supset Gx)$

We can do this since the two definite descriptions pick out the properties *F* and *G* as well as the particular *a*. Clearly, it is absurd to think the two causally relevant properties to be

>*x* is described on p. *N* of the *Times* of date *d*

and

>*x* is described on p. *M* of the *Times* of date *d'*.

EXPLANATIONS AND EXPLAININGS 313

Once all this is recognized we see Davidson has in no way rendered it implausible that, when a singular causal judgment "*a* caused *b*" is justified, not only is there a law, but there is a *law present in and part of that judgment*. What Davidson does is confuse issues by failing to distinguish the properties mentioned in the explanatory law from the properties mentioned by a definite description used to pick out the particular in the event to be explained. These may indeed coincide, but they need not, and, since they need not, Davidson's account of singular causal judgments, quoted above, is wrong. He has not shown that we must accept "that we may correctly give the cause without saying enough about it to demonstrate that it was sufficient . . . ".[274]

It is, *of course*, true that when we mention the explaining event, we may not explicitly mention all the properties needed to demonstrate this event is sufficient for the occurrence of the event to be explained. But they should be implicit in the context. This is what led Mill[275] — and, indeed, ourselves — to hold that most singular causal judgments are elliptical, a claim which Davidson naturally denies:

> . . . There is not, as Mill and others have maintained, anything elliptical in the claim that a certain man's death was caused by his eating a particular dish, even though death resulted only because the man had a particular bodily constitution, a particular state of health, and so on.[276]

Elsewhere he says that

> What is partial . . . is the *description* of the cause; as we add to the description of the cause, we may approach the point where we can deduce, from this description and laws, that an effect of the kind described would follow.[277]

We must, as Davidson says, "distinguish firmly between causes and the features we hit upon for describing them",[278] but we must also, as he does not add, distinguish between indefinite descriptions and definite descriptions and, in the latter, those that mention causally relevant properties and those that do not. Does the fact that the description, presumably the definite description, is partial in the properties it mentions force us to say the unmentioned things are *not* (implicitly) part of the explanation? Are we forced to withdraw Mill's claim that what is explicitly said is elliptical for an implicit and fuller explanation?

We have

(k_1)   The *F* caused the *G*

where this is taken to explain that the *G* is a *G* on the basis of the *F* being an *F*. To say that this is elliptical for not mentioning explicitly other relevant factors is to say that these other relevant factors are implicit in the context. Instead of (k_1) we have (e.g.)

(k_2)   The F caused the G but did so only because it was not H

(k_2), and so also (k_1) when it is made fully explicit, is thus to be understood as a collapsed form of

(k_3) $(x) [Fx \supset (Gx \lor Hx)]$
$\underline{F(\imath x) (Fx) \mathbin{\&} {\sim}H(\imath x) (Fx)}$
$G(\imath x) (Gx)$

where we take

(k_3*) $(\exists! x) (Fx)$
$(\exists! x) (Gx)$
$(\imath x) (Fx) = (\imath x) (Gx)$

to be further implicit premisses.[279] Now, if we treat definite descriptions in the fashion of Russell then (if 'the F' is successful then) it follows from

(k_4)   the F is ${\sim}H$

that

(k_5)   the F = the thing which is $F \mathbin{\&} {\sim}H$

and conversely. This enables us to think of ourselves as explanding (k_1) to

(k_6)   The $(F \mathbin{\&} {\sim}H)$ caused the G

when we discover the relevance of H and the necessity of asserting that it is absent from the F, i.e., of asserting (k_4). In this sense, that of (k_5) following from (k_4), we can speak of the description being partial, and being made less partial as more relevant factors are explicitly stated. (k_6) would then expand into the collapsed argument

(k_7) $(x) [Fx \supset (Gx \lor Hx)]$
$\underline{F(\imath x) (Fx \mathbin{\&} {\sim}Hx) \mathbin{\&} {\sim}H(\imath x) (Fx \mathbin{\&} {\sim}Hx)}$
$G(\imath x) (Gx)$

with some sort of further implicit premisses as (k_3). (k_7) explains *the same event* as (k_3) and in both *the explaining events are the same*. It is just that the events are picked out by different definite descriptions in the two cases.

Thus, given (k_4), which is part of (k_3), the explanation (k_1) = (k_3) is the same as the explanation (k_6) = (k_7). It is just that (k_6) = (k_7) uses a definite description mentioning more causally relevant properties than does the definite description in (k_1) = (k_3). None of this means, however, these other factors, those causally relevant but not *explicitly* mentioned, cannot be taken to be *implicitly* mentioned. Davidson has not shown we *must not* construe such explanations as (k_1) as elliptical for (k_2) and (k_3); the distinction he draws is not sufficient for this.

If Davidson's distinction is to do the work he requires of it, then there must be some further thought involved in it. For our case to hold what we have to argue is that when a person offers an explanation such as (k_1)

The F caused the G

then (supposing the judgment to be rational) we can find implicit in that assertion a law. This means, roughly, that if we question why this person thinks the causal relation obtains then he will assert a law enabling us to deduce the one event from the other. And if this law turns out to be something like

(k_8) $(x) [Fx \supset (Gx \lor Hx)]$

then we must also suppose he is implicitly asserting such necessary additional conditions as

(k_9)   The F is not H

or, what is the same (since (k_4) and (k_5) entail each other), that the definite description

(k_{10})   The F

used in (k_1) is implicitly the definite description

(k_{11})   The (F & $\sim H$)

where the mentioned properties are, since (k_8) is logical equivalent to

(k_{12}) $(x) [(Fx \& \sim Hx) \supset Gx]$

sufficient for the deduction of the event to be explained. Davidson raises the question, "Can we ... analyze 'a caused b' as meaning that a and G may be described in such a way that the existence of each could be demonstrated, in the light of causal laws, to be a necessary and sufficient condition for the existence of the other?", where we also have "the condition that either

316   CHAPTER 3

a or *b*, as described, exists", i.e., that the definite descriptions are successful. He argues for a negative answer:

> Then on the proposed analysis one can show that the causal relation holds between any two events. To apply the point in the direction of sufficiency, imagine some description '($\imath x$) (Fx)' under which the existence of an event *a* may be shown sufficient for the existence of *b*. Then the existence of an arbitrary event *c* may equally be shown sufficient for the existence of *b*: just take as the description of *c* the following: '($\imath y$) ($y = c$ & ($\exists ! x$) (Fx))'. It seems unlikely that any simple and natural restrictions on the form of the allowable descriptions would meet this difficulty . . . .[280]

To see whether this negative answer is justified, let us first examine the deductions in question. We have '($\imath x$) (Fx)' describing *a* and, let us suppose, '($\imath x$) (Gx)' describing *b*. *a* and *b*, as described, must exist:

(m_1) ($\exists ! x$) (Fx)
(m_2) ($\exists ! x$) (Gx)

And the existence of *a*, so described, is sufficient for the existence of *b*, so described. Thus, the laws are such that (m_2) is deducible from (m_1). Davidson is here assuming we wish to explain not only why *the G is a G* but also why there is exactly one *G*; we wondered about this just above, but here let us grant him the assumption. We now pick *c*, and form this description of it

(m_3)   ($\imath y$) [$y = c$ & ($\exists ! x$) (Fx)]

c, so described, exists:

(m_4)   ($\exists ! y$) [$y = c$ & ($\exists ! x$) (Fx)]

Finally, the existence of *c*, so described, is sufficient for the existence of *b*, described as before; that is, (m_2) is deducible from (m_4). All this is quite true. It is the nature of the deductions that it is important to notice. What (m_4) asserts is

(m_5)   ($\exists y$) [$y = c$ & ($\exists ! x$) (Fx) & (z) [$z = c$ & ($\exists ! x$) (Fx): $\supset z = y$]]

But

(y) [$p \supset Fy$]

is logically equivalent to

$p \supset (y) (Fy)$

and

$$(\exists x)(Fx \mathbin{\&} p)$$

is logically equivalent to

$$p \mathbin{\&} (\exists x)(Fx)$$

So (m_5) is logically equivalent to

(m_6) $(\exists ! x)(Fx) \mathbin{\&} (\exists y)[y = c \mathbin{\&} (z)(z = c \supset z = y)]$

If we assume the usual rule for interpreting individual constants, *that each name names exactly one thing*, then the right-hand conjunct of (m_6) is a necessary truth. But a conjunction, one conjunct of which is a necessary truth, is logically equivalent to the remaining conjunct. (m_6) is logically equivalent to (m_5); (m_5) is what (m_4) asserts; (m_6) is logically equivalent to its left-hand conjunct, i.e., (m_1). So *(m_4) is logically equivalent to (m_1)*. In other other words, to say (m_2) is deducible from (m_4) is to say nothing more nor less than to say (m_2) is deducible from (m_1). To say c exists under the description (m_3) is to say nothing more than that a exists and that (m_1) obtains. Since (m_2) is deducible from (m_1) we should not be surprised it is deducible from the fact that c, as described by (m_3), exists. The only thing that is relevant to the deduction of (m_2) is the existence of a as given by (m_1); the existence of c is irrelevant in just the way that in

$$\frac{(p \vee q) \mathbin{\&} (r \vee \sim r)}{\sim q}$$
$$p$$

'$r \vee \sim r$' is irrelevant to the deduction of p. It is not just that irrelevant information is introduced; the point is that no factual information at all is introduced (all that is added is the linguistic point that 'c' is a name). We may conclude, then, that although (m_2) is deducible from (m_4), c is in fact explanatorily irrelevant to the existence of b as described. Of course, the definite description (m_3) does denote c. That is,

(m_7) $c = (\imath y)[y = c \mathbin{\&} (\exists ! x)(Fx)]$

is true. And in fact, (m_7) is true just in case it is both true that 'c' names exactly one thing and true that (m_1). But the former is a necessary truth. So (m_7) holds if and only if (m_1) holds. But this hardly means c is of explanatory relevance to the existence of b as described. Davidson thinks otherwise, but is mistaken. Very likely he is led into this mistake by confusing the properties used to explain an event with the property mentioned in the definite description used to pick out the explaining event; the latter in no way needs to be a property causally relevant to the event to be explained.

The point may be put in another way. (m_7) and (m_4) are logically equivalent to (m_1) and the fact that c exists. c, as described by (m_3), does exist; but to say the latter, i.e., to say (m_4), is to say no more than that (m_1) is true and c exists. (m_2) is deducible from (m_1). Hence, even if c were not to exist, (m_2) would still obtain and would still be explained by (m_1). The existence of c is thus causally irrelevant to the existence (m_2) of b as described. And this yields a "simple and natural restriction on the form of allowable descriptions": an event c, no matter how described, is causally irrelevant to an event b under some description just in case that b would exist under that description even if c were not to exist. The causal irrelevance of c is shown by the irrelevance of its existence to the deduction of (m_2) from (m_4).

Kim has introduced considerations similar to those of Davidson but perhaps more interesting. He calls the problem the problem of "piggy-back" laws, or "the problem of parasitic constant conjunctions."[281] He considers, following Foster,[282] that

a's being F is causally sufficient for a's being G

might be analyzed somewhat as follows:

there is a relation R such that

(i) Fa, Gb, and $R(a, b)$
(ii) $(x)\,[Fx \supset (\exists y)\,(R(x, y)\,\&\,Gy)]$
(iii) $(x)(y)\,[R(x, y) \supset (z)\,(R(x, z) \supset z = y)]$
       $(x)(y)\,[R(x, y) \supset (z)\,(R(z, y) \supset z = x)]$

From

Fa

and (ii) it follows that

$(\exists y)\,(R(a, y)\,\&\,Gy)$

From this and (iii), the existence and uniqueness conditions, we obtain

$E!\,(\imath y)\,(R(a, y))$

and hence we have

$$G(\imath y)(R(a, y))$$

From (i) and the existence and uniqueness conditions we have

$$b = (\imath y)(R(a, y))$$

Thus, we explain b's being G in terms of a's being F, on the basis, first, of the law (ii), and on the basis, second, of b's being the thing R-ed by a, where this last depends upon, among other things, (iii) ensuring the uniqueness condition is fulfilled and (ii) the existence condition. Furthermore, we know from (i) and (iii) that

$$a = (\imath x)(R(x, b))$$

Thus, we also explain b's being G in terms of the thing, that R's b, being F.

Kim now essays the argument for "piggy-back" laws:

If a's being F is causally sufficient for b's being G, then for any object c there exists a property H such that c's being H is causally sufficient for b's being G. For let R_1 be the spatiotemporal relation between c and a, and let R_2 be the spatiotemporal relation between c and b. And we set H to be the property denoted by the expression '$(\exists y)$ $(Fy \& R_1(x, y))$'. Then, the law '$(x)(Hx \supset (\exists y)(Gy \& R_2(x, y))$' holds; and the other conditions are obviously satisfied. To make this more concrete, consider this case: the object b's being heated is causally sufficient for its expanding (here $a = b$ and the relation R can be taken as identity). Let c be an object exactly fifty miles due north of the object that is being heated. The property H in this case is the property an object has in virtue of there being another object 50 miles due north that is being heated. Moreover, given the law that all objects expand when heated, we have the law that for any object x if x has the property H, then there exists an object 50 miles due south which is expanding. From this it follows that c's having property H is causally sufficient for b's expanding.[283]

Instead of law (ii) we now have the following law, which, as Kim says, rides "piggy-back" on (ii):

(ii*) $(x)[(\exists v)(Fv \& R_1(x, v)) \supset (\exists y)(R_2(x, y) \& Gy)]$

The relation R_2 is the relative product of relations R_1 and R. It is defined as

$$R_2(x, z) = \text{df.} \ (\exists y)(R_1(x, y) \& R(y, z))$$

Given that (iii) holds for R_1 and R, it can be proved that R_2 also satisfies (iii). With the definition

'Hx' is short for '$(\exists v)(Fv \& R_1(x, y))$'

then (ii*) is

$$(x)\,[Hx \supset (\exists y)\,(R_2(x, y)\ \&\ Gy)]$$

If it is true that

$$Hc$$

then from (ii*) we obtain

$$(\exists y)\,(R_2(c, y)\ \&\ Gy)$$

Using existential instantiation we have

$$R_2(c, y)\ \&\ Gy$$

By definition of R_2 the first conjunct is the same as

$$(\exists v)\,(R_1(c, v)\ \&\ R(v, y))$$

and existential instantiation again gives us

(*) $R_1(c, v)\ \&\ R(v, y)$

The first conjunct establishes existence and (iii) uniqueness for the following definite description

$$E!\,(\imath x)\,(R_1(c, x))$$

and it follows that

$$v = (\imath x)\,(R_1(c, x))$$

Independently we know, and may use as an additional premiss, the fact that

$$a = (\imath x)\,(R_1(c, x))$$

Since, from the second conjunct of (*), we have

$$R(v, y)$$

then we also have, by substituting identicals,

$$R(a, y)$$

As before,

$$E!\,(\imath y)\,(R(a, y))$$

and, again as before, it follows that

$$y = (\imath y)(R(a, y))$$

But, since

$$Gy$$

it follows that

$$G(\imath y)(R(a, y))$$

Thus, using the (ii*) we have explained the same fact as before, using a truth about c, an arbitrarily selected individual, rather than one about a. Clearly, the deduction need not proceed by way of the additional premiss, by way of introducing 'a'. We could have used, instead of 'a', simply the definite description '$(\imath x)(R_1(c, x))$'. We would then have had

$$y = (\imath y)\{R[(\imath x)(R_1(c, x)); y]\}$$

which, given the definition of 'R_2' abbreviates to

$$y = (\imath y)(R_2(c, x))$$

The conclusion therefore is

$$G(\imath y)(R_2(c, y))$$

Since R_2 is by Kim's assumption the relation between b and c, and since (iii) guarantees uniqueness, we have

$$b = (\imath y)(R_2(c, y))$$

and hence

$$Gb$$

If we assume that it is irrelevant how b is picked out, irrelevant what name or definite description is used to do that, then we have succeeded in explaining the same fact, using two different laws, and two different initial conditions. This by itself is not paradoxical, but it becomes so once it is recognized that, provided the relations involved have the structure (iii), when we explain a fact using a law of the sort (ii) we can explain the same fact using any arbitrarily selected individual.

The basic law (ii) entails the new law (ii*) that rides piggy-back upon it. We can see this as follows. Since

$$(v)\,[Fv \supset Q] \equiv (\exists v)\,(Fv) \supset Q$$

is a logical equivalence, (ii*) is logically equivalent to

$$(x)\,(v)[(Fv \,\&\, R_1(x, v)) \supset (\exists y)\,(R_2(x, y) \,\&\, Gy)]$$

Universally instantiate twice, and we obtain what is to be proved. To begin, assume (ii) as the only premiss. Now make the assumption for a conditional proof:

$$Fv \,\&\, R_1(x, v)$$

Universally instantiate from (ii):

$$Fv \supset (\exists y)\,(R(v, y) \,\&\, Gy)$$

Modus ponens yields

$$(\exists y)\,(R(v, y) \,\&\, Gy)$$

and we existentially instantiate:

$$R(v, y) \,\&\, Gy$$

We also have

$$R_1(x, v)$$

so that we obtain

$$R_1(x, v) \,\&\, R(v, y)$$

and, by existential generalization, that

$$(\exists z)\,[R_1(x, z) \,\&\, R(z, y)]$$

By the definition of 'R_2' this is the same as

$$R_2(x, y)$$

Hence, we have

$$R_2(x, y) \,\&\, Gy$$

and existential generalization yields

$$(\exists y)\,(R_2(x, y) \,\&\, Gy)$$

We now discharge the assumption of the existential instantiation with this as our conclusion and then discharge the conditional proof. Universally generalizing twice yields (ii*) as the conclusion we desired. So (ii) entails (ii*). But the converse does not hold.

Originally, we explained Gb by citing Fa. For this, we used (ii). We then explained $G(\imath y)\,(R_2(c, y))$ by citing Hc. But to explain $G(\imath y)\,(R_2(c, y))$ is simply another way of explaining Gb; it is the same fact, only b is referred to by means of a definite description and not by its name. In effect, the same is true of the initial conditions. 'Hc' expands into

$$(\exists v)\,(Fv \,\&\, R_1(c, v))$$

We can existentially instantiate this to obtain

$$Fv \,\&\, R_1(c, v)$$

The second conjunct yields the existence condition and (iii) the uniqueness condition, so that we have

$$v = (\imath x)\,(R_1(c, x))$$

With the additional information that

$$a = (\imath x)\,(R_1(c, x))$$

we immediately obtain

$$Fa$$

From here we deduce '$G(\imath y)\,(R(a, y))$' as before, using (ii) as the premiss. Thus, to assert that Hc is really nothing more than to assert that Fa; it is just that one refers to a in a more complicated way.

What (ii) asserts is that

> objects of such and such a sort cause events of so and so a type

What (ii*) asserts is that

> objects of such and such a sort when described in a certain way cause events of so and so a type when the latter are re-described in the correspondingly appropriate way.

We should not really be surprised, then, that (ii) entails (ii*). It is really no more surprising than the fact that the

(iv) $(x)\,(Fx \supset Gx)$

entails the law

$$(x) \, [(Fx \, \& \, Hx) \supset (Gx \, \& \, Hx)]$$

or, what is logically equivalent,

(iv*) $(x) \, [(Fx \, \& \, Hx) \supset Gx]$

Note that the latter does not entail (iv), just as (ii*) does not entail (ii). It is important to recognize this, that the inference does *not* go the other way.

On the basis of (iv*) we can say of an object a that is F and H that it will be G. On the basis of (iv) we can also say of a that even if it had not been H it would still have been G. So, even though (iv*) is true, (iv) permits us to say of a that its being H is causally irrelevant to its being G. On the other hand, (iv*) might be true and (iv) false. In the latter case we should have no reason for supposing that a's being H was causally irrelevant to its being G. Similar remarks hold for (ii) and (ii*). On the basis of (ii*) we can deduce that b is G by citing c's being H. But on the basis of (ii) we can also say that even if c had not existed b would still have been G. What is crucial to b's being G is the fact that a is F. If c had not existed, then a would not have been the object R_1-ed by c, for there would have been no such object. But from the latter, neither the non-existence of a nor a's not being F can be inferred. The CN Tower will continue to exist and have the height it has even if the definite description "the tallest free-standing structure in the world" comes to be false of it. Since (ii) tells us b would have been G even if c had not existed, we can infer the causal irrelevance to b's being G of c, and of c's being R_1 to a. On the other hand, had (ii*) been true and (ii) false, we could not infer the causal irrelevance of c and of a's being R_1 to c; to the contrary, in that case we would have to hold they were in fact causally relevant.

I conclude that there is no real problem with respect to the laws that ride piggy-back upon such laws as (ii). If (ii) enables us to explain b's being G by citing a's being F, then there will be a law piggy-back upon (ii) that enables us to explain b's being G by citing any arbitrarily chosen object c; there will always be a property of c such that the piggy-back law and c's having that property will suffice to explain b's being G. But the same (ii) will enable us to infer that b would have been G even if c had not existed, in other words, that c is causally irrelevant to b's being G. So long as we know these arbitrarily chosen individuals are all causally irrelevant to b's being G

EXPLANATIONS AND EXPLAININGS 325

then there is no real problem. To say this is, of course, to say that (i), (ii) and (iii) must be supplemented if they are taken to define causal sufficiency. But there is nothing wrong with that.

Kim comes fairly close to making this point, but then shies away.

> We feel that for an object to have this sort of artificially concocted property (recall the case of H above) is not always for it to undergo, or be disposed to undergo, a "real change"; my being 50 miles east of a burning barn is hardly an event that happens to me. But it would be a mistake to ban all such properties; my being in spatial contact with a burning barn is very much an event that happens to me.[284]

My standing in a spatial relation to something else is an event that "happens to me" only when that fact is causally relevant to later events that happen to me: that seems to be the import of what Kim says. But he does not notice that the law (ii) provides him with a criterion for excluding as not being involved in "real changes" those facts consisting of my being related spatio-temporally to various arbitrarily chosen individuals. In other words, he does not see that the piggy-back problem is easily dissolved.

We do not, of course, deny (ii*) is a law. It (supplemented by other information) perfectly reasonably supports the counterfactual, "If d were H, then b would be G". Nor need we deny that (ii*) offers an explanation. Indeed, we should not deny that (ii*) and the fact that c is H explains b being G — for, it satisfies our criterion for being an explanation: the principle of predictability. But to explain is not to explain causally. In order for a law to causally explain, along the lines suggested by the conjunction of (i), (ii), and (iii), then it cannot be one riding piggy-back on another, in the way Kim has so cleverly shown some laws have others parasitic upon them. The criterion for a law being causal thus becomes contextual: (ii*) is causal *unless* there is another law (ii) upon which it rides piggy-back. There remains the question, however, *why are causal explanations in this sense are preferable to the non-causal explanations riding piggy-back upon them*? To assert that one is causal, the other is not, does not answer this question. Kim does not address himself to this issue, presumably because he does not recognize that, in spite of (ii*), (ii) renders the c of which H is true causally irrelevant to b being G. For us, however, the answer is clear enough. '$p \vee q$' has less content than 'p', since the latter entails but it not entailed by the former; (iv*) has less content than (iv), since the latter entails but is not entailed by the former; (ii*) has less content than (ii), since the latter entails but is not entailed by the former. An explanation based on (ii*) has less explanatory content than one based on (ii). *The principle of explanatory content directs us to explain* *a fact in terms of a law rather than those that ride Kim-style piggy-back upon it.*

Kim, as we know, tends to think in more formalistic terms than we do. So it likely would not occur to him to solve his problem of parasitic constant conjunctions in terms of our principle of explanatory content. So, instead he profers, tentatively to be sure, an alternative solution. He suggests that we impose stronger restrictions upon the "R" of (i), (ii) and (iii). Let 'loc(x)' represent the spatiotemporal location of an event, and let '$R \in Ct$' represent that R is a relation of spatiotemporal contiguity. Then we would first define "*direct contiguous cause*" by keeping (i) and (iii) and replacing (ii) by

(iia) $(x) [Fx \supset (\exists y) [Gy \& R(\text{loc}(x), \text{loc}(y)) \& R \in Ct]]$ [285]

And next we would define "*contiguous cause*" in terms of the ancestral of direct contiguous causation: If 'e' and 'e'' are events of the sort we are considering — a's being F and b's being G — and 'f' and 'g' are variables ranging over such events, then we would use the definition

e is a contiguous cause of e' if and only if $e \neq e'$ and e bears to e' the ancestral of the relation of direct contiguous causation — that is to say, $(S) (e' \in S \& (f) (g) (f \in S \& g$ is a direct contiguous cause of $f. \supset g \in S): \supset e \in S)$.[286]

Then a's being F is causally sufficient for b's being G just in case the former is a contiguous cause of the latter. And what this amounts to saying is that one event is causally sufficient for another just in case there is a causal chain of events going from the former to the latter, and in which each event is a direct contiguous cause of its successor.

This apparently solves the problem of parasitic constant conjunctions, for, as Kim remarks, "there seems to be no general argument to show that our definition of 'contiguous causation' succumbs generally to this sort of difficulty."[287] But this apparent solution is in fact quite unsatisfactory. For it requires there to be no action at a distance; this requirement means Newton discovered no causal laws when he discovered gravity. The "solution" also requires that non-spatial events cannot affect spatiotemporal events; this requirement means non-spatial mental events cannot affect our bodily states. Both these requirements are absurd, and render unsatisfactory this proposed solution to the problem of parasitic constant conjunctions. But that need not bother us, for the latter problem really is no problem at all, as we saw.

Kim saw himself as posing a problem for those who wish to analyze causation we do, though he was sympathetic to the latter sort of attempt. We conclude he has presented us with no real problem at all. Davidson was unsympathetic to our proposals. He saw his arguments not as presenting problems for our analysis but as presenting arguments against it. But we conclude that he has given us no reason why we should not construe (f_1)

> The short circuit caused the fire

as (f_2)

> The fact that there was a short circuit caused it to be the case that there was a fire

i.e., why we should not construe 'causes' as a non-truth-functional connective.

It still does not follow we *should* so construe it. Only if our arguments defending the Humean analysis are correct should it be so construed. Those arguments are to the effect that if 'causes' is taken to be a non-truth-functional connective analyzable along Humean lines then certain philosophical problems can be solved. Davidson offers a similar case for construing 'causes' as an ordinary relation between particulars.

His proposal [288] is that, just as we normally construe

(x_1)   Jack fell down *before* Jack broke his crown

as

(x_1')   There exist times t and t' such that Jack fell down at t, Jack broke his crown at t', and t preceded t'

in order to eliminate 'before' as a non-truth-functional connective in (x_1), replacing it by an ordinary two-place relation in (x_1'), so also we should construe

(x_2)   Jack fell down, *which caused it to be the case that* Jack broke his crown

as

(x_2')   There exist events e and e' such that e is a falling down of Jack, e' is a breaking of his crown by Jack, and e caused e'.

Here, 'Jack's fall' does not impute uniqueness, as does 'the short circuit' in (f_1). To handle such cases as involve uniqueness, we should proceed something as follows. We take

(x_3)   Jack's fall caused the breaking of Jack's crown

and construe it as

> The one and only one falling down of Jack caused the one and only one breaking of his crown by Jack

or, in symbols,

(x_3') ($\imath e$) F(Jack, e) caused ($\imath e$) B(Jack's crown, e)

But for our purposes (x_2') will suffice.

What we should concentrate upon is

(y_1) e_1 caused e_2

where 'caused' is a normal two-place relation. Compare this to another ordinary relational statement

(y_2) The chair is to the left of the table (at t_1)

If we move the table so that (y_2) comes to be false and

(y_3) The chair is to the right of the desk (at t_2)

becomes true, the chair still remains the same object, as does the table. If another object occupied the place of the chair at t_1, and (y_2) was also true, then that second object and the chair would be identical; "material objects are identical if and only if they occupy the same places at the same time."[289] But nonetheless, what a chair is spatially related to in a certain way can change while the chair remains the same. Is the same even *possibly* true of the 'causes' of (y_1)? Try replacing 'e_2' by 'e_3', where

(y_4) e_3 = the freezing over of Niagara Falls.

and where

(y_5) e_1 = Jack's fall

This yields

(y_6) e_1 caused e_3

or, in words, given (y_4) and (y_5) and principles of substitution acceptable to Davidson:

(y_7) Jack's fall caused the freezing over of Niagara Falls

The intuition is that 'Jack's fall' is quite out of place here: it seems quite implausible to say that the *same* event can fill the blank in

(y_8) x caused e_2

and also in

(y_9) x caused e_3

This is another way of putting the point that Davidson's position fails to account for what we above called the "third feature" of the nomological connective: in this context what it amounts to saying is that the causal relation of (y_1) can obtain between events only if those events are of *kinds* that are causally related. Spatial relations are the model of external relations. What we are saying, therefore, is that *'causes' cannot be an external relation*. Another model could be adopted, however, that of internal relations. A pitch of which

x is higher by a third than A

was true could not also be a pitch of which

x is higher by a third than A

It could not change such a relation and still be the same pitch. It is in the natures of pitches that they be related in certain ways to other pitches; change those relations and the very nature would be changed, one would have a different pitch. Thus, if Davidson is to make any case for his position he cannot construe 'causes' as a purely external relation; it must be construed as an internal relation. In that case, events must have natures. And if they do, certain descriptions of events are priviledged. Some will be essential — those that get at the nature of the event — the rest accidental.

Davidson simply ignores the "third feature" of causal connections. He leaves the impression that "causes" might well be a purely external relation. On the other hand, he has also said that " . . . it is not *events* that are necessary or sufficient as causes, but events as *described* in one way or another."[290] But Davidson has not developed the point. And certainly he has not developed any distinction between essential and accidental descriptions of events. It has been Kim who has pursued the problem more consistently than Davidson. Kim accepts[291] the basic framework of events, but attributes an internal structure to them.

He takes such phrases as 'the sinking of the Titanic', 'Jack's fall', and so on, to be singular terms. They refer to events. But they presuppose reference to other entities, specifically, a thing (Titanic, Jack), a property (sinking, fall) and a time. For simplicity, consider only monadic properties. Then, according to Kim, we form singular terms of the sort

(T$_1$) $[(x, t), P]$

to refer to events. 'x' refers to the "constitutive object", 't' to the "constitutive time", and 'P' to be constitutive property. We also have the "existence condition"

(T$_2$) $[(x, t), P]$ exists if and only if the object x exemplifies the property P at the time t

that is

$[(x, t), P]$ exists if and only if x has P at t

We may take 'P' to be a primitive empirical predicate.[292] Kim also lays down the "identity condition"[293]

(T$_3$) $[(x, t), P] = [(y, t'), Q]$ if and only if $x = y$ and $t = t'$ and $P = Q$

This yields the crucial distinction between descriptions that are essential and those that are accidental:

It is important to notice the distinction drawn by our analysis between properties *constitutive* of events and properties *exemplified* by them. An example should make this clear: The property of dying is a constitutive property of the event [(Socrates, t), dying], i.e., Socrates dying at t, but not a property exemplified by it; the property of occurring in a prison is a property this event exemplifies, but is not constitutive of it.[294]

It is in this context that Kim places the problem of causation:

(T$_4$)   "Under our account, ... if Socrates' drinking Hemlock (at t) was the cause of his dying (at t'), the two generic events, drinking hemlock and dying, must fulfill the requirement of lawlike constant conjunction."[295]

The crucial question is, what do the singular terms (T$_1$) refer to? (T$_4$), in one respect, reduces the problems about (y$_1$)

e_1 caused e_2

to the question of laws. In (y_1), 'e_1' and 'e_2' will be Kim's event referring terms. They will stand in the causal relation just in case their constitutive properties are causally related. So, using the notion of laws and of properties constitutive of events, Kim defines or analyzes the causal relation that Davidson more or less took as primitive, or at least refused to analyze. Lawful connections among properties are precisely what we ourselves find unobjectionable; that, after all, was Hume's point. The part we must question is the notation of '[...]' in (T_1).

Is (T_1) really just *shorthand* for something else, a notational device? In that case it is (T_2) which is crucial;

$[(x, t), P]$ exists

is then simply shorthand for

x has P at t

If this is so, then causation holds between facts; the appearance of it holding between particulars (events) is merely an appearance conveyed by the representation. In short, if we construe the terms (T_1) as eliminable then *Kim's account is not different from ours*. Its justification consists simply in the fact that it brings the formalism more into conformity with some of the grammatical nicities of ordinary English. This is like the inverse iota "operator": this has no real place in the *Principia Mathematica* formalism (which is also Hempel's L) that we are using; its only function is as a piece of short-hand to induce upon sentences, the logical form of which is quantificational, a grammatical form akin to that of English sentences in which phrases beginning with the definite article 'the' occur in the subject-place. The point remains, however, that so long as Kim's account of events is so construed, then, *logically speaking*, his account of the causal connective is not different from ours.

On the other hand, Kim may take terms like (T_1) to function as definite descriptions. Then (T_2) and (T_3) provide the existence and uniqueness conditions for these descriptions. But, to what entities do they refer? When I observe the fact that a is P at t, do I also observe the entity $[(a, t), P]$ over and above that fact? How does one distinguish it visually within the complex one is observing? The empiricist in one rebels at such entities. Moreover, they proliferate far too much; thus, Kim speaks of conjunctive events also.[296] And if he gives ontological status in this way to 'and', there seems no reason why he should not also give it to the other connectives. There is no reason

332 CHAPTER 3

why there should not be determinable, i.e., disjunctive, events, as well as determinate ones. Moreover, since to have length of 4 yards is a concept operationally defined using 'if ... then', and since having that property can be causally relevant to at least some "real changes", and since the latter is the criterion for constitutive properties, then one should have implicative also as well as conjunctive events. All this, surely, should give pause even to one who does not share our empiricist convictions. But there are more problems. Although the constitutive properties may be complex, still, upon this line, events themselves are not facts (complexes) but simples. They are simples which nonetheless have properties as "constitutive" of them; they are simples in the way in which the nominalist has natured simples or perfect particulars. They are in an ontology that takes functions (definite descriptions) to be unanalyzable notions (otherwise we could eliminate the definite description in (T_2) after the fashion of Russell); these functions map the events onto the more basic entities upon which they are ontologically dependent. In short, Kim's ontology, taken in the way just suggested, is thoroughly Meinongian. We are in no position to examine this sort of ontology more deeply; in any case, Bergmann has done so recently and very effectively.[297] Suffice it to say for here, that Kim gives us no reason for accepting this second alternative reading of his account of events. We should, therefore, stick with the other alternative, the alternative which amounts to our own position.

I therefore conclude that if one pushes Davidson's position so that it can account for the "third feature" of nomological connections, viz, the relevance of properties, then one is driven to a position like Kim's. But if one goes this way then the result is either our own position, or one that is bizarrely Meinongian. Davidson's attempt to take a third position on causation, different from ours and from that of Scriven and Collins fails. For ourselves and for Scriven and Collins, 'causes' represents a non-truth-functional connective. In order to solve the analytic/synthetic distinction, Davidson proposed construing it as an ordinary two-term relation, introducing an ontology of events so it might have entities other than facts to connect. As soon as we pushed him to account for the "third feature" of the nomological connective, events had to be construed after the fashion of Kim. Taken one way, events are mere *façons de parler*, and the position turns out to be ours, in which causal connections hold between facts and are analyzable. Taken in another way, events turn out to be entities in a manifestly implausible ontology, one certainly that neither Davidson nor Kim gives us any reason for accepting. There remains, too, at least for Davidson, if not for Kim, the second of the philosophical problems we have mentioned, that of how we know when the unanalyzable relation obtains. This is a difficulty that also faces the account of Scriven and Collins, that takes the relation to be a *primitive* non-truth-functional connective.

The difficulty with the nomological connective is that if it is taken as primitive, then what we called the two philosophical problems cannot be solved. That is, first, the usual explication of the analytic/synthetic distinction, and the only proposed explication which is at all plausible, is not available, so that distinction goes unexplicated. This problem Davidson solved by taking it to be a relation in the strict sense and not a connective. And, second, it has to be explained how we know when the causal connection obtains. To this latter issue we now turn.

Collins simply asserts we can know such causal connections.[298] This is his denial of (t_2), the thesis that if one knows A causes B then one knows a law associating A-like events with B-like events. This denial in turn is the basis for his denial of (T^*), the thesis that whenever one correctly makes a causal assertion one is *thereby* also making an assertion of the corresponding law. But Collins simply asserts. Davidson does the same,[299] or, rather, he simply cites the spurious authority of Ducasse.[300] They both write almost as if Hume had never written. There is no reason to think this view to be even plausible. Or, rather, one might suggest, the view becomes plausible only because it smuggles in an anthropomorphic Aristotelian view of causation. But that view, too, is false, however apparently plausible such anthropomorphism might be.

It is true that Collins attempts to make his position plausible by means of examples.[301] The examples are of the ink-bottle and poison and non-dying sort. It is claimed that in these cases there are no laws available for the defender of the deductive model to invoke to save his thesis. But we know this is not so. The laws Collins demands are perfect. It is true there are no perfect laws available in these examples. So much is correct. It does not follow there are no laws which are available. And, in fact, in these cases there are imperfect laws available, as we saw when we discussed these examples.[302] Because Collins ignores imperfect laws he finds in these examples a plausibility for his denial of (t_2), a plausibility which disappears upon closer examination. So Collins is simply reduced to asserting the denial of (t_2). This assertion we are given no good reason for accepting, while simultaneously we do know good reasons for rejecting it, viz, the positive case which the defender of the deductive model gives for analyzing (CE_1), "A caused B", as (CE_2):

334   CHAPTER 3

> Whenever an A-like event occurs, then a B-like event occurs;
> and that is an A-like event
> _____
> so, this is a B-like event

Scriven is somewhat more ambitious in the case that he makes. As with Collins, the nomological connective turns out to be a primitive connective. Scriven suggests we do in fact have knowledge when it obtains. This knowledge is not a knowledge of laws; Scriven in this way denies (t_2) and therefore (T*). This knowledge of causal connections is not to be further analysed. It is, however, philosophically non-problematic. At least, this is Scriven's claim. And because it is non-problematic, at least the second of the two philosophical problems is laid to rest. Scriven does not address himself to the first of these two problems about the nomological connective, so to that extent his position remains weaker than that of the defender of the deductive model. Still, he weakens the case of the latter if he can in fact establish there is a way of knowing causal connections to obtain which is not a knowledge of laws. Let us see if he does establish this point.

Scriven attempts to make his point in terms of the ink-bottle example.[303] In such a case we can, he holds, identify causes even though no true universal hypothesis is known.

> This capacity for identify causes is learnt, is better developed in some people than in others, can be tested, and is the basis for what we call judgements.[304]

Judgment thus yields knowledge of causal connections.

> Even if Hume was correct [about a law being involved in causation], it only follows that each particular causal statement is an instance of a law; it does not follow that the explanation of ... event E consists in a *deduction* of E from the laws *plus* antecedent conditions.[305]

The laws at best provide "role-justifying grounds" for explanations.[306] But where we typically rely on judgment to yield knowledge of causal connections, e.g., in the ink-bottle case, we do not even have laws available for "role-justifying". All we have available in these cases are truisms, rather than laws. We recognize causes easily and reliably,[307] and, for the typical cases, one's "*principles* of judgment are most nearly, though inadequately, expressed in the form of truisms,"[308] where a truism is, according to Scriven, as we know, a statement which is "trivial, though not empty," and it "tells us nothing new at all; [though] it says something, and it says something true, even if vague and dull."[309] These truisms Scriven also calls normic statements:

The normic statement says that everything falls into a certain category except those to which certain special conditions apply. And, although the normic statement itself does not explicitly list what count as exceptional conditions, it employs a vocabulary which reminds us to our knowledge of this, our trained judgement of exceptions.

If exceptions were few in number and readily described, one could convert a normic statement into an exact generalization by listing them. Normic statements are useful where the system of exceptions, although perfectly comprehensible in the sense that one can learn how to judge their relevance, is exceedingly complex.[310]

We have, of course, been over a good deal of this above.[311] That permits us to be relatively brief in our discussion at this point.

We know that Scriven's truisms or normic statements can all be interpreted as laws, that is, as imperfect laws. Thus, in those cases where we rely on "judgment" we do indeed have laws available. Identification of causes by "judgment" is not independent of knowledge of laws. Scriven thinks so only because he identifies all knowledge of laws with process knowledge and then proceeds to say that therefore the knowledge we express in truisms is not knowledge of laws. We know by now that what is required is the rejection of that initial identification. In the case of truisms or normic statements, there are *two* possibilities. *First*, there is *ex post facto* explanation. On this possibility, the initial conditions are known on the basis of inference from the law and the event to be explained. As we know, there is nothing special about these cases, and in particular no reliance on anything other than knowledge of individual facts or events and of laws, which means no reliance on some special faculty of "judgment" that operates in the absence of such knowledge of laws. There is no reliance on such a faculty of "judgment" that operates in the absence of such knowledge of laws. There is no reliance on such a faculty because we do in fact rely upon knowledge of laws. In these cases, there is explanation where the possibility of prediction cannot be actual. The *second* possibility also involves these examples, where the known (imperfect) laws permit explanation but not prediction. We also know that in these cases a trained scientist (e.g., a medical doctor) can antecedently predict, often with great success, even though he does not know the initial conditions, in the sense of *knowing that* such and such are the initial conditions and knowing (prior to the occurrence of the predicted event) that these initial conditions obtain. In these cases, as we know, it makes sense to speak of the scientist relying upon his (trained) "judgment" when he predicts. Only, such "judgment" is not a matter of *knowledge*, i.e., knowledge that, but rather is only a skill, a bit of acquired "know how", which is *not* a substitute for the knowledge ("knowledge that") which science desires. The predictions in this case are non-reasoned predictions, skillfully made, but nonetheless made in ignorance, and specifically in ignorance of the initial conditions. We discussed this in Section 2.6. What Scriven does with these examples is to take the "know how" which they involve and transform it into a "knowledge that " his primitive nomological connective obtains. Recall that what Scriven wants to argue is that there is a non-problematic knowledge that his primitive nomological connective obtains. He makes his case for such non-problematic knowledge on the basis of examples, of the ink-bottle or poison and death sort, where it is possible to explain but not predict, i.e., not predict on the basis of known laws. In these cases we can, he says, identify causes and predict without laws on the basis of a non-problematic sort of "judgment". This much is true. Only that "judgment" is a sort of "know how". What he directs our attention to is this non-problematic "know how". He does not establish what he set out to establish, namely, that there is a non-problematic knowledge (knowledge that) of the primitive nomological connective. He has therefore not succeeded in showing there is a non-problematic knowledge of causes independent of our knowledge of laws. Indeed, in all his examples, such knowledge of laws is present – though the laws are imperfect, their imperfection leads him to overlook their presence. It therefore follows that he has *in no way* weakened the case of the defender of the deductive model that knowledge of causes is to be analyzed in terms of knowledge of laws.

We might note that Scriven blurs together the two possibilities we just mentioned. The first possibility was that of *ex post facto* explanation. This involves only "knowledge that". The second possibility was that of non-reasoned prediction of events which one can later *ex post facto* explain. This involves "know how" which is not "knowledge that". By confusing the two possibilities, the fact that "know how" as well as "knowledge that" is involved in the latter case comes to be obscured by the fact that only "knowledge that" is involved in the former case. Thus, the "know how" can be made to appear to be the "knowledge that" which Scriven requires for his argument to go through; that is, it appears as the basic "knowledge that" of his primitive nomological connective. We know why Scriven confuses the two possibilities. He notes that the non-reasoned prediction is not based on inference via laws. He identifies laws with perfect laws. He notes (correctly) that in those cases where only *ex post facto* explanation is possible there are no perfect laws. He concludes that in the latter case there are no laws. Hence, it also is not a matter of inference via laws. Whence the two possibilities are essentially similar, according to Scriven. The wrong move is – we are by now familiar with it — that of identifying laws with perfect laws, or, in other words, of ignoring imperfect laws.

We must resort to trained non-reasoned predictions where our knowledge is imperfect, in particular where some of the relevant variables are not known. We saw this in our discussion, in Sections 2.5 and 2.6, of explanation and judgment. In those Sections we considered such laws as

$(x) \ \{Kx \supset [(Px \ \& \ C^*x) \equiv Dx]\}$
$(x) \ \{Kx \supset [((Px \ \& \ C^*x) \vee F_1x \vee F_2x \vee F_3x) \equiv Dx]\}$
$(\exists! f) \ \{\mathscr{C}f \ \& \ (x) \ [Kx \supset (fx \equiv (Fx \equiv Gx))]\}$

(which were there labelled (5.46), (5.36) and (5.23) respectively). Such laws as these have the feature that they, on the one hand, permit *ex post facto* explanation, but, on the other hand, require antecedent prediction to be non-reasoned, based on skilled "know how." What gives these laws this feature is the fact that one of the factors is the one we labelled "C^*". "C^*" is a label which we use for the relevant variables that we believe exist but the specific nature of which we do not know. We have reason to believe *that* the variables exist but do not know *not what* they are. As we noted in our discussion of (5.23), this means "C^*" is not itself a concept but the *hope of a concept*. It is because "C^*" is but the hope of a concept that in order to predict in these cases we must rely on skilled "know how" and non-reasoned prediction. Skilled "know how" and "C^*", the hope of a concept, go together. Scriven illegitimately transforms this "know how" into a "knowledge that" his primitive nomological connective obtains in the conditions where the "know how" is successful. This means "C^*" is in effect transformed from the hope of a concept, from a non-concept, into the concept of his primitive nomological connective. Scriven so to speak fills in the non-concept with a spurious conceptual content, that of his primitive nomological connective.

But from the fact that Scriven mistakes "know how" for "knowledge that", it does not follow directly that he will make it into a "knowledge that" his primitive nomological connective obtains. Structurally, however, that is the only reasonable transition. There is, moreover, a stronger structural connection between the occurrence of the word 'causes' in explanations (which Scriven takes to be the name of his primitive nomological connective), and the non-concept, or, rather, hope of a concept, "C^*". Let me try to bring this connection out.

We saw that another way of looking at "C^*" was as the potential name of relevant variables for which we can give a definite description we have reason to believe is successful. To say it is a "potential name of the relevant variables" and to say it is the "hope of a concept" is to say what amounts to the same thing. Taking "C^*" as a potential name enables us to see exactly that Scriven overlooks its nature as a *potential* name, or as the *hope* of a concept, and in effect takes it to be the concept of his primitive nomological connective. We can see this if we examine an argument Scriven offers[312] in an attempt to establish that where we must rely on explanations in terms of causes, i.e., those based on a statement like

(c) A causes B

then in those cases it is *logically impossible* to speak of a law, from which he concludes against the deductive model.

He suggests the defender of the deductive model might hold that the causal assertion (c) could be replaced by the generality

(d)   Whenever an A-like event occurs a B-like event follows.

Scriven argues this cannot succeed. He considers two alternative readings of (d) and concludes both are inadequate for what the defender of the deductive model wants. The two alternatives depend upon two ways in which the 'A-like' of (d) may be interpreted. Upon the *first reading*, 'A-like' means like A in having all the features I can presently describe. This will not do since there may be causally relevant features which enter into my "judgment" that (a) holds but which I cannot describe. Upon this interpretation, (d) will omit some causally relevant variables and will therefore not be the universal law required by the defender of the deductive model. Upon the *second reading*, 'A-like' means like A in all respects, presently describable or not. This will not do, since in this case the antecedent of (d) will mention not just the causally relevant features but *all* the features of A. By the Identity of Indiscernibles the only event which has *all* the features of A is A itself. Thus, upon the reading 'A-like' simply picks out A in its uniqueness and (d) can *logically* apply to this one occasion only. Laws (by definition) are generalities for which it is logically possible that they apply to more than one event: "A generalization must have more than one possible instance; a particular statement is no less particular because it applies to 'everything' logically identical to its subject."[313] Thus, it follows once again that the defender of the deductive model has not got the law he needs.

We know by now how to meet this sort of argument. Basically, what is wrong with it is that it omits the third alternative reading of (d) which understands it as an imperfect law. What is interesting is how Scriven here misses this alternative. When he provides his readings for 'A-like' he uses the term "describe", and he uses this in such a way that to describe an event is to apply a predicate to the event. To apply a predicate to an event is to assert that the event falls under the concept the predicate names. To describe an event thus involves the name of a concept. Scriven states his two alternatives in terms of this sense of 'describe'. The third alternative, that which involves imperfect knowledge, requires us to assert an event falls under certain concepts even when we do not know precisely what those concepts are. We must say the event is an instance of certain relevant variables without knowing precisely or specifically what those variables are. Since we do not know what those concepts or variables are, we cannot use their names in order to make predications of an event. In Scriven's terms, we cannot "describe" the event. Given his use of 'describe', it is therefore not a real alternative. The difficulty lies, of course, in that narrow use of 'describe'. We can "describe" the number of shoes in a pair as "two", by means of a name, or we can "describe" the number as "the sum of one and one", by means of a definite description. This is a broader use of 'describe' than that of Scriven. It is, however, a perfectly legitimate use. In this sense of 'describe', we can describe an event, assert that it falls under a concept, by using a definite description of the concept: the event is described as falling under *the* concept or variables which our imperfect law says exist. We may have such a definite description without knowing specifically what those concepts are. "C^*" is associated with that definite description, and functions for us as a potential name of the concept which the definite description denotes.

We know that 'causes' is a connective and that (c) is to be read along the lines of (a):

(e)   This event, which is A-like, causes that event, which is B-like.

Here the 'A-like' and 'B-like' have to be read in such a way that an event is A-like just in case it has all the features I can presently describe this event (= A) as having, where 'describe' is used in Scriven's narrow sense. We have argued that implicit in (c) and (e) is a law, an imperfect law. This law Scriven ignores, but when we are trying to diagnose his position *we* need not ignore it. The relevant concepts we can name are mentioned in the two connected sentences of (e). The relevant concepts, which we cannot name, we mention by means of a definite description, a definite description justified by the imperfect law implicit in (c) and (e). But since the law is only implicit, so too is that definite description. We may say that in any particular context the word 'causes' carries this definite description. Which definite description is carried varies from context to context, depending on which imperfect law is implict. "C^*" is the potential name associated with the definite description. Thus, 'causes' carries this potential name. Scriven now transforms this potential name into a real name. We saw that he does this by mistaking the skilled "know how" of "judgment" for a sort of "knowledge that". But what does it name? It cannot be any ordinary properties or variables, since on the one hand, we can use names of these to "describe" events, in Scriven's narrow sense of 'describe', and, on the other, the potential name is associated with a definite description of properties or variables, where the need for such a definite description prevents us from "describing" events in terms of these properties, in Scriven's narrow sense of 'describe'. The potential name thus comes to name some entity other than a property or variable we would mention in describing an event. So what does it name? The fact that this potential name is carried by 'causes' provides Scriven with the cue: it names his primitive nomological connective. In this way, the hope of a concept "C^*" comes to be invested with a spurious conceptual content, and to be thought of as naming what Scriven takes 'causes' to name, viz, his primitive nomological connective.

We may conclude, therefore, that Scriven completely fails to challenge the defender of the deductive model, and that he comes to think he has succeeded only because he mistakes the hope of a concept for the concept he wants, namely, the concept − the spurious concept − of his primitive nomological connective.

Let me end this discussion by noticing a great danger which is present if we start to accept Scriven's position.

"C^*" is the hope of a concept. That is to say that it marks our ignorance. It marks that we have but imperfect knowledge where we would prefer less imperfect knowledge. This specific ignorance is carried by the word 'causes'. Now, there is an ever-present tendency of the human mind that when it is ignorant of some part of the world it will project its own attributes into that part of the world. We "understand" ourselves, and therefore we "understand" the objects whose behaviour we are explaining because those objects *are* ourselves by virtue of our anthropomorphic projections. We fool ourselves that we have complete "understanding". By virtue of this tendency of anthropomorphic projection to spuriously fill in our areas of ignorance, to create for ourselves the illusion of knowledge, we also tend to rest content with such ignorance. "C^*" comes to be filled with anthropomorphic projections. This mark of ignorance, this non-concept, comes to be filled with a spurious content. Indeed, I would suggest that it is this spurious anthropomorphic content which is the essence of Scriven's primitive nomological connective. Be that as it may, the point remains that we tend to fill "C^*" with anthropomorphic strivings and thereby we remain content with, and accept as fully adequate, explanations based on the imperfect laws which involve "C^*". We rest content with (4.2)

$$(\exists f) \{\mathscr{C}f \& (x) [Kx \supset (fx \equiv (Fx \equiv Gx))]\}$$

rather than desiring to go on to the less imperfect (4.3)

$$(x) [Kx \supset (Cx \equiv (Fx \equiv Gx))]$$

(to use the formulae discussed in Section 1.4.) (4.2) does not provide as good an explantion as (4.3). Relative to what we can reasonably want to know, (4.3) being less imperfect, provides a better explanation than does (4.2). Yet we accept (4.2) as adequate. We accept it when we ought not. Our anthropomorphic projections lead us to accept as fully explanatory what is in fact less than what we want.[314] Resting content with what we have, we do not go on to engage in research designed to bring about the replacement of ignorance by knowledge, of the imperfect knowledge of (4.2) with the less imperfect knowledge of (4.3), of the hope of a concept "C^*" with the concept "C" which appears in (4.2). Our anthropomorphic projections thus come to inhibit the progress of science.

On the other hand, although there is a tendency on the part of the mind towards such anthropomorphism, it is a tendency which can be checked by a critical self-awareness, a critical consciousness with respect to what one does accept and what one is prepared to accept as providing explanations. The acquisition of such a critical self-awareness is, or ought to be, part of the training of any competent scientist.[315] Such a critical attitude requires, however, a set of norms in terms of which one can consciously evaluate the explanations one finds oneself tending to accept.

The difficulty with Scriven's behavioural approach is that it permits no formulation of such norms. It is descriptive and psychological rather than normative. His approach therefore renders it impossible to formulate the critical self-conscious attitude that alone can make it possible for scientists to resist the tendency to rest content with anthropomorphic projections, the critical attitude that alone can ensure the progress of science in spite of the tendency to stagnation that is a result of our anthropomorphizing. In this sense, *Scriven's approach is essentially anti-scientific*.

CONCLUSION

The deductive-nomological account of causation is both a descriptive and a normative thesis — it purports to describe how most attempts at scientific explanation in fact proceed, and it asserts that this is how they ought to proceed. This normative conception in the philosophy of science has its origins in the work of Galileo.[1] It has found its greatest defenders in Newton, Hume, John Stuart Mill, Mach, Russell, and the positivists of this century. The latter in particular introduced a variety of highly sophisticated analytical tools deriving from the work of modern logicians, and produced a powerful and persuasive conception of the scientific method.[2] Increasingly, however, this model has come under attack from various directions. Some of these, such as Feyerabend's and that of the so-called "Oxford philosophers" derive from a conception of logic radically different from that defended by the positivists and empiricists in general. We have not been able to examine the more general conceptions of these philosophers,[3] except where these touch very directly on the deductive-nomological model. Our task has been that of examining those positions that have argued that the deductive-nomological model must be supplemented or revised or qualified in some way. We have not been able to examine all the material available, but we have, I think, covered most of the major criticisms. In these discussions we concluded in each case that the criticism failed to touch the deductive-nomological model. We may therefore reasonably conclude, I think, that this aspect of the empiricist account of science has emerged unscathed from the barrage of attacks that has recently been launched against it. Recent tendencies to move away from empiricism towards a supposedly more adequate account of science have thus been shown to be unmotivated.

NOTES

NOTES TO CHAPTER 1

[1] *Logic of Scientific Discovery*, p. 59; his italics.
[2] *The Open Society and Its Enemies*, Vol. II, p. 362, n. 7.
[3] Book III, Ch. xii, Sec. 1.
[4] Cf. M. Scriven, 'Explanations, Predictions, and Laws', pp. 194–5.
[5] Cf. Wilson, 'Logical Necessity in Carnap's Later Philosophy', Chapter One.
[6] This is the essential core of the idea that statements of fact are objective; cf. G. Bergmann, 'Ideology'.
[7] Cf. Wilson, 'Logical Necessity in Carnap's Later Philosophy', Chapters One and Five.
[8] A. W. Collins, 'Explanation and Causality', p. 482ff.
[9] Hereafter, for the sake of brevity, I shall often speak simply of "deducing the explanandum" rather than the more tedious "deducing a sentence about the explanandum".
[10] Scriven, 'Explanations, Predictions, and Laws', p. 173ff.
[11] Collins, 'Explanation and Causality', p. 494ff.
[12] R. Chisholm, 'The Contrary-to-Fact Conditional'.
[13] Chisholm, 'Law Statements and Counterfactual Inference', emphasizes the connection between lawlikeness and subjunctive conditionals.
[14] That is, non-law *universal* statements.
[15] A. Pap, 'Disposition Concepts and Extensional Logic', and A. Burks, 'The Logic of Causal Propositions'.
[16] Cf. G. Bergmann, 'The Philosophical Significance of Modal Logic'.
[17] Cf. G. Frege, 'On the Foundations of Geometry'; F. Wilson, 'Implicit Definition Once Again'.
[18] Cf. S. Barker's comments on Pap's views in his review of *Minnesota Studies in the Philosophy of Science, Vol. II*, in which Pap's essay, 'Disposition Concepts and Extensional Logic', first appeared; and also the same point made against Barker himself in F. Wilson, 'Barker on Geometry as *a Priori*'.
[19] W. Kneale, 'Natural Laws and Contrary-to-Fact Conditionals', and *Probability and Induction*; and also A. C. Ewing, 'A Defence of Causality'.
[20] Cf. F. Wilson, 'Acquaintance, Ontology, and Knowledge', for a discussion of arguments to this effect.
[21] Kneale, *Probability and Induction*; and Ewing, 'A Defence of Causality', at least for causation in the extra-mental realm. The search for necessary connections thus generates a radical scepticism about whether we ever know any causal connections at all. But such radical scepticism is always consequent upon the introduction of entities that transcend sensible experience; cf. 'Acquaintance, Ontology, and Knowledge'.
[22] However, see also Wilson, 'The Lockean Revolution in the Theory of Science'.
[23] A. C. Ewing, *Fundamental Questions of Philosophy*, Chapter VIII.

NOTES TO CHAPTER 1

24 *Treatise*, p. 77.
25 Cf. Wilson, 'Acquaintance, Ontology, and Knowledge', and 'Hume's Theory of Mental Activity'.
26 *Treatise*, p. 139.
27 *Treatise*, p. 105.
28 As, for example, in R. A. Imlay, 'Hume on Intuitive and Demonstrative Inference'.
29 *Treatise*, p. 172.
30 *Treatise*, p. 139.
31 *Treatise*, p. 77.
32 *Treatise*, p. 155.
33 *Treatise*, p. 156.
34 *Treatise*, p. 172.
35 Chisholm, 'Law Statements and Counterfactual Inference', p. 230.
36 G. Bergmann has emphasized the role of context — and also the limitations on its role; see his discussion of the context theory of meaning in his 'Intentionality'.
37 W. Kneale, 'Natural Laws and Contrary-to-Fact Conditionals'.
38 K. Popper, 'A Note on Natural Laws and So-Called "Contrary-to-Fact Conditionals"'.
39 Kneale, 'Natural Laws and Contrary-to-Fact Conditionals', p. 124.
40 *Ibid.*, p. 124.
41 Cf. A. J. Ayer, *The Problem of Knowledge*, pp. 71–5; Tom Beauchamp and T. A. Mapples, 'Is Hume Really a Sceptic About Induction?'; and Tom Beauchamp and A. Rosenberg, *Hume and the Problem of Causation*.
42 *Treatise*, p. 112.
43 *Treatise*, p. 116.
44 *Treatise*, p. 123.
45 *Treatise*, p. 143ff.
46 *Treatise*, p. 173ff.
47 *Treatise*, I, III, iii.
48 *Treatise*, p. 139.
49 That is, we could go out and gather more evidence, but prior to such further evidence coming in we can rely only on the sample we already have.
50 Compare the treatment of subjective and objective justification, in the content of a discussion of utilitarianism, in G. E. Moore, *Ethics*, pp. 118–121.
51 C. J. Ducasse, 'Causality: A Critique of Hume's Analysis'.
52 *Ibid.*, p. 223.
53 D. Davidson, 'Causal Relations', p. 160.
54 *Ibid.*, p. 160.
55 In Chapter 3, below.
55a This is the terminology of J. L. Mackie, 'Causes and Conditions', p. 21. We discuss the relevant ideas further in Section 1.4, below, and in Chapter 3, and have some further remarks specifically on the notion of field in Section 1.4, below.
56 *Treatise*, pp. 131–5; p. 154.
57 *Treatise*, p. 153.
58 *Treatise*, pp. 152–3.
59 *Treatise*, I, III, xiii; cf. Wilson, 'Hume's Defence of Causal Inference'.
60 *Treatise*, p. 175.
61 *Treatise*, p. 134.

NOTES TO CHAPTER 1

62 *Treatise*, p. 193; cf. Wilson, 'Is There A Prussian Hume'.
63 J. S. Mill, *System of Logic*, Book III, Ch. III, Sections i and ii.
64 *Treatise*, p. 156.
65 This causal connection itself receives a Humean analysis; cf. Wilson, 'Hume's Theory of Mental Activity'.
66 We are not Hegelians; cf. Wilson, 'Acquaintance, Ontology, and Knowledge', and 'Meaning Is Use'.
67 Mill, *System of Logic*, p. 223.
68 Cf. the discussion of Kuhn in F. Wilson, *Reasons and Revolutions*; and also Wilson, 'Kuhn and Goodman: Revolutionary vs. Conservative Science'.
69 This presupposes one accepts both the so-called consequence and converse consequence conditions of confirmation. John Stuart Mill accepts both these not implausible conditions; cf. *System of Logic*, Bk. III, Chapter IV, Sec. 1. We comment further on these conditions below. See Note 81.
69a Cf. Wilson, 'Hume's Sceptical Argument against Reason'.
70 F. N. Kerlinger, *Foundations of Behavioural Research*, pp. 56–7.
71 See his notes to James Mill's *Analysis of the Phenomena of the Human Mind*, Vol. I, p. 350, p. 402ff.
72 *Ibid.*, p. 407.
73 *Ibid.*, p. 437.
74 *Ibid.*, p. 437–8.
75 E. Nagel, 'Carnap's Theory of Induction', Section VI.
76 Bk. III, Ch. VIII.
77 *Treatise*, pp. 174–5, Rules 5–8.
78 The note by John Stuart Mill in his father's *Analysis*, p. 436.
79 Vacuous occurrences of 'G' and any of the 'F's are not among the possibilities; cf. *System of Logic*, p. 63, p. 104.
80 Mill recognizes the need for these assumptions: see *System of Logic*, p. 369. Cf. also Hume, *Treatise*, pp. 173–4, Rule 4.
80a See *Reasons and Revolutions*, for more on these conditions.
81 See Note 69.
82 I refer to Russell's analysis of definite descriptions and of functions.
83 *System of Logic*, p. 373. Cf. Hume, *Treatise*, pp. 173–4, Rule 4.
84 Section 1.4, below.
85 See the selections from Russell in H. Feigl and M. Brodbeck, *Readings in The Philosophy of Science*.
86 Cf. *Treatise*, p. 173, Rules 1–3.
87 See *System of Logic*, Bk. III, Ch. V, Secs. 2, 6, 7, 8, and Ch. VI.
88 *Ibid.*, p. 374.
89 *Ibid.*, p. 372.
90 *Ibid.*, p. 373.
91 *Ibid.*, p. 373.
92 *Ibid.*, pp. 275–6.
93 *Ibid.*, p. 374.
94 *Ibid.*, p. 200ff, p. 371.
95 *Ibid.*, p. 188ff.
96 *The Logical Problem of Induction*.

NOTES TO CHAPTER 1

[97] *System of Logic*, p. 369.
[98] *Ibid.*, p. 373.
[99] *Ibid.*, p. 376.
[100] *Ibid.*, p. 375.
[101] C. S. Peirce, *Collected Papers*, 5.419 (= Vol. 5, paragraph 419).
[102] T. A. Goudge, *The Ascent of Life*, p. 207.
[103] T. Kuhn, *The Structure of Scientific Revolutions*. Kuhn's thought is analyzed in my *Reasons and Revolutions*.
[104] T. A. Goudge, *The Thought of C. S. Pierce*.
[105] See Note 101.
[106] Goudge, *The Ascent of Life*, p. 207.
[107] Goudge, *The Thought of C. S. Peirce*, pp. 189–90; his italics. Compare E. H. Madden's chapter on Peirce in R. Blake, C. J. Ducasse, and E. H. Madden, *Theories of Scientific Method*.
[108] I. Levi, 'Hacking, Salmon on Induction'.
[109] Goudge, *The Thought of C. S. Peirce*, p. 193.
[109a] Hume attempts a kind of vindication of inductive inference, but Hume's argument does not fall victim to this criticism; cf. Wilson, 'Hume's Defence of Causal Inference'.
[110] Peirce, *Collected Papers*, 5.590.
[111] E. H. Madden, 'Peirce and Contemporary Issues in the Philosophy of Science', p. 40; Goudge, *The Thought of C. S. Peirce*, pp. 197–8; Peirce, *Collected Papers*, 2.776.
[112] Goudge, *The Thought of C. S. Peirce*, pp. 200–1.
[113] Peirce, *Collected Papers*, 2.786.
[114] *Ibid.*, 2.786.
[115] *Ibid.*, 2.634.
[116] *Ibid.*, 5.60; Goudge, *The Thought of C. S. Peirce*, pp. 200–1.
[117] Peirce, *Collected Papers*, 6.477.
[118] *Ibid.*, 5.598.
[119] *Ibid.*, 5.591–92.
[120] Goudge, 'Pragmatism's Contribution to an Evolutionary View of Mind', p. 137.
[121] Goudge, *The Thought of C. S. Peirce*, pp. 209–11.
[122] Madden, 'Peirce and Contemporary Issues in the Philosophy of Science', p. 42.
[122a] See Chapter 3, Note 208, below.
[123] S. Toulmin, *Foresight and Understanding*, Chapter VI, has pursued the evolutionary analogy with some insight, but, alas, couples it with a version of the thesis that "all concepts are theory-laden" (cf. *Reasons and Revolutions*), so that he ends up with a radically distorted view of science, one that is essentially Aristotelian; see Wilson, 'Explanation in Aristotle, Newton, and Toulmin'.
[124] Goudge, *The Ascent of Life*, p. 208.
[125] *Ibid.*, p. 209.
[126] Cf. Kuhn, 'Reflections on My Critics', Section VI; cf. Wilson, 'Kuhn and Goodman: Revolutionary vs. Conservative Science'.
[127] Kuhn, *The Structure of Scientific Revolutions*, Second Edition, p. 76; *Reasons and Revolutions*, Section IX.
[128] *Structure*, Chapter VIII; *Reasons and Revolutions*, Section X.
[129] *Structure*, Chapter IX; *Reasons and Revolutions*, Section X. Cf. Wilson, 'Kuhn and Goodman: Revolutionary vs. Conservative Science'.

NOTES TO CHAPTER 1

130 Cf. Note 101.
131 Cf. Goudge, *Ascent of Life*, p. 209.
132 See the quoted passage cited in Note 109.
132a Kuhn accepts that there are such over-arching theory-independent standards: "Finally, at a still higher level [the highest], there is another set of commitments without which no man is a scientist. The scientist must, for example, be concerned to understand the world and to extend the precision and scope with which it has been ordered. That commitment must, in turn, lead him to scrutinize, either for himself or through colleagues, some aspect of nature in great empirical detail. And, if that scrutiny displays pockets of apparent disorder, then these must challenge him to a new refinement of his observational techniques or to a further articulation of his theories. Undoubtedly there are still other rules like these, ones which have held for scientists at all times" (Kuhn, *Structure of Scientific Revolutions*, p. 42).
133 Cf. C. Kordig, *The Justification of Scientific Change*; Wilson, *Reasons and Revolutions*.
134 *Reasons and Revolutions*, Sections IX and X. Also 'Logical Necessity in Carnap's Later Philosophy', Chapter Five.
135 Cf. F. Cunningham, *Objectivity in Social Science*.
136 Cf. C. Kordig, *The Justification of Scientific Change*.
137 Cf. Wilson, 'Hume's Theory of Mental Activity'.
138 *Treatise*, p. 448.
139 *Ibid.*, p. 610.
140 *Ibid.*, p. 611.
141 *Ibid.*, p. 611.
142 In his discussion of reduction sentences, G. Bergmann, 'Comments on Professor Hempel's "The Concept of Cognitive Significance",' p. 260ff, fails to distinguish adequately the idea of psychological context providing the criterion of lawlikeness, from the idea of an evidential context establishing the subjective worthiness of a law-assertion.
143 Cf. J. W. N. Watkins, 'The Paradoxes of Confirmation'.
144 Cf. R. Chisholm, 'Law Statements and Counterfactual Inference'.
145 Cf. N. Goodman, *Fact, Fiction, and Forecast*.
146 For example, justifying asserting the ideal gas law as an instance of the empirically more adequate van der Waal's law.
147 Scriven makes much of this point, in his 'Explanations, Predictions, and Laws', p. 208ff, but the next comment meets it sufficiently.
148 Cf. Moore, *Ethics*, pp. 118–121.
149 That is, the evidence should have been acquired through the use of the scientific method.
150 Scriven, 'Explanations, Predictions, and Laws', pp. 190–91, tries to use these points against Hempel and Oppenheim, 'Studies in the Logic of Explanation'. What I say here is sufficient comment.
151 See Chapter 3, below.
152 Cf. the title of Toulmin's book: *Foresight and Understanding*, i.e., foresight *vs.* understanding.
153 We discuss "mere forecasting" in much more detail in Chapter 2, below.
154 Collins, 'Explanation and Causality', pp. 485–86.
155 *Ibid.*, p. 482.

156 *Ibid.*, p. 485.
157 See G. Bergmann, *Philosophy of Science*, Chapter Two, for a detailed discussion of these ideas. For their role in the on-going process of science, see *Reasons and Revolutions*.
158 Cf. Bergmann, *Philosophy of Science*, Chapter Two; M. Brodbeck, 'Explanation, Prediction, and "Imperfect Knowledge"'.
159 Cf. G. Bergmann, 'Frequencies, Probabilities, and Statistics'; also Wilson, 'Hume's Sceptical Argument against Reason', and 'Is Hume a Sceptic with regard to Reason?'.
160 This is slightly inaccurate; we correct it when we discuss the examples (4.1), etc., below.
161 Cf. W. Salmon, *Logic*, First Edition, pp. 75–6.
161a Cf. *Ibid.*, p. 75.
161b However, in periods of "revolutionary science" this rule is, for sound reasons, relaxed somewhat; cf. Wilson, 'Kuhn and Goodman: Revolutionary vs. Conservative Science'.
162 T. Reid, *Essays on the Intellectual Powers of Man*, Essay III, Ch. iv, p. 253.
163 *System of Logic*, Bk. III, Ch. V, Sec. 6.
164 *Ibid.*, p. 222.
165 C. D. Broad, *Mind and Its Place in Nature*, p. 455.
166 C. G. Hempel, 'The Function of General Laws in History'.
167 Hempel, 'The Function of General Laws in History', pp. 345–48.
168 J. L. Mackie, 'Causes and Conditions'.
169 *Ibid.*, p. 16.
170 *Ibid.*, pp. 27–30.
171 J. Anderson, 'The Problem of Causality'.
172 Mackie, 'Causes and Conditions', p. 21ff.
173 *Ibid.*, p. 22, pp. 30–2.
174 *Ibid.*, p. 24.
175 Chapters 2 and 3, below.
176 Mackie, 'Causes and Conditions', pp. 30–32.
177 See the Russell selections in H. Feigl and M. Brodbeck, *Readings in The Philosophy of Science*.
178 Mackie, 'Causes and Conditions', pp. 30–32.
179 A similar point is made in M. Mandelbaum, *The Anatomy of Historical Knowledge*, pp. 200–202.
180 I discuss theories in greater detail in *Reasons and Revolutions*.
181 Cf. W. Sellars, 'Scientific Realism or Irenic Instrumentalism?'
182 See 'Explanation in Aristotle, Newton, and Toulmin', for more detail.
183 See G. Bergmann, *Philosophy of Science*, Chapter Three. Also, *Reasons and Revolutions*.
184 See *Reasons and Revolutions*.
185 Collins, 'Explanation and Causality', p. 488.
186 These examples have certain features that account for certain aspects of the process of concept formation. See Wilson, 'Definition and Discovery' and 'Is Operationism Unjust to Temperature?'
186a A similar point is made in B. C. van Fraasen, *The Scientific Image*, pp. 144–5.
187 Scriven, 'Explanation, Prediction, and Laws', p. 213.
188 Feyerabend, 'Explanation, Reduction, and Empiricism', p. 46ff. For another

NOTES TO CHAPTER 2

discussion of the Galileo example, cf. R. Yoshida, *Reduction in the Physical Sciences*, p. 31f.

[188a] I. Cohen ('Newton's Theory vs. Kepler's Theory and Galileo's Theory') agrees with our, rather than the Scriven-Feyerabend, position on the explanation of Galileo's Law by Newton.

[189] See also *Reasons and Revolutions*.

[190] Cf. the Boyle excerpt on his gas law, in M. Boas Hall, *Robert Boyle on Natural Philosophy*, p. 341.

[191] As, for example, does Achinstein, in his *Concepts of Science*. See my 'Discussion' of that work.

[192] Cf. G. Bergmann, 'Comments on Professor Hempel's "Concept of Cognitive Significance"'; F. Wilson, 'Acquaintance, Ontology, and Knowledge', and 'Logical Necessity in Carnap's Later Philosophy', Chapter One.

[193] Popper, 'Science: Conjectures and Refutations', pp. 62–3.

[194] Popper, 'Three Views Concerning Human Knowledge', p. 111ff.

[195] Popper, 'The Demarcation Between Science and Metaphysics', p. 257.

[196] *Ibid.*

[197] Popper, 'Three Views Concerning Human Knowledge', p. 111.

[198] *Ibid.*, p. 114.

NOTES TO CHAPTER 2

[1] Collins, 'Explanation and Causality', p. 487.

[2] As Hume points out, the idea of *cause* involves, essentially, the idea of *necessary connection*; cf. *Treatise*, pp. 77: "An object may be contiguous and prior to another, without being considered as its cause. There is NECESSARY CONNEXION to be taken into consideration; and that relation is of much greater importance, than any of the other two above-mentioned".

[3] Cf. Brodbeck, 'Explanation, Prediction, and "Imperfect" Knowledge'.

[4] Cf. Wilson, 'Logical Necessity in Canap's Later Philosophy', Chapters One and Two.

[5] Cf. Bergmann, 'On Non-Perceptional Intuition'.

[6] Cf. Wilson, *Reasons and Revolutions*.

[7] Cf. Wilson, *Reasons and Revolutions*.

[8] Ackoff, *Scientific Method*, p. 430; cf. pp. 3–4.

[9] *Ibid.*, p. 117.

[10] Cf. Bergmann, 'The Logic of Quanta', pp. 477–82.

[11] Cf. Bergmann, *Philosophy of Science*, Chapter Two; Wilson, *Reasons and Revolutions*.

[12] Scriven, 'Explanations, Predictions, and Laws'; Collins, 'Explanation and Causality'.

[12a] Compare J. S. Mill, *System of Logic*, p. 201: "Now it has been well pointed out ... that ... Time, in its modifications of past, present, and future, has no concern either with the belief itself, or with the grounds of it. We believe that fire will burn to-morrow, because it burned to-day and yesterday; but we believe, on precisely the same grounds, that it burned before we were born, and that it burns this very day in Cochin-China. It is not from the past to the future, as past and future, that we infer, but from the known to the unknown; from facts observed to facts unobserved; from what we have perceived, or been directly conscious of, to what has not come within our experience. In this last

predicament is the whole region of the future; but also the vastly greater portion of the present and of the past".

[13] Scriven, 'Explanations, Predictions, and Laws', p. 179f.

[14] *Ibid.*, p. 177, p. 181.

[15] 'Truisms as the Grounds for Historical Explanations', p. 468; also 'Explanations, Predictions, and Laws', p. 176.

[16] Scriven, 'Explanation and Prediction in Evolutionary Theory', p. 480.

[17] Bromberger, 'Why-Questions', p. 83. The points we shall make about this example will also apply to other examples of his presented on p. 72.

[18] Cf. S. Bromberger, 'Why-Questions'; Scriven, 'Explanations, Predictions, and Laws'.

[19] Presented at the Dalhousie Working Conference on Causality, 1973.

[20] For more details on the role of such *ex post facto* explanations in science, see *Reasons and Revolutions*.

[21] Also presented at Dalhousie Working Conference on Causality.

[22] Cf. Symon, *Mechanics*, p. 182ff.

[23] Scriven, 'Explanations, Predictions, and Laws'.

[24] This derives from Bromberger; see Hempel, 'Deductive-Nomological vs. Statistical Explanation', p. 109. Bromberger presents it in his 'Why-Questions', p. 71.

[25] Cf. E. Mach, *The Principles of Physical Optics*.

[26] In fact, however, it is hard to find a contemporary text that presents geometrical optics uncontaminated by wave optics.

[27] H. M. Blalock, *Causal Inferences in Nonexperimental Research*, pp. 11–21, pp. 38–51.

[28] Cf. Wilson, 'Review of Mandelbaum's *The Anatomy of Historical Knowledge*'.

[29] As, for example, by Scriven in his 'Cause, Connections, and Conditions', p. 240ff.

[29a] But cf. Wilson, 'Explanation in Aristotle, Newton, and Toulmin'.

[30] For example, von Wright, 'On the Logic and Epistemology of the Causal Relation', p. 97.

[31] Cf. Addis, 'Ryle's Ontology of Mind'.

[32] Collingwood, 'On the So-Called Idea of Causation', p. 86.

[33] *Ibid.*, p. 87.

[34] Cf. Sartre, *Being and Nothingness*, p. 180ff. Compare the discussions of motive in Ryle, *The Concept of Mind* and Peters, *The Concept of Motivation*; and, in criticism, Addis, 'Ryle's Ontology of Mind' and Wilson, *'Meaning Is Use'*.

[35] Cf. Addis, 'Ryle's Ontology of Mind'. See also Chapter 3, below.

[36] Cf. Bergmann, 'Purpose, Function, and Scientific Explanation'.

[37] *Ibid.*; also Grünbaum, 'Causality and the Science of Human Behavior'.

[38] Cf. Addis, *The Logic of Society*, Ch. III.

[39] Collingwood, 'On the So-Called Idea of Causation', p. 89.

[40] Gasking, 'Causation and Recipes', p. 483.

[41] Rosenberg, 'Causation and Recipes: The Mixture as Before?'

[42] Gasking, 'Causation and Recipes', p. 482.

[43] *Ibid.*

[44] Simon and Rescher, 'Cause and Counterfactual'.

[45] Collingwood, 'On the So-Called Idea of Causation', p. 90.

[46] Gasking, 'Causation and Recipes', p. 484.

[47] von Wright, 'On the Logic and Epistemology of the Causal Relation'.

⁴⁸ *Ibid.*, pp. 140–5.
⁴⁹ *Ibid.*, p. 107.
⁵⁰ *Ibid.*; cf. p. 110, p. 111.
⁵¹ *Ibid.*, pp. 105–6.
⁵² *Ibid.*, p. 104.
⁵³ *Ibid.*
⁵⁴ *Ibid.*, p. 105.
⁵⁵ *Ibid.*, p. 106.
⁵⁶ *Ibid.*, p. 107; his italics.
⁵⁷ *Ibid.*, p. 105.
⁵⁸ *Ibid.*
⁵⁹ Taylor, 'The Metaphysics of Causation', pp. 39–40.
⁶⁰ Mill, *System of Logic*, p. 216n (on p. 217).
⁶¹ Hart and Honoré, 'Causal Judgment in History and in the Law', p. 222.
⁶² *Ibid.*, p. 221.
⁶³ *Ibid.*, p. 218ff.
⁶⁴ Cf. Davidson, 'Causal Relations', p. 150ff; see Chapter 3, below.
⁶⁵ Mill, *System of Logic*, p. 216n (on p. 217).
⁶⁶ Toulmin, *Foresight and Understanding*, p. 59.
⁶⁷ *Ibid.*, pp. 45–6.
⁶⁷ᵃ I have argued this in detail in 'Explanation in Aristotle, Newton, and Toulmin'.
⁶⁸ Scriven, 'The Key Property of Laws–Inaccuracy', p. 101.
⁶⁸ᵃ See Section 3.4, below.
⁶⁹ Scriven, 'Truisms as the Grounds for Historical Explanations', p. 468.
⁷⁰ Hempel and Oppenheim add that relevant 'could'; see 'Studies in the Logic of Explanation', p. 249, p. 279.
⁷¹ Scriven, 'Truisms as the Grounds for Historical Explanations', p. 456; cf. also his 'Explanations, Predictions, and Laws', p. 175.
⁷² Scriven, 'Truisms as the Grounds for Historical Explanation', p. 456.
⁷³ *Ibid.*, p. 462.
⁷⁴ Cf. Brodbeck, 'Explanation, Prediction, and "Imperfect" Knowledge'.
⁷⁵ Scriven, 'Explanations, Predictions, and Laws', p. 184; 'Truisms as the Grounds for Historical Explanations', pp. 456–7, p. 467; 'Cause, Connections, and Conditions', p. 240ff; and Collingwood, 'On the So-Called Idea of Causation', pp. 86–7. See also Ruddick, 'Causal Connections'.
⁷⁶ Scriven, 'Truisms as the Grounds for Historical Explanations'.
⁷⁷ But see also Wilson, *'Meaning Is Use'*.
⁷⁸ *Ibid.*; also Wilson, 'Logical Necessity in Carnap's Later Philosophy'.
⁷⁹ Scriven, 'Truisms as the Grounds for Historical Explanations", p. 466.
⁸⁰ *Ibid.*
⁸¹ Cf. Wilson, *Reason and Revolutions*, Section III.
⁸² *Reasons and Revolutions*, Sections III, and VII.
⁸³ *Reasons and Revolutions*, Section VIII; 'Definition and Discovery'; 'Is Operationism Unjust to Temperature?'
⁸⁴ Cf. *Reasons and Revolutions*, Section III.
⁸⁵ Collins, 'Explanation and Causality', pp. 495–6.
⁸⁶ *Ibid.*, p. 498.

87 *Ibid.*, p. 497.
88 Bromberger, 'Why-Questions', p. 76.
89 Hempel and Oppenheim recognize this point; see their 'Studies in the Logic of Explanation', p. 258.
90 Cf. Wilson, *Reasons and Revolutions*, Sections III and VIII.
91 *Ibid.*, Section III.
92 *Ibid.*, Section III and IV.
93 *Ibid.*, Sections III, IV, VIII and X.
94 Scriven, 'Truisms as the Grounds for Historical Explanations', p. 456.
95 *Ibid.*, p. 458.
96 But compare the important discussion in P. Meehl, 'Problems in the Actuarial Characterization of a Person'.
97 Cf. *Ibid.*
98 Cf. M. R. Westcott, *Toward a Contemporary Psychology of Intuition*.

NOTES TO CHAPTER 3

1 For a discussion of this way of characterizing the context of discovery, cf. Wilson, 'Definition and Discovery'.
2 In Section I of that Chapter.
3 Bromberger, 'Why-Questions', p. 69f.
4 *Ibid.*, p. 69.
4a Cf. J. Urmson, 'Parenthetical Verbs'.
5 Bromberger, 'Why-Questions', p. 66.
6 *Ibid.*, p. 67.
7 *Ibid.*
8 Cf. *Reasons and Revolutions*, Sec. VIII.
9 The following quotes are from Bromberger, 'Why-Questions', p. 70.
10 Bromberger, 'Why-Questions', p. 70. See also his 'Approach to Explanation'.
11 C^* is the hope of a concept, C is the concept that, after research, fulfills that hope. The reference is back to the discussion in Chapter 2 Sections 2.5 and 2.6. See also Wilson, 'Definition and Discovery'.
12 In Section 2.5 of that chapter.
13 Collins, 'Explanation and Causality', p. 483.
14 *Ibid.*
15 *Ibid.*, p. 486; italics added.
16 M. Scriven, 'Explanations, Predictions, and Laws', p. 225.
17 *Ibid.*
17a I should add that Scriven at least attempts to spell out the criteria in detail. Others, in contrast, do not even do that. For example, B. C. van Fraassen, *The Scientific Image*, introduces an elaborate apparatus for characterizing explanations as answers to questions, but fails to address the issues that Scriven raises. We are treated (p. 141ff.) to a discussion of why-questions as ordered triples — $Q = \langle P_k, X, R \rangle$ — and so on, a discussion in which the now fashionable jargon of set theory is deployed with little effect beyond the creation of a spurious aura of exactitude. We are told (pp. 144–5) that B is an answer to Q just in the case where it expresses a proposition A such that A stands in the

relation R to $\langle P_k, X \rangle$, where the latter picks out the facts to be explained. The relation R is that of *explanatory relevance*: thus, what we are told is that the answer must explain what the questioner is concerned to understand. *And this is about all that we are told about R!* In effect, then, van Fraassen takes this to be a *primitive term*, incapable of analysis, but (one presumes) intrinsically normative. It is, perhaps, a simple non-natural normative relation, possibly analogous to the simple non-natural relation of "fittingness" that Ewing and others in the 'thirties introduced into moral philosophy. To be sure, van Fraassen does give us some commentary about his R. We are told (p. 104) that to be relevant is to afford grounds for believing that what is to be explained actually occurred, i.e., the answer must be such that the phenomena to be explained could have been predicted if the proposition the answer expresses had been known antecedently; we are introduced (p. 106ff.) to some ideas of statistical relevance; we are informed through some examples (p. 123ff.) that background information is often relevant to determining relevance; and are shown (p. 147ff.) how background knowledge and statistics are interconnected. The point is that none of this is to develop a systematic defence of a model of what explanation *ought to be*, that is, ought to be relative to the cognitive interests of the explainee; nor is it to articulate and defend a set of criteria for better and worse explanation, relative again to the same cognitive interests. This would require, *one*: a detailed explication of the notion of explanatory relevance, and, *two*: an argument justifying the claim that the concept thus explicated is, relative to our cognitive interests, the notion of explanatory relevance that we *ought* to adopt. Van Fraassen does neither of these things. To that extent one can, I think, justifiably charge him with introducing, like Ewing, a primitive, unanalyzed normative relation. Scriven, in contrast, is sensitive to these points. However wrong his account is, we may say that he at least addresses the significant issues, and does so in sufficient detail to make his position worth the critic's systematic examination.

[18] This and the following quotes are from M. Scriven, 'Truisms as the Grounds for Historical Explanations', pp. 446–7; his italics throughout. Cf. also his 'Explanations, Predictions, and Laws', p. 200ff.

[19] Scriven, 'Explanations, Predictions, and Laws', p. 200; 'Truisms as the Grounds for Historical Explanations', p. 446.

[20] Scriven, 'Explanations, Predictions, and Laws', p. 202, 205, and 207.

[21] Remarks of Collins, 'Explanation and Causality', p. 484, and of Scriven, 'Explanations Predictions, and Laws', p. 202, are quite irrelevant, it seems to me.

[22] Chapter 1, Section 1.4.

[23] See 'Explanation in Aristotle, Newton, and Toulmin'.

[24] Cf. Scriven, 'Explanations, Predictions and Laws', p. 202.

[25] Scriven, 'Explanations, Predictions, and Laws', p. 201.

[26] *Idem.*

[27] Cf. *Reasons and Revolutions*, Sections II, and XI.

[28] Hempel and Oppenheim, 'Studies in the Logic of Explanation'.

[29] Part I, Section 4 and Part II intervene between the initial statement of the model and the attempt in Part III to state formal criteria for being an explanatory argument. In conducting the argument defending the idea of scientific explanation in such areas as psychology and biology, Hempel and Oppenheim use only the initial statements, not the formal criteria. I suggest below that this is of some importance in getting the criteria of Part III in the correct perspective.

30 These are laid out in Part III, Section 7, of 'Studies in the Logic of Explanation', pp. 270–8.

30a *Ibid.*, p. 248.

31 Davidson, 'Causal Relations', pp. 152–3. Davidson regularly treats definite descriptions like '$(\imath y)(Hy)$' as a special sort of singular term. They are *not* a sort of *name*, however. For, if 'a' is *any* name, then the argument

(a) $$\frac{(x)(\phi x)}{\phi a}$$

is valid. In contrast,

(b) $$\frac{(x)(\phi x)}{\phi(\imath y)(Hy)}$$

is *invalid*. A valid argument is *truth-preserving*. (a) is in this way valid. But (b) is not, since the premiss may be true and the conclusion not true, e.g., when there is *nothing* that satisfies the propositional function 'Hy', i.e., when the definite description '$(\imath y)(Hy)$' *is unsucc*essful. What *is* valid is the argument

(c) $$\frac{E!\,(\imath y)(HY)}{(x)(\phi x)} \\ \phi(\imath y)(Hy)$$

where the first premiss asserts that the definite description is successful, i.e., that there is one and only one individual that is H. *Provided* that the definite description is successful, *then* any argument form that is valid for a name (e.g., (a)) is also valid for the definite description.

Russell, of course, proposed an analysis of statements containing definite descriptions. The statement

$$\phi(\imath y)(Hy)$$

ought, he argued, be construed as

$$(\exists y)\,[Hy\,\&\,(x)(Hy \supset x = y)\,\&\,\phi y]$$

Under this analysis, definite descriptions do *not* appear as singular terms in the language. Rather, they are simply pieces of *eliminable short-hand*, and the only *genuine singular terms* are *names*.

The language L that Hempel and Oppenheim propose for the language of science is of this Russellian sort, and this is of some relevance for the cogency of Davidson-type criticisms, as we shall see.

32 Kim, 'Causation, Nomic Subsumption, and the Concept of Event'.

33 *Ibid.*, p. 220; I have made some notational changes which, however, do not affect the sense.

34 *Ibid.*, p. 221.

35 Ackermann, 'Deductive Scientific Explanation', p. 163.

36 *Ibid.*, p. 162.

37 Hempel, 'Postscript (1964)', p. 295, makes the same points, and same plea for a set of criteria rationalized by the concept of scientific explanation itself.

38 'Studies in the Logic of Explanation', p. 246.
39 *Ibid.*, pp. 247–9.
40 *Ibid.*, p. 247.
41 *Ibid.*, p. 257.
42 *Ibid.*, p. 248.
43 *Ibid.*, p. 246.
44 *Ibid.*, p. 249.
45 Hempel, 'Explanation in Science and in History', p. 10; italics added.
46 Morgan, 'Archaeology and Explanation', pp. 268, 273, and 275.
47 See Watson, Leblanc and Redman, 'The Covering Law Model in Archaeology: Practical Uses and Formal Interpretations', which is commenting on Morgan, 'Archaeology an Explanation'.
48 Eberle, Kaplan and Montague, pp. 419–20.
49 Morgan, 'Omer on Scientific Explanation'.
49a Cf. W. E. Johnson, *Logic*, Part I, pp. 38–49.
50 Ackermann, 'Deductive Scientific Explanation', p. 164.
51 Thorpe, 'The Quartercentary Model of D-N Explanation', p. 189. Thorpe derives it from Hempel's 'Postscript (1964)', p. 294.
52 This example derives from Ackermann and Stenner, 'A Corrected Model of Explanation', p. 166. It is considered by Kim, 'On the Logical Conditions of Deductive Explanation', p. 287.
53 Kim, 'On the Logical Conditions of Deductive Explanation', p. 289.
54 Morgan, 'Kim on Deductive Explanation', p. 438.
55 This example is from Kim, 'On the Logical Conditions of Deductive Explanation', p. 288.
56 This example is considered by Hempel and Oppenheim, 'Studies in the Logic of Explanation', p. 275.
57 *Ibid.*
58 See also the discussion of Ryle in Section 3.5, below.
59 Kim, 'On the Logical Conditions of Deductive Explanation'.
60 *Ibid.*, p. 287.
61 *Ibid.*, p. 289n.
62 *Ibid.*, p. 289.
63 Thorpe, 'The Quartercentary Model of D-N Explanation', pp. 194–5.
64 *Ibid.*, pp. 190–1.
65 *Ibid.*, p. 192.
66 *Ibid.*
67 *Ibid.*
67a *Ibid.*, 194.
67b *Ibid.*
68 Omer, 'On the D-N Model of Scientific Explanation', p. 419.
69 *Ibid.*, pp. 420–1.
70 *Ibid.*, p. 422.
71 *Ibid.*
72 *Ibid.*
73 Kim, 'On the Logical Conditions of Deductive Explanation', p. 288.
74 Ackermann and Stenner, 'A Corrected Model of Explanation', p. 169.
75 Omer, 'On the D-N Model of Scientific Explanation', p. 424.

76 *Ibid.*
77 *Ibid.*, p. 425.
78 Ackermann, 'Deductive Scientific Explanation', pp. 160–1.
79 Morgan, 'Omer on Scientific Explanation', p. 115.
80 This reading of the argument is Morgan's.
81 Omer, 'On the D-N Model of Scientific Explanation', p. 426.
82 *Ibid.*
83 Morgan, 'Omer on Scientific Explanation', p. 111.
84 *Ibid.*, pp. 111–2.
85 The reference is to Hempel, 'Aspects of Scientific Explanation', p. 347n.
86 Omer, 'On the D-N Model of Scientific Explanation', p. 433.
87 On the notion of "imperfect knowledge", cf. Bergmann, *Philosophy of Science*, Chapter Two; Brodbeck, 'Explanation, Prediction, and "Imperfect" Knowledge'; and Chapters 1 and 2, above.
88 See the quoted passage cited in Note 76, above.
89 Cf. *Reasons and Revolutions*, Sections IX and X.
90 W. Dray, *Laws and Explanations in History*, p. 158ff.
91 T. A. Goudge, *Ascent of Life*.
92 'Philosophical Literature', pp. 100–1.
93 Cf. *Reasons and Revolutions*, Section VI.
94 Goudge, *The Ascent of Life*, p. 123 and 75.
95 *Ibid.*, p. 123.
96 *Ibid.*
97 *Ibid.*, p. 126.
98 The example is drawn from M. Scriven, 'Explanation and Prediction in Evolutionary Theory'.
99 "($\exists ! x$)" represents that "there is at least one and at most one x such that ...".
100 As we know from Chapter 2, Section 2.5, Scriven has wrongly concluded that, since the symmetry of explanation and prediction sometimes thus breaks down, since (in other words) we can sometimes explain *ex post facto* where we cannot predict, therefore explanation is not by deduction from laws.
101 For example, as presented in Hempel and Oppenheim, 'Studies in the Logic of Explanation'.
102 Cf. Scriven 'Explanation and Prediction in Evolutionary Theory', 'Explanations, Predictions and Laws', and 'Truisms as the Grounds for Historical Explanations'.
103 Goudge, *The Ascent of Life*, pp. 125–6.
104 *Ibid.*, p. 65.
105 Goudge quotes this passage from G. G. Simpson, 'Evolution'.
106 Goudge, *The Ascent of Life*, p. 66.
107 *The Ascent of Life*, p. 68.
108 Goudge points out (*The Ascent of Life*, p. 61) that such *ex post facto* explanations are more than common in biology; and that their basis is the existence of mixed quantificational laws of the sort (E). Goudge speaks of "reading back into the historical record" rather than of "*ex post facto* explanations" but the point is the same. And he makes the point that the laws like (E) can be taken to be taken to be instantiations of a more general "uniformitarian principle".
109 *The Ascent of Life*, p. 68.

¹¹⁰ *Ibid.*, p. 71ff.
¹¹¹ *Ibid.*, p. 71; italics added.
¹¹² *Ibid.*
¹¹³ Compare the 'paradoxical' emphasized in the quoted passage cited in Note 111.
¹¹⁴ Cf. W. Dray, *Laws and Explanation in History*, pp. 164–6.
¹¹⁵ Cf. C. Hempel, 'Aspects of Scientific Explanation', p. 428–30.
¹¹⁶ Cf. M. Scriven, 'Appendix' to 'Truisms as th Grounds for Historical Explanations', pp. 471–5.
¹¹⁷ Goudge, *The Ascent of Life*, p. 77. M. Ruse, *The Philosophy of Biology*, pp. 89–92, suggests, wrongly I think, Goudge to be here arguing that *all* generalities involved in narrative explanations should be construed as "inference-tickets" and that therefore narrative explanations do not involve deduction from general premisses. I do not think Goudge is rejecting law-deduction as a condition of explanation, but rather only the idea that when we explain "*E* because *s*" the relevant law for the deduction is the generality "Whenever an event exactly like *s* occurs then an event exactly like *E* occurs". For, as Scriven points out (see Note 116), *E* and *s* are unique: only they are *exactly* like *E* and *s*. In which case the generality *is* tautological, and not a law. But it does not follow from this – nor (I think) does Goudge suggest it follows – that no law and no deduction from a law is involved. Rather, what Goudge suggests that we need an account of how the law-deduction occurs which is both more subtle and more complicated. (By the way, we discuss Scriven's point here at length at the end of Section 3.6, below.)
¹¹⁸ Goudge, *The Ascent of Life*, p. 74.
¹¹⁹ *Ibid.*, p. 123.
¹¹⁹ᵃ Compare the 'seem' emphasized in the quoted passage cited in Note 111.
¹²⁰ Goudge, *The Ascent of Life*, p. 74; his italics.
¹²¹ Cf. the discussion of various senses of 'hypothesis' in G. Bergmann, 'The Logic of Psychological Concepts'.
¹²² Goudge, *The Ascent of Life*, p. 74.
¹²³ Cf. Dray, *Laws and Explanations in History*, p. 158ff.
¹²⁴ Goudge, *Ascent of Life*, p. 75.
¹²⁵ *Ibid.*
¹²⁶ *Ibid.*, pp. 73–4.
¹²⁷ For this point in connection with integrating explanations, cf. *The Ascent of Life*, p. 68.
¹²⁸ Two theories can always be put into one axiomatic system simply by conjoining their separate axioms into one by meas of 'and'. The requirement that the two theories have a shared form or generic structure eliminates this trivial case. It is also necessary, however, to exclude genera of the "gruesome" sort designed by Goodman. Eliminating these "arbitrary" predicates is not so easy. But I think Goodman himself has given the essential ingredients for a solution, with his notion of a predicate being "non-arbitrary" just in the case where it has become *entrenched*. Cf. Wilson, 'Kuhn and Goodman: Revolutionary vs. Conservative Science'.
¹²⁹ Ruse, *The Philosophy of Biology*, pp. 66–7, does not take seriously the idea that the axiomatic model is a model, minimally, for the generic unification of several laws. If Ruse is correct, then simply to make a deductive inference is to be involved in axiomatics. Upon that criterion, *all* science is in fact at present axiomatized. Yet we can still distinguish classical mechanics and evolutionary theory: as Goudge asserts, in the former a

real generic unity of laws is achieved, in the latter such a unity is still a goal, and in fact a distant goal.

[130] For the importance of clearly distinguishing definite descriptions from definitions, compare Wilson, 'Dispositions: Defined or Reduced?' and 'A Note on Hempel on the Logic of Reduction'.

[131] Goudge, *The Ascent of Life*, p. 75.

[132] For an excellent discussion of historical laws, see G. Bergmann, *Philosophy of Science*, Chapter II, Section 5.

[133] Cf. E. Nagel, *The Structure of Science*, pp. 288–90.

[134] Goudge, *The Ascent of Life*, p. 61.

[135] *Ibid.*, p. 62.

[136] *Ibid.*, pp. 122–3.

[137] *Ibid.*, p. 175.

[138] *Ibid.*, p. 16. Ruse, *The Philosophy of Biology*, p. 65, fails to see how historicity complicates the task of axiomatization, that is, complicates it *in fact*, though not in principle — but that is all that Goudge wishes to argue.

[139] *The Ascent of Life*, pp. 33–4.

[140] *Ibid.*, p. 16. Ruse, *The Philosophy of Biology*, p. 62, misses this *de facto* complexity also.

[141] *The Ascent of Life*, p. 125.

[142] *Poetics*, Chapter 10.

[143] Ruse, *The Philosophy of Biology*, pp. 88–9, argues that truth is necessary to the success of a narrative explanation: "If we do not know which are the true conditions, then I fail to see how we can claim to have a narrative explanation either". Goudge's point is that a narrative explanation can do the job expected of it even if the truth of (some part of) the explanation is not known. (See, for example, Goudge's emphasis in the quotation cited in Note 120.) Narrative explanations are in this way distinguished from ordinary causal explanations, which are acceptable in an explaining situation only if the causal statement is known to be true, or, at least (given the limits of induction), if there is reason to beleive it to be true.

[144] Cf. Lakatos, 'Falsification and the Methodology of Scientific Research Programmes'.

[145] For a discussion of how the location of an hypothesis in the context of research affects the logical status of the concepts appearing in it, see F. Wilson, 'Definition and Discovery'.

[146] There is a tendency too quickly to deploy the "context of discovery/context of justification" dichotomy to dismiss the process and concentrate on the product; see *Reasons and Revolutions*.

[147] Kuhn, *The Structure of Scientific Revolutions*, Ch. III.

[148] *Ibid.*, Ch. V.

[149] The relevance of context for providing laws not explicitly stated is indicated in Goudge, *The Ascent of Life*, p. 76.

[150] M. Scriven, 'Truisms as the Grounds for Historical Explanations', p. 446; cf. Collins, 'Explanation and Causality', p. 489f.

[151] Cf. M. Brodbeck, 'Explanation, Prediction, and "Imperfect Knowledge"'; also G. Bergmann, 'The Revolt Against Logical Atomism'.

[152] Or rather: a verbal stimulus which is an instance of the subject-predicate form and where the predicate is of the type 'F'. Each instance of each of the types 'Fa', 'Fb', 'Fc',

NOTES TO CHAPTER 3

etc., is one of the relevant verbal stimuli. The idea is clear enough; the rest can be filled in by the reader himself.

153 Cf. Chapter 1, Section 1.4, above.
154 Scriven, 'Truisms as the Grounds for Historical Explanations', pp. 446–447, and 462–463; Collins, 'Explanation and Causality', p. 483.
155 Scriven, 'Explanations, Predictions, and Laws', p. 225; Collins, 'Explanation and Causality', p. 491.
156 Scriven, 'Explanations, Predictions, and Laws', p. 225; 'Truisms as the Grounds for Historical Explanations', p. 448ff.
157 Scriven, 'Explanations, Predictions, and Laws', p. 197ff.; 'Truisms as the Grounds for Historical Explanations', p. 449. Cf. Collins, 'Explanation and Causality', pp. 490–491.
158 See Sec. 3.1 of this Chapter.
159 Scriven, 'Truisms as the Grounds for Historical Explanations', pp. 449–50.
160 *Ibid.*, pp. 449–450; his italics.
161 Cf. G. Ryle, ' "If", "So", and "Because" '; and *The Concept of Mind*, p. 300. Also Toulmin, *The Uses of Argument*, p. 121.
162 S. Toulmin, *Philosophy of Science*.
163 Cf. Lewis Carroll, 'What the Tortoise Said to Achilles'.
164 Scriven accepts the Rylean analysis of inference and in particular that of "because"-statements; see 'Explanations, Predictions, and Laws', p. 200.
165 Cf. N. Colburn, 'Logic and Professor Ryle'. Also L. Addis, 'Ryle's Ontology of Mind', p. 34ff.
166 Cf. Toulmin, *Philosophy of Science*, pp. 49, 95, and 105.
167 *Ibid.*, p. 49.
168 *Ibid.*, p. 58ff, and 78ff.
169 *Ibid.*, p. 78.
170 This example is taken from H. G. Alexander, 'General Statements as Rules of Inference', p. 321.
171 Chapter 1, Sec. 1.5, above.
172 Alexander, 'General Statements on Rules of Inference', p. 320.
173 Cf. *Reasons and Revolutions*, Section VII.
174 Toulmin, *Philosophy of Science*, pp. 52–3.
175 *Ibid.*, p. 51.
176 *Ibid.*, pp. 28–30.
177 See *Reasons and Revolutions*, Section VII, for a dissection of another of Toulmin's arguments.
178 Toulmin, *Philosophy of Science*, p. 80.
179 *Ibid.*, p. 88; italics added.
180 *Ibid.*, p. 64; his italics.
181 Cf. J. S. Mill's discussion of colligation in his *System of Logic*, Bk. III, Ch. II, Secs. 4,5; also Bk. III, Ch. II, Sec. 6.
182 See, for example, *Philosophy of Science*, p. 45, first full paragraph.
183 'Explanation in Aristotle, Newton, and Toulmin'.
184 Toulmin, *Philosophy of Science*, p. 25.
185 *Ibid.*, p. 27.
186 *Ibid.*, p. 29.

NOTES TO CHAPTER 3

187 Cf. E. Mach, *The Principles of Physical Optics*.
188 Passage cited in Note 186 above.
189 Toulmin, *Philosophy of Science*, p. 30; cf. also p. 41.
190 *Ibid.*, pp. 93–5.
191 *Ibid.*, p. 25.
192 *Ibid.*, p. 41.
193 *Ibid.*, pp. 63–64.
194 Cf. *Reasons and Revolutions*, Sections III and X.
195 Cf. Mach, *The Principles of Physical Optics*, for a clear discussion of geometric optics and how it provides a good approximation to wave optics, i.e., how the latter is related to the former as the less imperfect to the more imperfect.
196 Toulmin, *Philosophy of Science*, pp. 24, and 28–30.
197 *Ibid.*, p. 36; also pp. 26, 30, and 63–4.
198 *Ibid.*, p. 37.
199 *Ibid*, pp. 36–7.
200 Goudge, *The Ascent of Life*, pp. 97–8.
201 *Ibid.*, p. 194.
202 *Ibid.*; his italics.
203 *Ibid.*, p. 196ff.
204 *Ibid.*, p. 80ff.
205 Cf. C. D. Broad, *Mind and Its Place in Nature*, pp. 82–3. Goudge refers to this section of Broad in *The Ascent of Life*, p. 195.
206 *Mind and Its Place in Nature*, p. 83.
207 In Section 3.4, above.
208 David Hume, in his *Dialogues Concerning Natural Religion*, has Philo ask the atheist (Part XII) " ... if it be not probable, that the principle which first arranged, and still maintains, order in this universe, bears not some remote inconceivable analogy to the other operations of nature, and among the rest to the oeconomy of human mind and thought". This is indeed the best that can be got from the argument from design. And to this the atheist agrees: "The atheist allows, that the original principle of order bears some remote analogy to it". This is surely correct; this conclusion follows *soundly* from the Argument from Design. Only, we now *know* (since Darwin) what it is that bears this analogy to the human mind: it is no transcendent God, but simple natural selection. Nor is the analogy anything more than the fact that natural selection and reasoning are processes in which problems of adaptation are solved; nothing, in short, sufficiently strong (as Hume knew) to establish anything about morals and politics, about praying or about aborting.
209 *The Ascent of Life*, p. 195.
210 W. Sellars, 'Scientific Realism or Irenic Instrumentalism'.
211 N. R. Campbell, *Foundations of Science* [formerly, *Physics: The Elements*], Chapter VI.
212 E. Nagel, *The Structure of Science*, Chapter VI, Section i.
213 Cf. Lakatos, 'Falsification and the Methodology of Scientific Research Programmes', Sections 3(a), 3(b). Cf. *Reasons and Revolutions*, Section VI.
214 M. B. Hesse, *Models and Analogies in Science*. Cf. Sellars, 'Scientific Realism or Irenic Instrumentalism', p. 344ff.
215 *Models and Analogies in Science*, pp. 39–40.

NOTES TO CHAPTER 3

[216] Cf. Lakatos, 'Falsification and the Methodology of Scientific Research Programmes', Section 3(c); G. Buchdahl, 'History of Science and Criteria of Choice'.

[217] On the role of the composition law, see G. Bergmann, *Philosophy of Science*, Chapter III. Also F. Wilson, 'Explanation in Aristotle, Newton and Toulmin'; and 'Discussion of Achinstein's *Concepts of Science*'.

[218] *The Structure of Scientific Revolutions* (Second Edition), p. 153ff; cf. Wilson, 'Kuhn and Goodman: Revolutionary vs. Conservative Science'. Campbell, *Foundations of Science*, pp. 129–37, suggests an account along the same lines.

[219] Cf. *Reasons and Revolutions*, Sections VI, IX and X.

[220] Cf. Chapter 1, Section 1.3, above.

[221] Cf. F. Wilson, 'Goudge's Contribution to the Philosophy of Science'.

[222] Cf. *Reasons and Revolution*, Sections I and VII; also G. Bergmann, 'The Revolt Against Logical Atomism'.

[223] Goudge, *The Ascent of Life*, p. 76.

[224] See the remark made above on the quoted passage cited in Note 125.

[225] Cf. Dray's discussion of "principles of action" in *Laws and Explanations in History*, Chapter V.

[226] Cf. Hempel, 'Aspects of Scientific Explanation', Section 10; L. Addis, *The Logic of Society*, Chapters III, VI, and IX; A. Rosenberg, *Microeconomic Laws*, Chapter V; M. Brodbeck, 'Meaning and Action'.

[227] Cf. T. Abel, 'The Operation Called *Verstehen*'.

[228] *Ars Poetica*, 333.

[229] T. A. Goudge, 'Pragmatism's Contribution to an Evolutionary View of Mind'; *The Ascent of Life*, pp. 205–11.

[230] 'Pragmatism's Contribution to an Evolutionary View of Mind', p. 139.

[231] *Ibid.*, p. 140.

[232] Cf. G. H. Mead, *Mind, Self and Society*, Sections I(4)–I(6).

[233] Cf. Bergmann, *Philosophy of Science*, Chapter III.

[234] Goudge, 'Pragmatism's Contribution to an Evolutionary View of Mind', pp. 142 and 146.

[235] M. Scriven, 'Explanations, Predictions, and Laws', p. 204; 'Truisms as the Grounds for Historical Explanations', pp. 453 and 456.

[236] Collins, 'Explanation and Causality', p. 491.

[237] *Ibid.*, p. 491.

[238] *Ibid.*, p. 491.

[239] *Ibid.*, p. 493.

[240] *Ibid.*, p. 492.

[241] *Ibid.*, pp. 494–496.

[242] *Ibid.*, p. 493.

[243] *Ibid.*, p. 496.

[244] *Ibid.*, p. 495.

[245] *Ibid.*, p. 494.

[246] *Ibid.*, p. 494ff.

[247] Cf. Wilson, 'Logical Necessity in Carnap's Later Philosophy', Chapter One.

[248] Cf. Wilson, 'Acquaintance, Ontology, and Knowledge'.

[249] Cf. Wilson, 'Logical Necessity in Carnap's Later Philosophy', Chapter Two.

[250] Cf. D. Davidson, 'Truth and Meaning'.

[251] D. Davidson, 'Causal Relations', pp. 159–160.
[252] *Ibid.*, p. 161.
[253] *Ibid.*, p. 151.
[254] *Ibid.*
[255] A. Pap, 'Disposition Concepts and Extensional Logic', p. 212.
[256] Davidson, 'Causal Relations', p. 152.
[257] A. Burks, 'The Logic of Causal Propositions', p. 369.
[258] Davidson, 'Causal Relations', p. 152.
[259] Cf. B. Tapscott, *Elementary Applied Symbolic Logic*, p. 373.
[260] *Ibid.*, p. 373.
[261] Davidson, 'Causal Relations', pp. 151–2.
[262] *Ibid.*, p. 152.
[263] *Ibid.*, p. 153.
[264] *Ibid.*, pp. 152–3.
[265] *Ibid.*, p. 160.
[266] *Ibid.*, p. 160.
[267] *Ibid.*, pp. 159–160.
[268] Cf. J. Kim, 'Causation, Nomic Subsumption, and the Concept of Event', p. 230.
[269] D. Davidson, 'The Individuation of Events', p. 171; cf. also his 'Causal Relations', p. 152.
[270] However, Davidson ('Causal Relations', p. 158) – unreasonably – imposes this as a condition laws must fulfill if they are to be causal.
[271] Cf. M. Mandelbaum, *The Anatomy of Historical Knowledge*, p. 110ff.
[272] *Ibid.*, p. 74 and p. 118ff.
[273] Cited in Note 267 above.
[274] Davidson, 'Causal Relations', p. 157.
[275] J. S. Mill, *System of Logic*, Bk. III, Ch. V, Sec. 3.
[276] Davidson, 'Causal Relations', pp. 155–6; cf. also p. 156.
[277] *Ibid.*, p. 156.
[278] *Ibid.*, p. 155.
[279] Since we assume (i_6): $(\exists! x) (Fx)$, the '$\sim$' in '$\sim H(\imath x) (Fx)$' has throughout what follows the following scope:

$$(\exists x) [Fx \,\&\, (y) (Fy \supset y = x) \,\&\, \sim Hx]$$

[280] Davidson, 'Causal Relations', p. 158.
[281] Kim, 'Causation, Nomic Subsumption, and the Concept of Event', p. 231.
[282] J. A. Foster, 'Psychophysical Causal Relations', pp. 65–66.
[283] Kim, 'Causation, Nomic Subsumption, and the Concept of Event', p. 231.
[284] *Ibid.*, p. 231.
[285] *Ibid.*, p. 232; cf. also Davidson, 'Causal Relations', p. 158.
[286] Kim, 'Causation, Nomic Subsumption, and the Concept of Event', pp. 232–3.
[287] *Ibid.*, p. 234.
[288] Davidson, 'Causal Relations', pp. 153–4.
[289] Davidson, 'The Individuation of Events', p. 172.
[290] *Ibid.*, p. 172.
[291] Kim, 'Causation, Nomic Subsumption, and the Concept of Event'.

292 *Ibid.*, p. 222.
293 *Ibid.*, p. 223.
294 *Ibid.*, p. 226.
295 *Ibid.*
296 Cf. Kim's discussion of compound events, *ibid.*, pp. 233–4.
297 G. Bergmann, *Realism*.
298 Collins, 'Explanation and Causality', p. 494ff.
299 Davidson, 'Causal Relations', p. 160.
300 C. J. Ducasse, 'Critique of Hume's Conception of Causality'; we have discussed Ducasse's case, above, Chapter 1, Section 1.3.
301 Collins, 'Explanation and Causality', p. 494ff.
302 Chapter 2, above.
303 Scriven, 'Explanations, Predictions, and Laws', p. 456ff.
304 *Ibid.*, p. 456.
305 *Ibid.*, p. 453.
306 *Ibid.*, p. 446.
307 *Ibid.*, p. 458.
308 *Ibid.*, p. 458.
309 *Ibid.*, p. 458.
310 *Ibid.*, p. 466.
311 Chapter 2, above.
312 Scriven, 'Explanations, Predictions and Laws', pp. 473–76.
313 *Ibid.*, p. 475.
314 Cf. Wilson, 'Explanation in Aristotle, Newton, and Toulmin'.
315 Cf. Chapter 2, Section 2.6 above.

NOTES TO CONCLUSION

1 Cf. S. Drake, *Galileo*, p. 33f., p. 39 and 52.
2 There is a standard exposition in the anthology of Feigl and Brodbeck, *Readings in the Philosophy of Science*.
3 For Feyerabend, see Wilson, *Reasons and Revolutions*. For "Oxford Philosophy", see Wilson, *'Meaning Is Use'*. See also Wilson, 'Acquaintance, Ontology, and Knowledge', and 'Logical Necessity in Carnap's Later Philosophy'.

BIBLIOGRAPHY

Abel, T. (1953): 'The Operation Called *Verstehen*', in H. Feigl and M. Brodbeck: *Readings in the Philosophy of Science*.
Ackermann, R. (1965): 'Deductive Scientific Explanation', *Philosophy of Science* 32.
Ackermann, R. and Stenner, A. (1966): 'A Corrected Model of Explanation', *Philosophy of Science* 33.
Achinstein, P. (1970): *The Concepts of Science*, Baltimore, Md.
Ackoff, R. (1962): *Scientific Method*, New York.
Addis, L. (1965): 'Ryle's Ontology of Mind', in L. Addis and D. Lewis: *Moore and Ryle: Two Ontologists*.
Addis, L. (1975): *The Logic of Society*, Minneapolis.
Addis, L. and Lewis, D. (1965): *Moore and Ryle: Two Ontologists*, The Hague.
Alexander, H. G. (1958): 'General Statements as Rules of Inference', in H. Feigl, M. Scriven, and G. Maxwell: *Minnesota Studies in the Philosophy of Science*, Vol. II.
Anderson, J. (1938): 'The Problem of Causality', *Australasian Journal of Philosophy* 16.
Anscombe, G. E. M. (1975): 'Causality and Determination', in E. Sosa: *Causation and Conditionals*.
Armstrong, D. (1978): *Universals and Scientific Realism*, Cambridge.
Ayer, A. J. (1956): *The Problem of Knowledge*, Harmondsworth, Middlesex.
Barker, S. (1959): 'Review of *Minnesota Studies in the Philosophy of Science*, Vol. II, *Philosophical Review* 58.
Beauchamp, T. and Mappes, T. A. (1975): 'Is Hume Really a Sceptic about Induction?' *American Philosophical Quarterly* 12.
Beauchamp, T. and Rosenberg, A. (1981): *Hume and the Problem of Causation*, Oxford.
Bergmann, G. (1951): 'Comments on Prof. Hempel's "The Concept of Cognitive Significance",' *Proceedings of the American Academy of Arts and Sciences* 80. Reprinted in his *The Metaphysics of Logical Positivism*.
Bergmann, G. (1945): 'Frequencies, Probabilities, and Positivism', *Philosophy and Phenomenological Research* 6.
Bergmann, G. (1954): 'Ideology', in his *The Metaphysics of Logical Positivism*.
Bergmann, G. (1959): *Meaning and Existence*, Madison, Wisc.
Bergmann, G. (1954): 'On Non-Perceptual Intuition', in his *The Metaphysics of Logical Positivism*.
Bergmann, G. (1957): *Philosophy of Science*, Madison, Wisconsin.
Bergmann, G. (1962): 'Purpose, Function, and Scientific Explanation', *Acta Sociologica* 5.
Bergmann, G. (1964): *Realism*, Madison, Wisc.
Bergmann, G. (1954): *The Metaphysics of Logical Positivism*, New York.
Bergmann, G. (1960): 'The Philosophical Significance of Modal Logic', *Mind* 69.
Bergmann, G. (1959): 'The Revolt Against Logical Atomism', in his (1959) *Meaning and Existence*.

Black, M. (ed.; 1950): *Philosophical Analysis*, Ithaca, N.Y.
Blake, R. M., Ducasse, C. J., and Madden, E. H. (1960): *Theories of Scientific Method: The Renaissance Through the Nineteenth Century*, Seattle.
Blalock, H. M. (1964): *Causal Inferences in Nonexperimental Research*, New York.
Broad, C. D. (1925): *Mind and Its Place in Nature*, London.
Brodbeck, M. (1962): 'Explanation, Prediction and "Imperfect Knowledge"', in H. Feigl and G. Maxwell: *Minnesota Studies in the Philosophy of Science*, Vol. III.
Brody, B. (ed.; 1970): *Readings in the Philosophy of Science*, Englewood Cliffs, N.J.
Bromberger, S. (1968): 'Approach to Explanation', in R. J. Butler, *Analytic Philosophy*, Second Series.
Bromberger, S. (1970): 'Why-Questions', in B. Brody (ed.; 1970): *Readings in the Philosophy of Science*.
Burks, A. W. (1951): 'The Logic of Causal Propositions', *Mind*, n.s. 60.
Butler, R. J. (ed.; 1968): *Analytic Philosophy*, Second Series Oxford.
Campbell, N. R. (1957): *The Foundations of Science* (Formerly *Physics: The Elements*), New York.
Carroll, L. (1895): 'What the Tortoise Said to Achilles', *Mind*, n.s. 4.
Chisholm, R. (1960): 'Law Statements and Counterfactual Inference', in E. H. Madden (1960): *The Structure of Scientific Thought*.
Chisholm, R. (1949): 'The Contrary-to-Fact Conditional', in Feigl and Sellars: *Readings in Philosophical Analysis*.
Cohen, I. B. (1974): 'Newton's Theory vs. Kepler's Theory and Galileo's Theory', in Elkana, Y.: *The Interaction Between Science and Philosophy*.
Colburn, N. (1954): 'Logic and Professor Ryle', *Philosophy of Science* 21.
Collingwood, R. (1938): 'On the So-Called Idea of Causation', *Proceedings of the Aristotelian Society* 38.
Collins, A. W. (1966): 'Explanation and Causality', *Mind*, n.s. 75.
Colodny, R. G. (1962): *Frontiers in Science and Philosophy*, Pittsburgh.
Cunningham, F. (1973): *Objectivity in Social Science*, Toronto.
Davidson, D. (1975): 'Causal Relations', in his *Essays on Actions and Events*.
Davidson, D. (1982): *Essays on Actions and Events*, Oxford.
Davidson, D. (1969): 'The Individuation of Events', in his *Essays on Action and Events*.
Davidson, D. (1984): *Inquiries into Truth and Meaning*, Oxford.
Davidson, D. (1967): 'Truth and Meaning', in his *Inquiries into Truth and Meaning*.
Drake, S. (1980): *Galileo*, Oxford.
Drake, S. (1978): *Galileo at Work*, Chicago.
Dray, W. (1957): *Laws and Explanation in History*, London.
Dray, W. (ed.; 1966): *Philosophical Analysis and History*, New York.
Ducasse, C. J. (1960): 'Causality: A Critique of Hume's Analysis', in E. H. Madden: *The Structure of Scientific Thought*.
Eberle, R., Kaplan, D., and Montague, R. (1961): 'Hempel and Oppenheim on Explanation', *Philosophy of Science* 28.
Elkana, Y. (ed.: 1974): *The Interaction Between Science and Philosophy*, Atlantic Highlands, New Jersey.
Ewing, A. C. (1932–3): 'A Defence of Causality', *Proceedings of the Aristotelian Society* 33.
Ewing, A. C. (1951): *Fundamental Questions of Philosophy*, London.

Feigl, H. and Brodbeck, M. (eds.; 1953): *Readings in the Philosophy of Science*, New York.
Feigl, H. and Maxwell, G. (eds.; 1961): *Current Issues in the Philosophy of Science*, New York.
Feigl, H. and Maxwell, G. (eds.; 1962): *Minnesota Studies in the Philosophy of Science*, Vol. III, Minneapolis.
Feigl, H. and Scriven, M. (eds.; 1956): *Minnesota Studies in the Philosophy of Science*, Vol. I, Minneapolis.
Feigl, H., Scriven, M., and Maxwell, G. (eds.; 1958): *Minnesota Studies in the Philosophy of Science*, Vol. II, Minneapolis.
Feigl, H. and Sellars, W. (eds.; 1949): *Readings in Philosophical Analysis*, New York.
Feyerabend, P. (1962): 'Explanation, Reduction, and Empiricism', in H. Feigl and G. Maxwell: *Minnesota Studies in the Philosophy of Science*, Vol. III.
Flew, A. (ed.; 1960): *Essays in Conceptual Analysis*, London.
Foster, J. A. (1968): 'Psychophysical Causal Relations', *American Philosophical Quarterly* 5.
Frege, G. (1960): 'On the Foundations of Geometry', *Philosophical Review* 69.
Friedman, M. (1974): 'Explanation and Scientific Understanding', *Journal of Philosophy* 71.
Gardiner, P. (ed.; 1959): *Theories of History*, Glencoe, Ill.
Gasking, D. (1955): 'Causation and Recipes', *Mind* 64.
Goodman, N. (1979^3): *Fact, Fiction, and Forecast*, Third Edition, Indianapolis.
Goudge, T. A. (1973): 'Pragmatism's Contribution to an Evolutionary View of Mind', *Monist* 57.
Goudge, T. A. ($1960-1975^2$): 'Philosophical Literature', Chapter 5 of C. Klink (ed.): *Literary History of Canada*.
Goudge, T. A. (1961): *The Ascent of Life*, Toronto.
Goudge, T. A. (1950): *The Thought of C. S. Peirce*, Toronto.
Grünbaum, A. (1953): 'Causality and the Science of Human Behavior', in H. Feigl and M. Brodbeck: *Readings in the Philosophy of Science*.
Hall, Marie Boas (1965): *Robert Boyle on Natural Philosophy*, Bloomington, Indiana.
Hart, H. L. and Honoré, A. M. (1966): 'Causal Judgment in History and in the Law', in W. Dray: *Philosophical Analysis and History*.
Hempel, C. G. (1965): 'Aspects of Scientific Explanation', in his *Aspects of Scientific Explanation and Other Essays*.
Hempel, C. G. (1965): *Aspects of Scientific Explanation and Other Essays*, New York.
Hempel, C. G. (1962): 'Deductive-Nomological vs. Statistical Explanation', in Feigl and Maxwell: *Minnesota Studies in the Philosophy of Science*, Vol. III.
Hempel, C. G. (1962): 'Explanation in Science and in History', in R. G. Colodny: *Frontiers in Science and Philosophy*.
Hempel, C. G. (1965): 'Postscript (1964)' to 'Studies in the Logic of Explanation', in his *Aspects of Scientific Explanation and Other Essays*.
Hempel, C. G. (1965): 'Studies in the Logic of Confirmation', in his *Aspects of Scientific Explanaton*.
Hempel, C. G. (1959): 'The Function of General Laws in History', in P. Gardiner: *Theories of History*.
Hempel, C. and Oppenheim, P. (1965): 'Studies in the Logic of Explanation', in C. Hempel, *Aspects of Scientific Explanation and Other Essays*.

Heese, M. B. (1963): *Models and Analogies in Science*, London.
Hume, D. (1947): *Dialogues on Natural Religion*, edited by N. K. Smith, Edinburgh.
Hume, D. (1888): *A Treatise on Human Nature*, edited by L. A. Selby-Bigge, Oxford.
Imlay, R. A. (1975): 'Hume on Intuitive and Demonstrative Inference', *Hume Studies* 1.
Johnson, W. E. (1921, 1922, 1924): *Logic* (in three volumes), London.
Kerlinger, F. N. (1965): *Foundations of Behavioural Research*, New York.
Kim, J. (1975): 'Causes and Counterfactuals', in E. Sosa: *Causation and Conditionals*.
Kim J. (1973): 'Causation, Nomic Subsumption, and the Concept of Event', *Journal of Philosophy* 70.
Kim J. (1963): 'On the Logical Conditions of Deductive Explanation', *Philosophy of Science* 30.
Kitcher, P. (1976): 'Explanation, Conjunction and Unification', *Journal of Philosophy* 73.
Kitcher, P. (1981): 'Explanatory Unification', *Philosophy of Science* 48.
Klibansky, R. (ed.; 1968): *Contemporary Philosophy*, Florence.
Klink, C. (ed.; 1960–1975^2): *Literary History of Canada*, Second Edition, Vol. III, Toronto.
Kneale, W. (1950): 'Natural Laws and Contrary-to-Fact Conditionals', *Analysis* 10.
Kneale, W. (1949): *Probability and Induction*, Oxford.
Kordig, C. (1971): *The Justification of Scientific Change*, Dordrecht.
Kuhn, T. (1970): 'Reflections on My Critics', in I. Lakatos and A. Musgrave: *Criticism and the Growth of Knowledge*.
Kuhn, T. (1962^1; 1970^2): *The Structure of Scientific Revolutions*, Chicago (1962; second revised edition, 1970).
Lakatos, I. (1970): 'Falsification and the Methodology of Scientific Research Programmes', in I. Lakatos and A. Musgrave: *Criticism and the Growth of Knowledge*.
Lakatos, I., and Musgrave, A. (eds.: 1970): *Criticism and the Growth of Knowledge*, London.
Lewis, D. (1973): *Counterfactuals*, Cambridge, Mass.
Mach, E. (1959^2): *The Principles of Physical Optics*, Second Edition, trans. by J. Anderson and A. Young, New York.
Mackie, J. L. (1975): 'Causes and Conditions', in E. Sosa: *Causation and Conditionals*.
Madden, E. H. (1960): 'Charles Sanders Peirce's Search for a Method', in R. M. Blake, C. J. Ducasse, and E. H. Madden: *Theories of Scientific Method: The Renaissance Through the Nineteenth Century*.
Madden, E. H. (1968): 'Peirce and Contemporary Issues in the Philosophy of Science', in R. Klibansky (ed.): *Contemporary Philosophy*.
Madden, E. H. (ed.; 1960): *The Structure of Scientific Thought*, Boston.
Mandelbaum, M. (1977): *The Anatomy of Historical Knowledge*, Baltimore.
Mead, G. H. (1934): *Mind, Self and Society*, Chicago.
Meehl, P. (1956): 'Problems in the Actuarial Characterization of a Person', in H. Feigl and M. Scriven: *Minnesota Studies in the Philosophy of Science*, Vol. I.
Mill, James (1878^2): *Analysis of the Phenomena of the Human Mind*, Second Edition, with Notes by J. S. Mill, London.
Mill, J. S. (1961^8): *System of Logic*, Eighth Edition, London.
Moore, G. E. (n.d.): *Ethics*, London.
Morgan, C. G. (1973): 'Archaeology and Explanation', *World Archaeology* **4**.

Morgan, C. G. (1973): 'Omer on Scientific Explanation', *Philosophy of Science* **40**.
Nagel, E. (1963): 'Carnap's Theory of Induction', in P. Schilpp, *The Philosophy of Rudolf Carnap*.
Nagel, E. (1961): *The Structure of Science*, New York.
Norton, D. F., Robison, W., and Capaldi, N. (eds.; 1979): *McGill Hume Studies*, San Diego.
Omer, I. (1970): 'On the D-N Model of Scientific Explanation', *Philosophy of Science* **37**.
Pap, A. (1956): 'Disposition Concepts and Extensional Logic', in H. Feigl and M. Scriven: *Minnesota Studies in the Philosophy of Science*, Vol. I.
Peirce, C. S. (1931–5): *Collected Papers*, ed. by C. Hartshorne and P. Weiss, Cambridge, Mass.
Peirce, C. S. (1955): 'How to Make Our ideas Clear', in J. Buchler (ed.): *Philosophical Writings of Peirce*, New York.
Peters, R. S. (1953): *The Concept of Motivation*, London.
Popper, K. (1949): 'A Note on Natural Laws and So-called "Contrary-to-Fact Conditionals",' *Mind* **58**.
Popper, K. (1969^3): *Conjectures and Refutations*, Third Edition, London.
Popper, K. (1969): 'Science: Conjectures and Refutations', in his *Conjectures and Refutations*.
Popper, K. (1969): 'The Demarcation Between Science and Metaphysics', in his *Conjectures and Refutations*.
Popper, K. (1961): *The Logic of Scientific Discovery*, New York.
Popper, K. (1969): 'Three Views Concerning Human Knowledge', in his *Conjectures and Refutations*.
Popper, K. (1950): *The Open Society and Its Enemies*, Princeton.
Reid, T. (1863): *Essays on the Intellectual Powers of Man*, in Vol. I of his *Works*, ed. by W. Hamilton, Edinburgh.
Rescher, N. (ed.; 1959): *Essays in Honor of Carl G. Hempel*, Dordrecht, Holland.
Rosenberg, A. (1973): 'Causation and Recipes: The Mixture as Before', *Philosophical Studies* **24**.
Rosenberg, A. (1976): *Microeconomic Laws*, Pittsburgh.
Ruddick, W. (1968): 'Causal Connection', *Synthese* **18**.
Ruse, M. (1973): *The Philosophy of Biology*, London.
Ryle, G. (1950): ' "If", "So," and "Because",' in M. Black: *Philosophical Analysis*.
Ryle. G. (1949): *The Concept of Mind*, London.
Salmon, W. (1963^1): *Logic*, First Edition, Englewood Cliffs, New Jersey.
Sartre, J.-P. (1966): *Being and Nothingness*, trans. by H. Barnes, New York.
Schilpp, P. A. (ed.; 1963): *The Philosophy of Rudolf Carnap*, LaSalle, Ill.
Scriven, M. (1966): 'Causes, Connections, and Conditions in History', in W. Dray: *Philosophical Analysis and History*.
Scriven, M. (1959): 'Explanation and Prediction in Evolutionary Theory', *Science* **130**.
Scriven, M. (1962): 'Explanations, Predictions, and Laws', in H. Feigl and G. Maxwell: *Minnesota Studies in the Philosophy of Science*, Vol. III.
Scriven, M. (1961): 'The Key Property of Physical Laws – Inaccuracy', in H. Feigl and G. Maxwell: *Current Issues in the Philosophy of Science*.
Scriven, M. (1959): 'Truisms as the Grounds for Historical Explanations', in P. Gardiner: *Theories of History*.

Sellars, W. (1967): *Philosophical Perspectives*, Springfield, Ill.
Sellars, W. (1967): 'Scientific Realism or Irenic Instrumentalism', in his *Philosophical Perspectives*.
Simon, H. and Rescher, N. (1936): 'Cause and Counterfactual', *Philosophy of Science* 33.
Simpson, G. G. (1950): 'Evolution', article in *Chamber's Encyclopaedia*, London.
Sosa, E. (ed.; 1975): *Causation and Conditionals*, London.
Sumner, L. W., Slater, J., and Wilson, F. (1980): *Pragmatism and Purpose*, Toronto.
Symon, K. R. (1953): *Mechanics*, Reading, Mass.
Tapscott, B. L. (1976): *Elementary Applied Symbolic Logic*, Englewood Cliffs, New Jersey.
Taylor, R. (1975): 'The Metaphysics of Causation', in E. Sosa: *Causation and Conditionals*.
Thorpe, D. A. (1974): 'The Quartercentenary Model of D-N Explanation', *Philosophy of Science* 41.
Toulmin, S. (1961): *Foresight and Understanding*, London.
Toulmin, S. (1960): *Philosophy of Science* (Torchbook Edition), New York.
Toulmin, S. (1958): *The Uses of Argument*, Cambridge.
Urmson, J. O. (1960): 'Parenthetical Verbs', in A. Flew: *Essays in Conceptual Analysis*.
van Fraassen, B. C. (1980): *The Scientific Image*, Oxford.
von Wright, G. H. (1975): 'On the Logic and Epistemology of the Causal Relation', in E Sosa: *Causation and Conditionals*.
von Wright, G. H. (1965^2): *The Logical Problem of Induction*, Second Edition, Oxford.
Watkins, J. W. N. (1970): 'The Paradoxes of Confirmation', in B. Brody: *Readings in the Philosophy of Science*.
Watson, P. J., Leblanc, S. A., and Redman, C. L. (1974): 'The Covering Law Model in Archaeology: Practical Uses and Formal Interpretations', *World Archaeology* 6.
Westcott, M. R. (1968): *Toward a Contemporary Psychology of Intuition*, New York.
Wilson, F. (1970): 'Acquaintance, Ontology, and Knowledge', *New Scholasticism* 54.
Wilson, F. (1982): 'A Note on Hempel on the Logic of Reduction', *International Logic Review* 13.
Wilson, F. (1968): 'A Note on Operationism', *Critica* 2.
Wilson, F. (1969): 'Barker on Geometry as *a Priori*', *Philosophical Studies* 20.
Wilson, F. (1967): 'Definition and Discovery', *British Journal for the Philosophy of Science* 18 and 19.
Wilson, F. (1971): 'Discussion of Achinstein, *The Concepts of Science*', *Philosophy of Science* 38.
Wilson, F. (1969): 'Dispositions: Defined or Reduced?' *Australasian Journal of Philosophy* 47.
Wilson, F. (in preparation): *Empiricism and Darwin's Science*.
Wilson, F. (1969): 'Explanation in Aristotle, Newton and Toulmin', *Philosophy of Science* 36.
Wilson, F. (1980): 'Goudge's Contribution to Philosophy of Science', in L. W. Summer et al.: *Pragmatism and Purpose*.
Wilson, F. (1984): 'Is Hume a Sceptic with regard to Reason', *Philosophy Research Archives* 10.
Wilson, F. (1982): 'Is There a Prussian Hume?', *Hume Studies* 8.

Wilson, F. (1983): 'Hume's Sceptical Argument against Reason', *Hume Studies* 9.
Wilson, F. (1983): 'Hume's Defence of Causal Inference', *Dialogue* 22.
Wilson, F. (1979): 'Hume's Theory of Mental Activity', in D. F. Norton, N. Capaldi, and W. Robison: *McGill Hume Studies*.
Wilson, F. (1965): 'Implicit Definition Once Again', *Journal of Philosophy* 60.
Wilson, F. (1968): 'Is Operationism Unjust to Temperature?' *Synthese* 18.
Wilson, F. (1983): 'Kuhn and Goodman: Revolutionary vs. Conservative Science', *Philosophical Studies* 44.
Wilson, F. (1967): 'Logical Necessity in Carnap's Later Philosophy', in A. Hausman and F. Wilson: *Carnap and Goodman: Two Formalists*, The Hague.
Wilson, F. (in preparation): *'Meaning Is Use'*.
Wilson, F. (in preparation): *Reasons and Revolutions*.
Wilson, F. (1979): 'Review of Mandelbaum, *The Anatomy of Historical Knowledge*', *Philosophical Review* 88.
Wilson, F. (1972): 'Review of Rescher, *Scientific Explanation*', *Dialogue* 11.
Wilson, F. (forthcoming): 'The Lockean Revolution in the Theory of Science', forthcoming in a Festschrift for R. F. McRae.
Yoshida, R. (1977): *Reduction in the Physical Sciences*, Halifax.

NAME INDEX

Ackermann, R., 200–202, 206, 209, 212–213, 215–216, 227, 229, 236–7
Ackoff, R., 81, 83
Alexander, H. G., 252, 275
Anderson, J., 56
Aristotle, x, 9, 11, 36, 67, 75, 86, 110, 184, 261, 333
Armstrong, D., x

Bacon, F., 30, 81, 277
Bergmann, G., x, xiii–xiv, 13n., 34, 332
Bergson, H., 286
Blalock, H. M., 104n., 105
Boyle, R., 49, 65, 294
Braybrooke, D., xviii
Brett, G. S., 252
Broad, C. D., 53, 252, 286–287
Bromberger, S., 91n., 155, 176–179
Burks, A., 9n., 299

Campbell, N. R., 287
Carnap, R., 29, 219
Carroll, L., 272
Charles, J. A. C., 294
Chisholm, R., 7n., 8n., 13
Cohen, I. B., 63n.
Collingwood, R., 109–112, 115, 121, 131
Collins, A. W., 4–5, 46n., 58, 75, 87, 89, 148, 151–155, 159, 175–176, 179–181, 183, 266–267, 290–293, 296–298, 304, 307, 332–334

Darwin, C., xvii, 37, 92
Davidson, D., xvii, 17–18, 187, 191, 195, 296–300, 303–308, 311–313, 315–318, 326–329, 331–333
de Retz, Cardinal, 19
Descartes, R., 35, 81
Dray, W., 251, 258n., 289

Ducasse, C. J., xviii, 17–19, 21, 25, 170, 333

Eberle, R., 187, 209, 211, 226–227, 229
Ewing, A. C., 9n., 10

Feyerabend, P., 62–63, 190, 241–242, 251, 289, 343
Forster, E. M., xv
Fortescue, Sir John, 109
Foster, J. A., 318
Frege, G., 9n.
Friedman, M., xv–xvi

Galileo, 1, 49–50, 62–64, 81, 86, 97, 343
Gasking, D., 111–116, 121
Goodman, N., xvii, 199
Goudge, T. A., xviii, 37–42, 251–252, 254, 256–263, 285–293

Hempel, C. G., xvii, 54–55, 163, 176, 187–189, 190–192, 195–207, 209, 215–216, 221, 223, 225–226, 229, 231, 236–237, 241–242, 248–249, 269, 331
Hesse, M. B., 287–288
Hooke, R., 253
Horace, 289
Hume, D., ix–xi, xiii–xiv, xvii, 2, 10–20, 29–30, 34–35, 37, 39, 42–43, 45, 52–53, 73, 82, 170, 250, 295, 327, 331, 333–334, 343

Kant, I., 36, 101
Kaplan, D., 187, 209, 211, 226–227, 229
Kepler, J., 1, 49–50, 93–94, 277, 284
Kerlinger, F. N., 27n.

NAME INDEX

Kim, J., xvii, 187, 195, 197–198, 209, 216, 218, 221, 226–229, 231, 235–236, 244, 297, 318–321, 325–326, 329–332
Kitcher, P., xv–xvi
Kneale, W., 9n., 13–14
Kuhn, T., xiii, xvii, 37, 41–43, 263, 288–289

Lakatos, I., 252, 287
Laplace, P. S., Marquis de, 101
Levi, I., 39n.
Lewis, D., x

Mach, E., 103, 343
Mackie, J. L., 55–57, 84
Madden, E. H., 39n., 41n.
Mead, G. H., 289
Meinong, A., 297, 332
Mill, J. S., x, xiii–xiv, xvi, xviii, 2, 18, 20, 23–24, 29–30, 32, 34–37, 52–53, 129, 134, 313, 343
Montague, 187, 209, 211, 226–227, 229
Morgan, C. G., 187, 204, 210, 222, 228, 230, 240–241

Nagel, E., x, 29, 287–288
Newton, Sir Isaac, 1, 6, 48–49, 56–57, 59, 62–63, 68, 70, 73, 93–94, 100–101, 116, 121, 170, 179, 252–253, 257, 260–261, 277, 326, 343

Ockham, W of, 103
Omer, I., 187–188, 231–236, 238–243, 245–246, 249–250
Oppenheim, P., 187–189, 191–192, 195–207, 209, 215–216, 221, 223, 225–226, 229, 231, 237, 242, 248–249

Pap, A., 9n., 298
Peirce, C. S., xiii, 37–40, 42–43
Popper, K., 1–2, 13–14, 36, 67–71, 73, 163, 199, 275

Prichard, H. A., 11
Ptolemy, 73, 277–278

Reid, T., 52–53
Rescher, N., x, xvii, 113–117
Rosenberg, A., 112, 115
Ruddick, W., 92–93
Ruse, M., 260n., 261n., 262n.
Russell, B., 12, 33–34, 56, 106–107, 197, 225, 252, 260, 314, 332, 343
Ryle, G., 271–273, 276, 290

Scriven, M., 5, 9, 11, 45n., 57, 62–63, 66, 87–89, 101, 135–141, 148, 151, 155–157, 159–161, 163, 165–166, 168, 175–176, 179–187, 190, 256, 263, 266–271, 290–291, 293, 296–298, 304, 307, 332–342
Sellars, W., 287
Simon, H., 113–117
Skinner, B. F., 244
Slater, J. G., xviii
Snell, 274–278, 282, 284–285
Socrates, 330
Stenner, A., 236
Strawson, P. F., 148
Sumner, L. W., xviii

Taylor, R., 127–129, 131
Teilhard de Chardin, P., 252
Thorpe, D. A., 216, 227–231
Toulmin, S., 41n., 134–136, 271–285, 287, 289–290

van Fraassen, B. C., 180n.
Veblen, T., xii
Volterra, V., 260
von Wright, G. H., 35, 118–126, 130–133

Watson, P. J., 289
Whewell, W., 36
Wilson, F., xviii, 9n.
Wittgenstein, L., 11

SUBJECT INDEX

action: connected to idea of causation via that of production 111ff, 124ff; connected to idea of causation via that of strategies 109; does not define causation xiv, 118ff; most primitive notion of causation 109; to be explained by laws 110f; therefore not fundamental to notion of explanation 111

causal fallacy: 50

causal relations: *see* causation

causality, principle of: 34ff; assertion justified by enumerative induction 34ff; connection with process laws 34; empirical but not falsifiable 36; a generic law 34; guiding hypothesis for eliminative induction 33ff

causation: action vs passion as a proposed criterion for 128ff; asymmetries in, accounted for in terms of the forms of the laws involved 112–113, 113–115; can be a matter of omissions as well as actions 129ff; connected to idea of action via that of production 111ff, 125; connected to idea of action via that of strategies 109; contrast to conditions is contrast of unexpected to expected 134–136; contrasted to conditions 134; contrasted to correlation xiv, 74, 104f, 106ff; four crucial features of 76f, 294, 329; involves asymmetries not present in all explanations 97ff, 104ff, 300; involves laws xi, 108, 290–343; independent of action xiv, 116–117; and interference 113, 120ff, 127ff, 131ff; most primitive notion of is human action 109; and production 111, 113, 121; supports assertion of counterfactual conditionals 122

construed as an ordinary two-place relation: 296; conflicts with third crucial feature of causation 329; must be an internal relation 329; raises no problem with analytic/synthetic distinction 297; raises problem of knowledge 297; relates events 297, 327ff; relates natured events 329

construed as a primitive non-truth-functional connective: 295; raises problem of analytic/synthetic distinction 295; raises problem of how we know it 295, 333; purported cases of knowing it plausible only if ignore imperfect laws 333, 295ff; not a conditional 300ff; a top philosopher's arguments against it being a conditional 298ff; these arguments plausible only if ignore imperfect laws 307

construed as a non-truth-functional connective analyzable by DN model: 295; raises no problem with analytic/synthetic distinction 296; raises no problem of knowledge 296

Humean account of: 10ff, 290ff; Broad's alleged counterexample to 53f; Collingwood's alleged counterexample to 109ff; entails fallibility of all judgments of 16; Gasking's alleged counterexample to 111ff; Kneale's counterexamples to regularity view not counterexamples to 13ff; necessary connection a matter of imputation x, 13; not a regularity view 10; objectively, causation = accidental generality 12; Popper's non-Humean view 13; Reid's alleged counterexample to 52; requires

377

causal judgments to involve a generality 12, 18ff, 263ff, 290ff; Scriven's alleged counterexamples to 254ff, 263ff, 290ff; Simon and Rescher's alleged counterexample to 113; subjectively, causation distinguished from accidental generalities 13; takes account of necessary connection 10; von Wright's alleged counterexample to 118ff

language of: does not appear in ideal explanations 106; less preferable than language of laws 106; often irremediably anthropomorphic 107; often leaves law only implicit 225; the latter a defect 225

See also explanation, law

clarity: Popper's view on how to contribute to it 67, 68

completeness: unreasonable notions of 81ff

confirmation xii, 30ff; consequence and converse consequence conditions of 25n, 32, 347

See also: Induction

common cause, fallacy of: 50f; involves imperfect knowledge 51

concepts, scientific: if non-primitive, then explicitly defined 11; if primitive, then designate sensible properties and relations 11; not theory-laden 281; ought to satisfy empiricist meaning criterion 67

conversational conciseness, principle of: its role as a norm for explainings 222; its role in defending the DN model 222f, 224; motivates leaving law statement implicit in explanations 225; over-ruled by science to require law statements to be explicit 225

conversational content, principle of: its role as a norm for explainings 219, 224; its role in defending the DN model 221f; Omer wrongly takes to be basic in establishing criteria for explanation 231ff; principle of explanatory content prior to 244; principle of predictability prior to 241; requires one to stay on topic 217, 220

correlation: contrasted to causal relation xiv, 74, 104f, 106ff; does not lack explanatory power xiii, 73, 104f; a special case of laws xiii, 73, 104

counter examples: a weak form of argument 86, 226, 241; do not call into question DN model 204, 207, 241; often boring 88; often lead to unprincipled tinkering 204, 207, 226; often not conclusive 86

counterfactual conditionals: accidental generalizations do not support assertion of 7-8; causation can support the assertion of 122; imperfect laws can support the assertion of 96; laws support assertion of x, 7-8

deduction: accounts for the explanatory must 4, 76; essential to explanation x, 7, 73, 75ff, 267, 270; Feyerabend's alleged counterexample to the necessity of in explanation 63; not restricted to Barbara xii, 163, 280, 281

definite descriptions: often appear in explanations 308ff; their misuse in arguments against the DN model 193, 196, 277, 307ff, 316ff, 339ff, 355f; their special role in *ex post facto* explanations 148ff, 164, 255

definition, implicit: yields no philosophical insight 9

empiricism: moves away from shown to be unmotivated 1-343

events: causation construed by a top philosopher as a relation between 297, 327ff; logical independence of 11, 76, 198f; logical independence of implies logical independence of predicates that refer to them 12, 199; causation as a relation between requires natured events 329

SUBJECT INDEX 379

Kim's views on 329ff; their affinity to Meinong's views 332; their affinity to deductivist views 331
evidence: active search for 20, 38ff, 43, 80; not part of explanation 268; reasons for this no support for excluding laws from explanation 269; unless one wrongly approaches explanation through psychology and linguistics 271
explaining: Ackermann's principle of redundancy as a norm for 201, 215; Ackermann's principle of triviality as a norm for 201; act of explaining something to someone xii, 173; an activity 3; contrasted to explanation xii, 3, 173; criteria for evaluating success of 181ff; explanation useful in only if subjectively worthy 182, 190, 205; law often stated only implicitly 5, 224, 293, 315; other premises often stated only implicitly 194f, 313ff; principle of conversational conciseness as a norm for 222; principle of conversational content as a norm for 219; principle of explanatory content a norm for 244ff; relation to a question 91; relevance of cognitive interests to 183ff; a transaction between explainor and explainee 3, 91f, 173
context of: contrasted to that of research 247, 250; models have no role in 290
scientific explanation of: 264ff; pragmatic interest in that explanation 264, 267; used to support claim laws not part of explanation 264ff
explanation: as an answer to a question 91, 258; an argument 2; an argument that is an explanation in one context may not be in another 214, 217, 221, 225, 227, 234; being inappropriate for explainings contrasted to being a non-explanation xii; better if better at predicting 83; better and worse 46, 74, 83; by unification xv; can satisfy idle curiosity xiii, 173; can satisfy pragmatic interests xii, 173; competing candidates for 80, 241ff; contrasted to explainings xii, 3, 173; contrasted to forecasting 46, 73; distinguished from explanation sketch 54; employed in contexts other than explainings 247; employed in context of research 247; equivalent in explanatory power if equivalent in predictive power 236; how to establish criteria for 1; ideal of defined by cognitive interests 83ff; identified with prediction 79; of imperfect by process 59ff; intuitive judgment not involved xii, 165ff; logical form of ix, 76ff; never certain we have one 44; never needed for a false generalization 62; non-formal criteria for usefulness of 181ff, 205, 209; not a transaction 3, 4; not an activity 3; permits prediction and control xii, 5, 77f, 202; physics does not provide a paradigm 1, 74–75; and the principle of predictability 79, 202, 204; process knowledge constitutes ideal xiv, 47; a set of sentences 2, 181; sometimes useless in an explaining transaction 89; statistical xvii; subjective contrasted to objective acceptability xi, 7, 44ff; true premises essential for 7; useful in explainings only if subjectively worthy 182, 190; weak contrasted to non- xii, 46; weak contrasted to strong xii, 46
acceptability of: context dependent xi, 173, 214; not determined by formal rules alone 210; not determined by linguistic rules alone 177
approach through psychology and linguistics: 175, 263ff; dangerous to science 341–342; essentially antiscientific 343; not normative 174, 341; permits anthropomorphic projections that inhibit research 341

behavioural criteria for: 173ff; an approach through psychology and linguistics 175, 263ff; do not provide norms for worth of explanation 174, 341; a misleading approach 178, 180, 292; non-normative nature of behavioural criteria often not noticed by critics of DN model 175; used to support claim laws not part of explanations 264ff, 291f

causal: false claim that they are non-extensional 192; often involves asymmetries 97ff, 104ff, 108ff; but in a way compatible with symmetry thesis 100ff; none without law-deduction xi, 290–243; but law often only implicit 225

criteria for: Hempel and Oppenheim's: 188ff, 215; are criteria for objective worth 189; presuppose principle of predictability 198ff, 202ff

criteria for: Omer's: 237, 239f; based on principle of conversational content 231ff, 236

criteria for: Scriven's: 180ff; are really criteria for explainings 181; criticized 186f

criteria for: Thorpe's verificational approach: 229ff; connection with principle of predictability unrecognized by Thorpe 231; essentially syntactical 231; connected with principle of predictability 231

evidence not part of: 268ff; reasons for this no support for excluding laws from explanation 269; unless one wrongly approaches explanation through psychology and linguistics 271, 137, 143ff, 154, 251

ex post facto: 137, 143ff, 154, 271; common in biology 358; does not refute symmetry thesis 139ff, 163, 358; imperfect 145; integrating explanations a special case of 257; prediction possible where statistical laws available 256; role of definite descriptions in 148, 256

the false claim of Davidson that any fact explains any fact: 191, 316; Kim's version of this claim 195, 319

formalist criticisms of DN model: xiv, 187ff; Ackermann's 200, 213, 229; Davidson's 191ff; Eberle, Kaplan and Montague's 209ff, 229; Hempel and Oppenheim's 223; Kim's 197ff, 218ff, 221f, 226ff, 235; Morgan's 222, 228, 240; Omer's 231ff; Thorpe's 216ff, 228ff

integrating: 251, 256ff; based on imperfect generic laws 257; a case of *ex post facto* explanation 257; does not conform to axiomatic model 257; does not conform to process model 257; in biology 256ff

law premise: necessity of x, 77, 206, 212f, 263–290; often only implicit in context 224, 293

narrative: as an answer to a question 258; as an answer to a how-possibly-question 260; based on imperfect generic laws 254, 258; does not conform to axiomatic model 253; does not conform to process model 252; fills in gaps by hypotheses 258; hypothetical element in answers how-possibly-questions 262; hypothetical element in does not conflict with scientific nature of explanation 262; hypothetical element in initiates research 262; laws are part of and not inference tickets 358; proposes research rather than terminates it 289; in biology 253ff; in history 251; often contains historical laws 260; provides an explanation sketch 258; role of models in 285ff; truth of premisses of not necessary for success of in explaining contexts 360; in this respect contrasted to causal explanation 360

none without knowing a law: xiv, 290–343; wrong belief to the contrary

SUBJECT INDEX

often the result of an approach through psychology and linguistics 292

provides a: must 4, 76, 198, 235; the must explicated by deduction 4, 77, 199

sketch: 54; distinguished from explanation using imperfect laws 54; the notion tends to generate confusion 55; narrative explanation a case of 258; often implicitly contains imperfect laws 55, 258

symmetrical with prediction: ix, xi, 5, 79, 202; this thesis not refuted by *ex post facto* explanations 139ff, 163, 358; this thesis not refuted by non-reasoned predictions 87; this thesis not refuted by non-scientific explanations 87

theoretical xv, 57ff, 80; achieved by deduction 57ff, 65; achieved by instantiating generic laws 58, 65f; contrasted to explanations of individual facts xv; requires deduction 80

See also causation, explainings, explanation, law, relevance

explanatory content: principle of: directs us to prefer causal laws to their 'piggy back' laws 327; its rationale derives from cognitive interests 244; its role as a norm for explainings 244ff; prior to principle of conversational content 244

falling bodies, Galileo's law of: a case of imperfect knowledge 62ff; false but approximately true 63; since false, never explained anything 63; subjectively worthy of acceptance 63

fallibility: not a vice 16

falsifiability: generic laws not falsifiable 276; often achieved via confirmation of contrary hypotheses 276; Popper's ideas concerning, silliness of, 36

fashion: inadvisability of pandering to 155

forecasting: a case of explanation 46; contrasted to explanation 46; its weak explanatory power 46

General, Mrs.: tentatively compared to certain philosophers 176

generalization, accidental: 7, 12, 112; contrasted to law 12f; does not support the assertion of counterfactual conditionals 7–8; incorrectly contrasted to causation rather than laws 112

geometrical optics: alleged counterexample to symmetry thesis 102ff, 275ff, 279ff

induction: Humean and Peircean views compared 39ff; always fallible 39, 81

eliminative: requires guiding generic hypothesis 25, 29ff; supports assertion of laws 18ff, 122, 123ff, 132f; supports assertion of imperfect laws 22ff, 26ff

enumerative: Mill's claim that this justifies assertion of guiding hypothesis for eliminative induction 29ff

intelligence: human: analogous to natural selection 362

interest: cognitive: xii, 78f, 81, 93; Popper's claim that scientist's not that of technologist 68; determines the importance of the principle of explanatory context 244; determines the importance of the principle of predictability 199; in causal asymmetries 98ff, 105ff; leads one to prefer causal relations to correlations 106; motivates scientific research 183; requires one not to use a weaker explanation when a stronger is available 184ff, 212, 216, 219; role in explainings 183, 244; used to define ideal of explanation 81; used to define models of

explanation 78ff; often expressed by a question 92f
idle curiosity: xii, 78, 81, 157; aims at matter-of-fact knowledge only since Galileo 86; motivates research not motivated by pragmatic interests 47, 186, 244, 253, 265; pleasure on its satisfaction 78
pragmatic: xii, 50, 78, 81, 265; may be satisfied by imperfect knowledge 85, 93, 157, 244, 265; motivates technologist 69
intuitive judgement: its purported role in explanation xii, 166, 334ff; its role in classifying variables 177; its role in helping predict when knowledge is imperfect 166ff, 335, 336; does not refute symmetry thesis 167, 335; improves with training 168; occurs in ordinary contexts 168ff

knowledge: imperfect xiv, 47, 48ff; role in causal fallacies, 50ff
process: xiii, 47, 56; characterized 47; constitutes ideal of explanation xiv, 48; Newton's account of the solar system, an example of, 48
See also explanation, law

law: can be asserted on the basis of a single instance 17ff, 170; causal xiii, 106ff, 118ff, 136ff; contrasted to accidental generalization x, 7, 12f, 66ff, 73; correlation as case of xiii, 86ff; evidential basis always a sample 17; explains by unifying individual facts xv; falsely contrasted by Popper to technological computation rules 68; Humean analysis of xi, xiii, 10ff; its assertion always uncertain 16, 39; language of, preferable to language of causes 106; mistaken view that one is often not part of explanation, Ryle's 271ff, Toulmin's 273ff; not a 'conceptual truth' 276, 279, 283; often quantificationally complex in everyday contexts 163; often quantificationally complex in science 275ff; a statement about a population 17; supports assertion of counterfactual conditionals x, 7–8; a true matter-of-fact generalization x
explanation of: 57; by less imperfect laws 57; by theories 57ff
'gappy': (= imperfect): xv, 55ff, 258ff
generic: 24, 36, 254ff, 277; accounts for scope of theories 58; a case of imperfect law 24, 254ff; often not falsifiable 25, 36, 139, 276; not therefore truisms 139, 276; research guided by 31; used to argue wrongly that laws not part of explanations 277ff
historical: 260; necessity for sometimes confused with uniqueness of history 260–261; occur in biology, 260, 359; sometimes overlooked 359
Humean account of: *see* causation, Humean account of
imperfect (= 'gappy'): xv, 24ff, 47, 48ff, 55ff, 254ff; account for predictions purportedly not explanations 88ff; account for explanations purportedly not predictions 151ff; basis of narrative explanations 254; can support the assertion of counterfactual conditionals 96; less imperfect explanation the more imperfect 53; often ignored by critics of DN model 151, 277ff, 298, 307; role in integrating explanations 257; role in narrative explanations 257ff; sometimes ignored by defenders of DN model 253, 256; often yields only determinable, not determinate, predictions 84, 258; 'piggy-back' law a special case of 325
mistaken distinction between its 'rule' and its 'scope of application': 66ff, 273ff; used to argue laws not part of explanations 273
'piggy-back': 319; does not yield causal explanation 325; entailed by causal

SUBJECT INDEX

laws 319; imperfect 325; no problem for DN model 324; principle of explanatory content directs us to prefer causal laws 327; wrongly thought to enable any fact to explain any fact 319ff
process: xiii, 34, 47ff, 57f; characterized 47; explains all imperfect laws 53; provides the ideal scientific explanation 47; yields determinate predictions 84
statistical: 49, 64, 85, 184, 254; an imperfect law 64, 254; yields predictions in connection with *ex post facto* explanations 256
unanalyzed modality account: rejected 9, 76f; 7ff, 76ff; leads to radical scepticism 345
See also causation, explanation
lawfulness: a contextual or pragmatic feature x, 11ff; the Humean account xi; its imputation justified only relative to evidential context 15ff; not a synatical nor semantical feature x; objective conditions alone do not justify its imputation 15
linguistic theory: not useful in the philosophy of explanation 155

measurement: limits to cannot be overcome 81
mechanics, quantum: its capacity to obscure issues in philosophy of science 102; the fashionable tendency to introduce it too quickly 102
metaphysical non-sense leading to scientific sense: Popper's mistaken views on 67
models: are analogies 286; their role in science 279ff, 285ff; does not support claim that laws are not 'conceptual truths' 279, 283ff; located in context of research 290; not located in context of explainings 290; not merely heuristic 287; not absolutely essential to research 287; use in narrative explanations 285ff; useful in suggesting hypotheses 287

natural selection: analogous to human intelligence 362

Oxford: its provincial nature 11

paradigms, Kuhnian 41, 253; are generic theories that guide research 24n, 253, 263; function as generic hypotheses for eliminative induction 41; in a Humean account of laws 41ff; in a Peircean view of induction 41, 288; no commitment to 'theory-ladenness' of concepts thesis 42
pendulum: alleged counterexample to symmetry thesis 94ff
philosophers, certain: tentatively compared to Mrs. General 176; their anti-scientific conclusions 289, 343; their capacity to choose apparently apt examples without exploring them in detail 105; their capacity to get to the point 177; their love of varnish 176; their capacity to be fair to their opponents 63; their capacity to spend inordinate time on formal details 219; their tendency to use examples the complexity of which obscures relevant points 103; their tendency to use the appropriate blade of Ockham's razor 103
philosophy of science: not just descriptive but also normative 1
physics: does not provide a paradigm of explanation 1, 54
post hoc, ergo propter hoc Fallacy: 51ff; involves imperfect knowledge 51, 52
positivism, logical: conclusion as to its death shown to be too hastily drawn 1–343
predicates, logical independence of: 11; entails a logical gap between sample

and population 16; entails that inductive inference is always fallible 16; reflects logical independence of properties of events 12, 199

predictability, principle of: 79; connection with Goodman's notion of entrenchment 199; connection to Popper's notion of a test 199; connection to Thorpe's verificational approach to explanation 231; exhausts concept of explanation 79, 202, 204; failure by Thorpe to recognize connection to his approach 231; imposes non-formal as well as formal constraints 205, 209, 210, 217; imposes the non-formal constraint that an explanation used in explaining be subjectively worthy 205; its rationale derives from our cognitive interests 199, 201; its rationale contrasted to that of Ackermann's principle of redundancy 201; prior to principle of conversational content 241

prediction: cases purportedly not explanation 88ff; and control, requires deduction from laws 77; identified with explanation 79, 202; identified with reasoned prediction 87; inference from known to unknown 87, 137; not merely a statement about the future 87; reasoned vs non-reasoned ix, xiv, 5; and the principle of predictability 79

of laws 87; does not refute symmetry thesis 87; based on generic laws 41; based on laws about laws xvi

symmetrical with explanation: ix, xi, 77; this claim not refuted by non-reasoned predictions 87; this claim not refuted by non-scientific explanations 87

premiss, false: does not explain even if subjectively acceptable 62

premiss, true: essential to explanation 7, 62

production: idea of, connects ideas of action and causation 111ff; to be explained via laws 112; therefore not fundamental to notion of causation 113

question: expresses a cognitive interest 93
 how-possibly: answered by narrative explanation 262
 why: elaborate definition of 176; latter not really necessary 177; often requires explanation 91, 100, 354

redundany, Ackermann's principle of: 200, 213; its rationale contrasted to that of principle of predictability 20; its role as a norm for explainings 201; its too hasty use to exclude explanatory arguments as non-explanatory 214; no rationale save to exclude counterexamples 201; not a norm for explanation 201; used to reject alleged counterexamples to DN model 215

relevance, explanatory: 354; left unexplained 354; taken as primitive 354

research, context of: contrasted to context of explaining 247, 250; explanations employed in 247; role of models in 290

scientific: motivated by cognitive interests 186; often guided by generic theories 165; often initiated by hypotheses of narrative explanations 262; role of models in suggesting hypotheses 267; role of paradigms in guiding 288; role of verstehen in suggesting hypotheses 289

sobriety, virtue of: capacity of Toronto to encourage 252

technological computation rules: 68; are laws, 68ff; Popper's erroneous views on 68ff

SUBJECT INDEX

theories: as guides in research xvi, 165; explain by deduction 80; explain by deduction from axioms 57; explain by unifying laws xv; explain when generic by instantiation 58; scope determines their explanatory power 58

generic: can justify accepting imperfect laws prior to instantial evidence 165; refuted by confirmation of a contrary theory 284; scope achieved by containing generic laws 58

'theory-ladenness' of concepts: 42, 276, 281, 384; a false thesis 42; a form of relativism 42; leads to an anti-scientific Aristotelian account of science; 348; not implied by Kuhnian paradigms 42

Toronto: a good place to do philosophy 252; its capacity to encourage sobriety 252

trains: their behaviour held by some to refute the DN model 93, 127ff

triviality, Ackermann's principle of: 201, 212, 213; its role as a norm for explaining 201; no rationale save to exclude counterexamples 201, 212; its rationale contrasted to that of principle of predictability 201, 212; not a norm for explanation 201

unification: by composition law xvi, 58; by generic laws xvi, 58, 359; by reduction xvi, 58; degrees of xv; of individual facts by laws xv; of laws by theories xv; strongest unification of individual facts provided by process laws xv; trivially and irrelevantly achieved by conjunction 359

verstehen: a source for hypotheses to guide research 289